MODERN ASPECTS OF ELECTROCHEMISTRY

No. 29

LIST OF CONTRIBUTORS

SYMEON I. BEBELIS
Department of Chemical Engineering
University of Patras
Patras GR-26500, Greece

MILAN M. JAKSIC
Department of Chemical Engineering
University of Patras
Patras GR-26500, Greece

STYLIANOS G. NEOPHYTIDES
Department of Chemical Engineering
University of Patras
Patras GR-26500, Greece

ILEANA-HANIA PLONSKI
Institute of Atomic Physics
IFTM
RO-76900 Bucharest, Romania

SHARON G. ROSCOE
Department of Chemistry
Acadia University
Wolfville, Nova Scotia, Canada
B0P 1X0

DARKO B. ŠEPA
Faculty of Technology and Metallurgy
University of Belgrade
11000 Belgrade, Yugoslavia

CONSTANTINOS G. VAYENAS
Department of Chemical Engineering
University of Patras
Patras GR-26500, Greece

RONALD WOODS
CSIRO Division of Minerals
Port Melbourne, Victoria 3207,
Australia

A Continuation Order Plan is available for this series. A continuation order will bring delivery of each new volume immediately upon publication. Volumes are billed only upon actual shipment. For further information please contact the publisher.

MODERN ASPECTS OF ELECTROCHEMISTRY

No. 29

Edited by

J. O'M. BOCKRIS

Texas A&M University
College Station, Texas

B. E. CONWAY

University of Ottawa
Ottawa, Ontario, Canada

and

RALPH E. WHITE

University of South Carolina
Columbia, South Carolina

PLENUM PRESS • NEW YORK AND LONDON

The Library of Congress cataloged the first volume of this title as follows:

Modern aspects of electrochemistry. no. [1]
  Washington Butterworths, 1954–
  v. illus., 23 cm.
  No. 1–2 issued as Modern aspects series of chemistry.
  Editors: no 1–   J. Bockris (with B. E. Conway, No. 3–)
  Imprint varies: no. 1, New York, Academic Press.—No. 2, London, Butterworths.
  1. Electrochemistry—Collected works. I. Bockris, John O'M.ed. II. Conway, B.E. ed.
(Series: Modern aspects series of chemistry)

QD552.M6   54-12732 re

ISBN 0-306-45162-X

© 1996 Plenum Press, New York
A Division of Plenum Publishing Corporation
233 Spring Street, New York, N. Y. 10013

10 9 8 7 6 5 4 3 2 1

Printed in the United States of America

Preface

MODERN ASPECTS OF ELECTROCHEMISTRY
VOLUME 29

This volume of *Modern Aspects* contains a remarkable spread of interests. In respect to the first, Dr. Darko Šepa takes up a much discussed subject— how does one evaluate the fundamental quantity, energy of activation? He deals with this in very considerable detail and allows us to see the various pitfalls which pertain to the evaluation, particularly those that arise as a consequence of the variation of the symmetry factor with temperature.

One of the more remarkable advances in electrochemistry since 1950 is described by Vayenas *et al.* It is the effect of electrical potential changes on catalyst surfaces. The subject—though only a few years old—is fully described in a remarkable chapter.

Then, Dr. Ileana-Hania Plonski from Bucharest, Romania, brings out the question of the mechanism of the dissolution of iron, clearly one of the more important reactions in technological electrochemistry. Dr. Plonski's contributions differ from those of others in that she stresses the possible effects that arise from the presence of hydrogen on the surface, even under anodic conditions when it should dissolve, because it may be retained to kink sites and therefore influence the rate of iron dissolution and, eventually, corrosion.

Few of those who have to do with the industrially important subject of mineral recovery are aware that for many years a series of Australian chemists, starting with Wark and continuing with Nixon (who first applied the mixed potential concept), applied electrochemical ideas to mineral floatation. Ronald Woods' article represents a culmination of many years of study and is far more advanced in what it tells the reader than any other document to date in this field.

Sharon Roscoe's article is about protein surfaces and, in the views of the editors, is a classic in the sense that it works out some of the consequences of considering proteins as electrodes. This opens up, of course, an area of biochemistry which may allow an enhanced understanding of many phenomena which have proved difficult to rationalize.

Contents

Chapter 3

EFFECT OF SURFACE STRUCTURE AND ADSORPTION PHENOMENA ON THE ACTIVE DISSOLUTION OF IRON IN ACID MEDIA

Ileana-Hania Plonski

Chapter 4

ELECTROCHEMICAL INVESTIGATIONS OF THE INTERFACIAL BEHAVIOR OF PROTEINS

Sharon G. Roscoe

Chapter 5

CHEMISORPTION OF THIOLS ON METALS AND METAL SULFIDES

Ronald Woods

Energies of Activation of Electrode Reactions: A Revisited Problem

Darko B. Šepa

Faculty of Technology and Metallurgy, University of Belgrade, 11000 Belgrade, Yugoslavia

I. INTRODUCTION

A century has passed since Roszkowsky in 1894 indicated in *Zeitschrift für Physikalische Chemie* that the electrode potential of the hydrogen evolution reaction, taken at constant current density, is affected by temperature.[1] His paper was the first report on the effect of temperature on the kinetics of electrode reactions. During the next half century after the publication of Roszkowsky's paper, however, studies of temperature effects in electrode kinetics were scarce and involved mostly determinations of temperature coefficients of the potential or overpotential for a few electrode reactions.[2–18]

In 1929, on the basis of a purely formal analogy to the Arrhenius equation in chemical kinetics,[19] Bowden introduced the *energy of activation* as a parameter for evaluation of the temperature effect in electrode kinetics.[20] Mathematically, he proposed two alternative ways of defining the energy of activation. One was based on the temperature coefficient of the electrode potential taken at constant current density, and the other on the dependence of the logarithm of current density, taken at constant electrode potential, on the reciprocal of the temperature. Bowden was the first to emphasize that the energy of activation of electrode reactions is linearly dependent on electrode potential and also was the first to determine experimentally the energy of activation of, for example, the hydro-

Modern Aspects of Electrochemistry, Number 29, edited by John O'M. Bockris *et al.* Plenum Press, New York, 1996.

gen evolution reaction at mercury and the oxygen evolution reaction at platinum in acid solutions.[20]

Bowden's concept of the potential-dependent energy of activation of electrode reactions was subsequently promoted in theoretical considerations of the hydrogen evolution reaction by Erdey-Gruz and Volmer[21] and Frumkin.[22]

In 1939, Eyring, Glasstone, and Laidler[23] applied the formalism of the newly developed transition-state theory, or absolute reaction rate theory, to electrode reactions. Thus, the energy of activation, formally introduced by Bowden,[20] became identified theoretically as either heat of activation or enthalpy of activation.

Using a model of the potential energy surface and general ideas of transition-state theory, Horiuti and Polanyi for the first time calculated the value of the energy of activation for a simplified model of the elementary reaction of hydrogen-ion discharge, taken as the rate-determining step in the hydrogen evolution reaction in aqueous solutions.[24]

Building on the foundation laid by an earlier review article by Bowden and Agar,[25] two important papers on this subject were published—one by Agar[26] in 1947, and the other, independently, by Temkin[27] in 1948. In these papers Bowden's ideas[20] were elaborated, but within the frame of transition-state theory and using the overpotential as an experimental substitute for the change of the Galvani potential difference in the cell. Suggesting the use of overpotential in studies of temperature effects in electrode kinetics, Agar and Temkin introduced the equilibrium state of the electrode reaction studied as the electrical reference state for evaluation of the phenomenon. It is interesting to note that in the subsequent literature only Agar and/or Temkin were credited for the introduction of the heat of activation[26] or energy of activation[27] into experimental electrochemistry, which oversimplifies the facts. However, it is true that the articles by these two authors initiated worldwide interest in studies of temperature effects in electrode kinetics. The increased number of papers published on the subject in the succeeding years proves this, for prior to the articles by Agar and Temkin this topic had hardly been treated at all.[28–30] Thus, during the ten years between 1949 and 1959, the number of papers dealing with the various aspects of temperature effects on the rate of electrode reactions that were published in well-known scientific journals outnumbered all previously published papers on this topic.[31–78]

A century after the publication of Roszkowsky's paper, extensive evidence on the topic has accumulated in the literature. Full insight into all the published data is hard to obtain, partly because some of the data are published in local or less significant scientific journals and partly because data on the subject are frequently a minor part of broader investigations of electrode kinetics. Nevertheless, closer examination of the literature on the topic enables one to classify it as follows.

Theoretical treatments of the subject involve mostly a rephrasing of Agar's and Temkin's basic considerations and are found in books,[46,79–84] review articles,[75,85–87] and within review articles on other aspects of electrode kinetics.[42,45,88–94] A few attempts at more advanced or alternative treatments of the subject, together with considerations on the significance of the energy of activation in electrode kinetics and descriptions of theoretical models of electrode reactions,[95–115] and reports of theoretical calculations of the energy of activation for simplified models of the electrode reaction,[24,37,67,72,116–119] including discussions of the role of the energy of activation in electrocatalysis,[120–124] can be classified in this group of articles.

Another group of articles contains mainly data on experimental values of various energies of activation for electrode reactions. Within this group of articles, the majority simply contain values of the energy of activation determined at zero overpotential (i.e., at the reversible potential) for various electrode reactions at different electrode materials and mostly in aqueous solutions. The most frequently studied reaction is the hydrogen evolution reaction at different electrode materials (Refs. 20, 31, 34–36, 40, 58, 62, 63, 68, 70, 73, 76, and 125–152), followed by metal deposition and dissolution reactions (Refs. 52, 59, 65, 66, 74, 75, and 153–179), redox reactions (Refs. 39, 41, 78, and 180–188), oxygen reduction and evolution reactions (Refs. 20, 30, 71, and 189–196), and a number of other inorganic and organic electrode reactions (Refs. 29, 63, and 197–207). A few compilations of data exist.[208,209]

There is a significant number of articles in which the reported energies of activation are determined at two or more overpotentials (or potentials), mostly only for either the cathodic or the anodic direction of the electrode reaction (Refs. 48, 50, 51, 54, 61, 69–71, 75, and 210–238). These articles are classified separately because they contain additional information on the phenomenon, namely, the dependence of the energy of activation on overpotential, or potential.

Finally, there is a group of articles which deals principally with the various aspects of the experimental methods and procedures used for determination of energies of activation of electrode reactions.[239–246]

Reviewing the literature on the topic published during the past century, it is clear that significant progress in the elucidation of the phenomenon has been made since the work of Roszkowsky. However, basic scientific approaches leading to a better understanding of the phenomenon have been developed by only a few authors, namely, Roszkowsky, Bowden, Eyring, Glasstone, and Laidler, ending with the generalized formulations of Agar and Temkin. Since the work of Agar and Temkin, further conceptual advances toward a closer understanding of temperature effects in electrode kinetics have been lacking.

Electrode kinetics has not benefited appropriately from the body of evidence on the effect of temperature that has been accumulated to date. A careful review of the literature indicates that within approximately the last 20 years, interest in research on this topic has been fading. This may be interpreted as indicating that a saturation point has been reached in terms of what can be achieved with the theoretical and experimental methods that are presently available for elucidation of the effect of temperature on the rate of electrode reactions.

Some weak points, both theoretical and experimental, in the present treatment of the phenomenon will be discussed in this chapter. Some slightly different considerations will be proposed, offering additional information on the phenomenon, but it will be shown that there is still a need for further studies on this topic.

II. PRIOR ART

1. Basic Relationships

Let us consider a simple, reversible electrode reaction occurring at an inert metal electrode, in accordance with the stoichiometric equation

$$O \text{ (soln)} + e^- = R \text{ (soln)} \tag{1}$$

In electrode kinetics the rate of the reaction in Eq. (1) is usually evaluated in terms of the current density, j, which is affected by a number of physical variables, such as the electrical state of the interface (represented theoretically as the Galvani potential difference, $\Delta\phi$), temperature, concentration of reacting species, and pressure, as well as by more complex variables hidden within such phrases as the "nature of the electrode material" and the "nature of the solution":

$$j = J(\Delta\phi, T, c_i, p, \ldots) \tag{2}$$

Assuming that the kinetics of the reaction in Eq. (1) is characterized in both directions by rate laws of first order with respect to the corresponding reacting species, then in accordance with transition-state theory, the rate law is written as follows:[45]

$$j_c = F \cdot k_c \cdot c_O \cdot \exp(-\overline{\Delta G_c^{\#}}/RT) \tag{3a}$$

and

$$j_a = F \cdot k_a \cdot c_R \cdot \exp(-\overline{\Delta G_a^{\#}}/RT) \tag{3b}$$

where k is the rate constant, independent of the electrical state of the interface, c is concentration, $\overline{\Delta G^{\#}}$ is the electrochemical Gibbs energy of activation, and the subscripts c and a represent the cathodic and the anodic direction of the reaction in Eq. (1); F, R, and T have their usual meaning.

According to Guggenheim,[247] the electrochemical Gibbs energy of activation can be conceptually represented by

$$\overline{\Delta G^{\#}} = \Delta G_0^{\#} \pm \beta \cdot F \cdot \Delta\phi \tag{4}$$

where the plus sign holds for the cathodic direction and the minus sign for the anodic direction of the reaction in Eq. (1), β is the symmetry factor, satisfying the condition $\beta_c + \beta_a = 1$, and $\Delta G_0^{\#}$ is the ideal, or chemical, Gibbs energy of activation of the reaction in Eq. (1), that is, a constant part of the electrochemical Gibbs energy of activation that is independent of $\Delta\phi$ at the electrode/solution interface.[†]

When Eq. (4) is introduced into Eq. (3), one obtains

$$j_c = F \cdot k_c \cdot c_O \cdot \exp(-\Delta G_{0,c}^{\#}/RT) \cdot \exp(-\beta_c F \Delta\phi_c/RT)$$

$$= j_{z,c} \cdot \exp(-\beta_c F \Delta\phi_c/RT) \tag{5a}$$

and

[†]The Galvani potential difference (GPD) is defined as the difference of the inner electric potentials within the Helmholtz part of the double layer.[248] However, in solutions with ionic strengths above approximately 0.1 mol/dm³, the difference of the inner electric potentials between the outer plane of the Helmholtz layer and the bulk of the solution is close to zero. In such cases, the GPD may be approximated by the difference of the inner electric potentials of the electrode material and the solution, with the inner electric potential of the solution as the relative zero point in the evaluation of $\Delta\phi$.[249]

$$j_a = F \cdot k_a \cdot c_R \cdot \exp(-\Delta G_{0,a}^{\#}/RT) \cdot \exp(\beta_a F \Delta\phi_a/RT)$$

$$= j_{z,a} \cdot \exp(\beta_a F \Delta\phi_a/RT) \tag{5b}$$

where $\Delta\phi_c < \Delta\phi_e$ and $\Delta\phi_a > \Delta\phi_e$, and $\Delta\phi_e$ is the reversible or equilibrium Galvani potential difference (GPD) of the reaction in Eq. (1), and j_z is the characteristic current density at the reference GPD (in this case, the charge-free state of the interface, $\Delta\phi = 0$, for the reaction in the direction indicated by the subscript).

It is simplest to study the effect of temperature on the rate of the electrode reaction if all variables except temperature are held constant. Then, it follows from Eq. (2) that

$$\frac{dj}{dT} = \left(\frac{\partial J}{\partial T}\right)_{\Delta\phi, c_i, p, \ldots} \tag{6}$$

When derivatives of the logarithms of the rate laws given by Eqs. (5) are taken with respect to the reciprocal of the temperature, for $\Delta\phi = $ const and assuming $\beta = $ const[†], one obtains

$$\left(\frac{\partial \ln j_c}{\partial T^{-1}}\right)_{\Delta\phi, c_i, p, \ldots} = \left(\frac{\partial \ln j_{z,c}}{\partial T^{-1}}\right)_{\Delta\phi_c=0, c_i, p, \ldots} - (1/R) \cdot \beta_c \cdot F \cdot \Delta\phi_c \tag{7a}$$

and

$$\left(\frac{\partial \ln j_a}{\partial T^{-1}}\right)_{\Delta\phi_a, c_i, p, \ldots} = \left(\frac{\partial \ln j_{z,a}}{\partial T^{-1}}\right)_{\Delta\phi_a=0, c_i, p, \ldots} + (1/R) \cdot \beta_a \cdot F \cdot \Delta\phi_a \tag{7b}$$

If the analogy to the Arrhenius equation[19] is followed and symbols for energies of activation at a controlled GPD (indicated in parentheses) are introduced into Eqs. (7), one obtains[‡]

$$U^{\#}(\Delta\phi_c) = U^{\#}(\Delta\phi_c = 0) + \beta_c \cdot F \cdot \Delta\phi_c \tag{8a}$$

and

[†]There is positive evidence that in some cases β is a function of temperature,[95] but any sound theoretical explanation that questions the temperature independence of β in electrode kinetics is still missing. Hence, the classical approach is followed here.

[‡]The symbol $U^{\#}$ for the energy of activation of electrode reactions will be used in this chapter, in accordance with the IUPAC recommendations on nomenclature in electrochemistry.[250]

$$U^{\#}(\Delta\phi_a) = U^{\#}(\Delta\phi_a = 0) - \beta_a \cdot F \cdot \Delta\phi_a \qquad (8b)$$

where $U^{\#}(\Delta\phi = 0)$ is the energy of activation characterizing the reaction in Eq. (1) in the direction indicated by the subscript, at an electrical reference state of zero, that is, $\Delta\phi = 0$.

The rate laws given by Eqs. (5) can also be presented in the form

$$\Delta\phi_c = (RT/\beta_c F) \cdot \ln\left(\frac{j_{z,c}}{j_c}\right) \qquad (9a)$$

and

$$\Delta\phi_a = (RT/\beta_a F) \cdot \ln\left(\frac{j_a}{j_{z,a}}\right) \qquad (9b)$$

Equations (9) offer an alternative possibility for studying the effect of temperature on the kinetics of electrode reactions: the change of $\Delta\phi$ with temperature may be monitored while the current density, together with the other parameters defining the electrode and solution, are held constant. So, under these restrictions, taking the derivatives of Eqs. (9) with respect to temperature, one finally obtains

$$\beta_c \cdot F \cdot T \cdot \left(\frac{\partial\Delta\phi_c}{\partial T}\right)_{j_c, c_i, p, \ldots} = U^{\#}(\Delta\phi_c = 0) + \beta_c F \Delta\phi_c = U^{\#}(\Delta\phi_c) \qquad (10a)$$

and

$$\beta_a \cdot F \cdot T \cdot \left(\frac{\partial\Delta\phi_a}{\partial T}\right)_{j_a, c_i, p, \ldots} = U^{\#}(\Delta\phi_a = 0) - \beta_a F \Delta\phi_a = U^{\#}(\Delta\phi_a) \qquad (10b)$$

Hence, irrespective of the procedure followed, the same final result, namely, $U^{\#}(\Delta\phi)$, is obtained. In experimental electrode kinetics, however, the majority of the pertinent data in the literature have been obtained following procedures based principally on Eqs. (7) and (8), probably because, $U^{\#}(\Delta\phi)$, as defined by Eqs. (10), is dependent on continuous variation of $\Delta\phi$ with temperature, which additionally complicates the determination of $U^{\#}(\Delta\phi)$. In the following, only procedures based on Eqs. (7) will be discussed.

In Eqs, (4)–(10), the electrical state of the interface where the reaction in Eq. (1) occurs is evaluated on an operative or $\Delta\phi$ scale of potential, which is theoretically correct. However, the GPD is not open to experimental determination.[248] For instance, the electrical reference state $\Delta\phi = 0$, introduced when the derivatives of Eqs. (5) are taken with respect to the reciprocal of temperature, assumes the interface to be balanced with respect to all charged particles and dipoles present there, irrespective of the temperature. This clear definition of the nonelectrified state of the interface is not related to any particular state of any electrode reaction or pertinent charge-transfer process and cannot be experimentally realized. Since chemistry is not involved in any defined way in the establishment of the particular electrical state of the interface, the electrical reference point has a theoretical but not a practical significance. However, this important electrical reference state is a useless reference point in experimental electrochemistry, because the absolute value of any GPD at any real interface is not accessible experimentally. Hence, energies of activation related to any specified value of $\Delta\phi$ are experimentally indeterminate. Thus, another electrical reference state of the interface should be sought in evaluation of the effect of temperature on the rate of electrode reactions, in order to satisfy one of the basic requirements, namely, $\Delta\phi$ = const.

As is well known, only changes in the GPD are accessible experimentally:

$$\Delta(\Delta\phi) = \Delta\phi_{\text{final}} - \Delta\phi_{\text{initial}} = \Delta\phi_f - \Delta\phi_i \qquad (11)$$

At a number of electrodes, the equilibrium state of certain electrode reactions can be observed. The electrical state of these interfaces can be reproduced experimentally with a relatively high precision.[251] However, any absolute specification of the electrical state at these interfaces is inaccessible. On the other hand, the equilibrium state is thermodynamically well defined. Thus, under isothermal conditions it is possible to experimentally prepare interfaces where equilibrium of an electrode reaction can be assured, which is characterized by a constant and unknown GPD but is at the same time a well-defined thermodynamic state. As a consequence, $\Delta(\Delta\phi)$ can be determined experimentally in accordance with Eq. (11), if $\Delta\phi_i = \Delta\phi_e$. However, any experimental determination of $\Delta(\Delta\phi)$ in accordance with Eq. (11) assumes the use of at least two interfaces and the formation of an electrochemical cell that contains an experimental electrical reference interface. Then,

$$\Delta(\Delta\phi) = \Delta\phi - \Delta\phi_e \qquad (12)$$

Such an experimental electrical reference interface is termed in electrochemistry the reference electrode (RE), while the other interface in the cell is called the working electrode (WE).

Depending on the choice of the electrode reaction that is held in equilibrium at the RE, two basic scales of potential are encountered in isothermal electrochemistry as substitutes for $\Delta(\Delta\phi)$. If it is the same reaction as the one occurring at the WE, the scale of overpotential is established as

$$\eta = \Delta\phi - \Delta\phi_e \tag{13}$$

and if it is the hydrogen electrode reaction in equilibrium, held at standard conditions [standard hydrogen electrode (SHE)], the scale of relative electrode potential is established:[†]

$$E = \Delta\phi - \Delta\phi_{SHE} \tag{14}$$

The theoretical equations (5), (7), and (8) can be used in experimental electrode kinetics only if $\Delta\phi$ in these equations is replaced by an experimentally accessible substitute, such as overpotential (Eq. 13) or relative electrode potential (Eq. 14).

When Eq. (13) is introduced into Eqs. (5), it follows that

$$j_c = F \cdot k_c \cdot c_O \cdot \exp(-\Delta G_{0,c}^{\#}/RT) \cdot \exp(-\beta_c F\Delta\phi_e/RT) \cdot \exp(-\beta_c F\eta_c/RT)$$

$$= j_0 \cdot \exp(-\beta_c F\eta_c/RT) \tag{15a}$$

and

$$j_a = F \cdot k_a \cdot c_R \cdot \exp(-\Delta G_{0,a}^{\#}/RT) \cdot \exp(\beta_a F\Delta\phi_e/RT) \cdot \exp(\beta_a F\eta_a/RT)$$

$$= j_0 \cdot \exp(\beta_a F\eta_a/RT) \tag{15b}$$

where j_0 is a characteristic current density at the electrical reference state $\eta = 0$. It is equal for both directions of the reaction in Eq. (1) and is usually called the exchange current density. Then,

$$\left(\frac{\partial \ln j_c}{\partial T^{-1}}\right)_{\eta_c, c_i, p, \ldots} = \left(\frac{\partial \ln j_0}{\partial T^{-1}}\right)_{\eta=0, c_i, p, \ldots} - (1/R) \cdot \beta_c \cdot F \cdot \eta_c \tag{16a}$$

[†]The scale of relative electrode potentials, which is also known as the SHE scale of potential, is based on the convention that $\Delta\phi_{SHE} = 0.000$ V (SHE).

and

$$\left(\frac{\partial \ln j_a}{\partial T^{-1}}\right)_{\eta_a, c_i, p, \ldots} = \left(\frac{\partial \ln j_0}{\partial T^{-1}}\right)_{\eta=0, c_i, p, \ldots} + (1/R) \cdot \beta_a \cdot F \cdot \eta_a \quad (16b)$$

Finally, when symbols for the corresponding energies of activation are introduced into Eqs. (16), one obtains

$$U^{\#}(\eta_c) = U^{\#}(\eta = 0) + \beta_c \cdot F \cdot \eta_c \quad (17a)$$

and

$$U^{\#}(\eta_a) = U^{\#}(\eta = 0) - \beta_a \cdot F \cdot \eta_a \quad (17b)$$

Analogously, when Eq. (14) is introduced into Eq. (5), it follows that

$$j_c = F \cdot k_c \cdot c_O \cdot \exp(-\Delta G_{0,c}^{\#}/RT)$$

$$\cdot \exp(-\beta_c F \Delta\phi_{SHE}/RT) \cdot \exp(-\beta_c F E_c/RT)$$

$$= j_{r,c} \cdot \exp(-\beta_c F E_c/RT) \quad (18a)$$

and

$$j_a = F \cdot k_a \cdot c_R \cdot \exp(-\Delta G_{0,a}^{\#}/RT)$$

$$\cdot \exp(\beta_a F \Delta\phi_{SHE}/RT) \cdot \exp(\beta_a F E_a/RT)$$

$$= j_{r,a} \cdot \exp(\beta_a F E_a/RT) \quad (18b)$$

where $j_{r,c}$ and $j_{r,a}$ are characteristic current densities at the electrical reference states, for example, at $E_c = 0$ and $E_a = 0$, respectively. Then,

$$\left(\frac{\partial \ln j_c}{\partial T^{-1}}\right)_{E_c, c_i, p, \ldots} = \left(\frac{\partial \ln j_{r,c}}{\partial T^{-1}}\right)_{E_c=0, c_i, p, \ldots} - (1/R) \cdot \beta_c \cdot F \cdot E_c \quad (19a)$$

and

$$\left(\frac{\partial \ln j_a}{\partial T^{-1}}\right)_{E_a, c_i, p, \ldots} = \left(\frac{\partial \ln j_{r,a}}{\partial T^{-1}}\right)_{E_a=0, c_i, p, \ldots} + (1/R) \cdot \beta_a \cdot F \cdot E_a \quad (19b)$$

Finally, when the corresponding definitions for the energies of activation are introduced into Eqs. (19), one obtains

Table 1

Various Experimentally Accessible Energies of Activation Discriminated with Respect to Direction of Reaction and Scale and Value of Potential Used to Control the Electrical State of the Working Electrode

Control of electrical state of WE	Symbol and relationship	Remarks
$\eta = 0$	$U^{\#}(\eta=0) = -R \cdot (\partial \ln j_0/\partial T^{-1})_{\eta=0,\, c_i,\, p,\ldots}$	$U^{\#}(\eta_c=0) = U^{\#}(\eta_a=0) = U^{\#}(\eta=0)$
$\eta_c = \text{const}$	$U^{\#}(\eta_c) = -R \cdot (\partial \ln j_c/\partial T^{-1})_{\eta_c,\, c_i,\, p,\ldots}$	$U^{\#}(\eta_c=0) = U^{\#}(\eta=0) + \beta_c \cdot F \cdot \eta_c$
$\eta_a = \text{const}$	$U^{\#}(\eta_a) = -R \cdot (\partial \ln j_a/\partial T^{-1})_{\eta_a,\, c_i,\, p,\ldots}$	$U^{\#}(\eta_a=0) = U^{\#}(\eta=0) + \beta_a \cdot F \cdot \eta_a$
$E_c = 0$	$U^{\#}(E_c=0) = -R \cdot (\partial \ln j_{r,c}/\partial T^{-1})_{E_c=0,\, c_i,\, p,\ldots}$	$j_{r,c} = j(E_c=0)$
$E_a = 0$	$U^{\#}(E_a=0) = -R \cdot (\partial \ln j_{r,a}/\partial T^{-1})_{E_a=0,\, c_i,\, p,\ldots}$	$j_{r,a} = j(E_a=0)$
$E_c = \text{const}$	$U^{\#}(E_c=0) = -R \cdot (\partial \ln j_c/\partial T^{-1})_{E_c,\, c_i,\, p,\ldots}$	$U^{\#}(E_c) = U^{\#}(E_c=0) + \beta_c \cdot F \cdot E_c$
$E_a = \text{const}$	$U^{\#}(E_a=0) = -R \cdot (\partial \ln j_a/\partial T^{-1})_{E_a,\, c_i,\, p,\ldots}$	$U^{\#}(E_a) = U^{\#}(E_a=0) - \beta_a \cdot F \cdot E_a$

$$U^{\#}(E_c) = U^{\#}(E_c = 0) + \beta_c \cdot F \cdot E_c \tag{20a}$$

and

$$U^{\#}(E_a) = U^{\#}(E_a = 0) - \beta_a \cdot F \cdot E_a \tag{20b}$$

From the theoretical point of view, energies of activation, as defined either by Eqs. (16) and (17) or by Eqs. (19) and (20), are experimentally accessible. Various definitions of the experimental energies of activation are presented in Table 1.

2. Varieties of Terms

The various definitions of energies of activation in experimental electrode kinetics, as given in Table 1, make it necessary to have an adequate terminology that makes proper distinctions among them. When the literature is consulted in this respect, a rather confusing situation is encountered. Problems start with the basic term, energy of activation, since several other terms are used synonymously, as presented in Table 2.

An even more chaotic situation is found when energies of activation are referred to a specified electrical state of the interface. The whole spectrum of terms, from the simplest (those from Table 2), without any indication of the pertinent constant potential or overpotential, to various more complex terms can be found in the literature for the single definition of EA. The compilation of terms presented in Table 3 illustrates the variety of terminology for $U^{\#}(\eta = 0)$ and $U^{\#}(E)$ alone.

This free-for-all in the selection of terms must be abandoned in experimental electrode kinetics to eliminate confusion. In this regard, IUPAC has published a document[250] in which it is recommended that the

Table 2
Terms Used as Synonyms for the Energy of Activation (EA) and Their Abbreviations

Term	Abbreviation
Activation energy	AE
Heat of activation	HA
Activation heat	AH
Enthalpy of activation	EnA

terms for any EA of electrode reactions should be formulated as follows: As the basic term, energy of activation should be used, and a clear indication of the pertinent constant potential (overpotential or electrode potential) must follow the basic term. Thus, full information has to be provided by both the symbol and the term used for any experimentally accessible EA of electrode reactions. For instance, in the case of electrode reactions occurring in the cathodic direction at constant overpotential, the symbol is $U^{\#}(\eta_c)$, and the corresponding term is the energy of activation at constant cathodic overpotential. For the experimentally accessible EAs in Table 1, the corresponding symbols and terms, based on the IUPAC recommendations, are given in Table 4. These terms are rather descriptive, and Table 4 includes appropriate abbreviations that will be used in this chapter.

Accounting for the diversity of experimentally accessible EAs is not simply a matter of specifying the choice of potential scale and indicating the value of the potential and the direction of reaction. Further distinctions are made by specifying the experimental conditions under which the EAs are determined. These conditions are defined by:

- The type of WE–RE cell
- The temperature regime in the WE–RE cell

The WE–RE cell may be of two basic types: (i) the simplest cell, a cell without transference,[†] and (ii) a cell with transference.[‡]

When the effect of temperature on the rate of electrode reactions at a constant electrical state of the WE is determined, there are two basic experimental techniques which may be distinguished with respect to the temperature regime used to control the temperature of the electrodes in the WE–RE cell: (i) the technique of *isothermal* cells, in which both the WE and the RE are kept at a common temperature throughout the measurements, and (ii) the technique of *nonisothermal* cells, in which the RE is kept at an arbitrarily selected constant temperature during all

---

[†]The simplest cell in electrochemistry is the cell without transference, which is composed of a WE and an RE made of the same electrode material and both immersed in a common solution.

[‡]Cells with transference are generally composed of a WE and an RE made of different electrode materials, and each is immersed in its own solution, but these two solutions are in contact.

Table 3

Variety of Terms Encountered in the Literature for the Energy of Activation at a Single Controlled Electrical State of the Working Electrode

Control of electrical state of WE	Term	Reference(s)
$\eta = 0$	EA	18,53,78,105,180,202
	AE	57,89–91,109,133,191,193,205,206,216
	HA	34,36–38,41,46,67,77,92,116,130,163,184
	AH	88
	EnA	133,187
	Apparent EA	78,79,82,102,110,131,136,182,194
	Apparent AE	185
	Apparent HA	104,150,186
	Real HA	78,93,94,97
	Standard HA	133,141,153,166,172,187
	Standard EnA	86,104
	EA at the reversible potential	71,97,128,161,219
	AE at the reversible potential	126,127,190,193,195,197,201,239
	HA at the reversible potential	121,132,178,235
	EnA at the reversible potential	31,40,42,45,62,76,182,187,192
	EnA at zero polarization	176
	Real EA at the reversible potential	27
	Apparent HA at the reversible potential	125,138,139,189
	Virtual HA at the reversible potential	45

$E = \text{const}$	
EA	29,30,39,223
HA	26
Relative EA	66
Apparent AE	169,197,207,209,233–235
Arbitrary AE	246
AE at a given potential	75,217
AE at a given voltage	95,152,216
Arrhenius EA at the potential V	161,232
Effective EA at constant potential	61,160

Table 4
Recommended Symbols, Names, and Abbreviations for Various Experimentally Accessible Energies of Activation of Electrode Reactions

Potential scale	Value of potential	Symbol	Energy of activation	
			Recommended name	Abbreviation
Overpotential	$\eta = 0$	$U^{\#}(\eta = 0)$	Energy of activation at zero overpotential	EAZOP
	$\eta_c = \text{const}$	$U^{\#}(\eta_c)$	Energy of activation at constant cathodic overpotential	EACOP-c
	$\eta_a = \text{const}$	$U^{\#}(\eta_a)$	Energy of activation at constant anodic overpotential	EACOP-a
Relative electrode potential	$E_c = 0$	$U^{\#}(E_c = 0)$	Energy of activation at zero electrode potential for cathodic reaction	EAZEP-c
	$E_a = 0$	$U^{\#}(E_a = 0)$	Energy of activation at zero electrode potential for anodic reaction	EAZEP-a
	$E_c = \text{const}$	$U^{\#}(E_c)$	Energy of activation at constant cathodic electrode potential	EACEP-c
	$E_a = \text{const}$	$U^{\#}(E_a)$	Energy of activation at constant anodic electrode potential	EACEP-a

measurements, and the WE is thermostated at various constant temperatures within the range examined.

The various combinations of WE–RE cell type and temperature regime that may be employed in EA determinations are schematically represented in Fig. 1.

Assuming that the same electrode reaction, in the same direction and at the same constant electrical state of the interface, occurs at the WE in all cases presented in Fig. 1, let us consider how the cell type and temperature regime used directly affect the measured emf and consequently the corresponding EA. In the case presented in Fig. 1a, which corresponds to the simplest cell with an isothermal regime, the measured emf is equal to

$$\text{emf} = \Delta\phi_{WE} - \Delta\phi_{RE} = \eta \tag{21}$$

irrespective of the temperature of the cell.

However, if a cell with transference is used in an isothermal regime, as presented in Fig. 1b, the measured emf is more complex:

$$\text{emf} = \Delta\phi_{WE} - \Delta\phi_{RE} + \Delta\phi_D + \Delta\phi_C = \eta + \Delta\phi_D + \Delta\phi_C \tag{22}$$

where $\Delta\phi_D$ is the experimentally inaccessible difference between the inner potentials of two isothermal solutions of different composition (called the diffusion or junction potential), and $\Delta\phi_C$ is the experimentally inaccessible difference between the inner potentials of two isothermal electrode materials of different composition (called the contact potential).

When the simplest cell is used in a nonisothermal regime, as presented in Fig. 1c, the measured emf is composed of the following terms:

$$\text{emf} = \Delta\phi_{WE}(T) - \Delta\phi_{RE}(T_0) + \Delta\phi_T \tag{23}$$

where $\Delta\phi_T$ is the experimentally inaccessible GPD at the interface formed by two solutions of the same composition but different temperatures.

Finally, if the cell with transference is used in a nonisothermal regime (Fig. 1d), the measured emf is more complex:

$$\text{emf} = \Delta\phi_{WE}(T) - \Delta\phi_{RE}(T_0) + \Delta\phi_D(T_0) + \Delta\phi_C(T_0) + \Delta\phi_T \tag{24}$$

where $\Delta\phi_D(T_0)$, $\Delta\phi_C(T_0)$, and $\Delta\phi_T$ have the same meanings as above. In this case, only the terms $\Delta\phi_{WE}$ and $\Delta\phi_T$ are affected by the temperature regime in the cell, because all other terms are referred to the constant temperature, T_0, at which the RE is held.

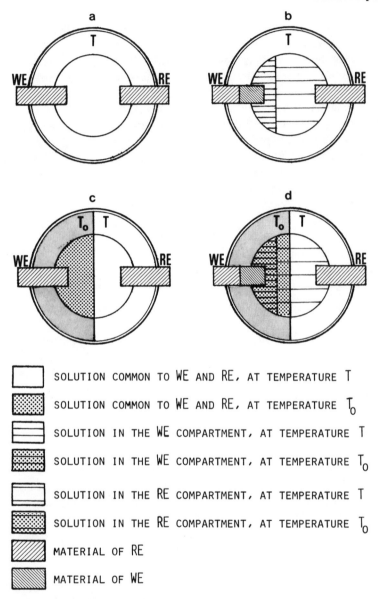

Figure 1. Various temperature regimes and compositions of the WE–RE cell. (a) Isothermal cell without transference; (b) isothermal cell with transference; (c) nonisothermal cell without transference; (d) nonisothermal cell with transference.

When expressions for $\Delta\phi_{WE}$ from Eqs. (21)–(24) are substituted for $\Delta\phi_c$ and $\Delta\phi_a$ in Eqs. (5), and derivatives taken with respect to the reciprocal of temperature, additional definitions of experimentally accessible EAs, beyond those in Table 4, are obtained. In Table 5 are listed all varieties of experimentally accessible EAs, where the choice of the cell type and of the temperature regime in the WE–RE cell is indicated by the left subscript to the corresponding symbol for EA. Due to the fact that each set of experimental conditions specified in Table 5 is characterized by a particular definition of EA, it is necessary to introduce appropriate additional symbols, names, and abbreviations, in order to distinguish among them. Therefore, experimental conditions pertinent to the cell type and temperature regime must be part of the symbol, name, and abbreviation for each of the experimentally accessible EAs. The following symbols are proposed:

i: isothermal regime in the WE–RE cell
n: nonisothermal regime in the WE–RE cell
s: simplest WE–RE cell, or cell without transference
t: WE–RE cell with transference

Further, it is proposed that these symbols precede the basic symbol of EA as a left subscript, in the manner presented in Table 5. Now, both the symbol and the full name of any particular experimentally accessible EA of electrode reactions must by rather extensive, to avoid any confusion in identification of the EA. For instance, $_{is}U^{\#}(E_c)$ (see Table 5) is named as follows: energy of activation at constant cathodic electrode potential, determined in the isothermal cell without transference. Because the use of such definitions to describe the pertinent EA for the reaction studied under specified experimental conditions is cumbersome, it is useful to introduce appropriate abbreviations. Thus, in the above-mentioned example, the abbreviation proposed is is-EACEP-c. Such abbreviations cannot be shortened further if full information concerning the EA is to be instantly conveyed.

Among the various definitions of experimentally accessible EAs listed in Table 5, the majority contain certain experimentally indeterminate parameters that have nothing to do with the reaction occurring at the WE. Therefore, the significance of some definitions can be questioned. This has been done already by Frumkin and Krishtalik,[252] who deny the physical significance of all EAs determined in cells with transference or in a nonisothermal regime. Although such a critical judgment is formally

Table 5

Recommended Symbols and Abbreviations for Experimentally Accessible Energies of Activation, Containing Information on the Potential Scale and Value of the Potential, Direction of Reaction and Temperature Regime in the WE–RE Cell, and Type of WE–RE Cell

Potential scale[a]	DR[b]	TR[c]	TC[d]	Symbol and relationship to electrical state of the WE	Abbreviation
Overpotential	c	i	s	$_{is}U^{\#}(\eta_c) = {}_{is}U^{\#}(\eta_c = 0) + \beta_c F\eta_c$	is-EACOP-c
	c	i	t	$_{it}U^{\#}(\eta_c) = {}_{it}U^{\#}(\eta_c = 0) + \beta_c F(\eta_c - \Delta\phi_C) + \beta_c FT[(d\Delta\phi_D/dT) + (d\Delta\phi_C/dT)]$	it-EACOP-c
	c	n	s	$_{ns}U^{\#}(\eta_c) = {}_{ns}U^{\#}(\eta_c = 0) + \beta_c F(\eta_c - \Delta\phi_T) + \beta_c FT[(d\Delta\phi_T/dT)]$	ns-EACOP-c
	c	n	t	$_{nt}U^{\#}(\eta_c) = {}_{nt}U^{\#}(\eta_c = 0) + \beta_c F[\eta_c - \Delta\phi_D(T_0) - \Delta\phi_C(T_0) - \Delta\phi_T] + \beta_c FT(d\Delta\phi_T/dT)$	nt-EACOP-c
	a	i	s	$_{is}U^{\#}(\eta_a) = {}_{is}U^{\#}(\eta_a = 0) - \beta_a F\eta_a$	is-EACOP-a
	a	i	t	$_{it}U^{\#}(\eta_a) = {}_{it}U^{\#}(\eta_a = 0) + \beta_a F[\eta_a - \Delta\phi_D - \Delta\phi_C) + \beta_a FT[(d\Delta\phi_D/dT) + (d\Delta\phi_C/dT)]$	it-EACOP-a
	a	n	s	$_{ns}U^{\#}(\eta_a) = {}_{ns}U^{\#}(\eta_a = 0) + \beta_a F[\eta_a - \Delta\phi_T) + \beta_a FT[(d\Delta\phi_T/dT)]$	ns-EACOP-a
	a	n	t	$_{nt}U^{\#}(\eta_a) = {}_{nt}U^{\#}(\eta_a = 0) + \beta_a F[\eta_a - \Delta\phi_D(T_0) - \Delta\phi_C(T_0) - \Delta\phi_T] + \beta_a FT(d\Delta\phi_T/dT)$	nt-EACOP-a
Relative electrode potential	c	i	s	$_{is}U^{\#}(E_c) = {}_{is}U^{\#}(E_c = 0) + \beta_c FE_c$	is-EACEP-c
	c	i	t	$_{it}U^{\#}(E_c) = {}_{it}U^{\#}(E_c = 0) + \beta_c F(E_c - \Delta\phi_D - \Delta\phi_C) + \beta_c FT[(d\Delta\phi_D/dT) + (d\Delta\phi_C/dT)]$	it-EACEP-c
	c	n	s	$_{ns}U^{\#}(E_c) = {}_{ns}U^{\#}(E_c = 0) + \beta_c F(E_c - \Delta\phi_T) + \beta_c FT[(d\Delta\phi_T/dT)]$	ns-EACEP-c
	c	n	t	$_{nt}U^{\#}(E_c) = {}_{nt}U^{\#}(E_c = 0) + \beta_c F[E_c - \Delta\phi_D(T_0) - \Delta\phi_C(T_0) - \Delta\phi_T] + \beta_c FT(d\Delta\phi_T/dT)$	nt-EACEP-c
	a	i	s	$_{is}U^{\#}(E_a) = {}_{is}U^{\#}(E_a = 0) - \beta_a FE_a$	is-EACEP-a
	a	i	t	$_{it}U^{\#}(E_a) = {}_{it}U^{\#}(E_a = 0) + \beta_a F[E_a - \Delta\phi_D - \Delta\phi_C) + \beta_a FT[d\Delta\phi_D/dT) + (d\Delta\phi_C/dT)]$	it-EACEP-a
	a	n	s	$_{ns}U^{\#}(E_a) = {}_{ns}U^{\#}(E_a = 0) - \beta_a F[E_a - \Delta\phi_T) + \beta_a FT(d\Delta\phi_T/dT)$	ns-EACEP-a
	a	n	t	$_{nt}U^{\#}(E_a) = {}_{nt}U^{\#}(E_a = 0) - \beta_a F[E_a - \Delta\phi_D(T_0) - \Delta\phi_C)(T_0) - \Delta\phi_T] + \beta_a FT(d\Delta\phi_T/dT)$	nt-EACEP-a

[a]Choice of experimentally accessible potential scale for the WE.
[b]Direction of the reaction occurring at the WE: c, cathodic; a, anodic.
[c]Temperature regime in the WE–RE cell: i, isothermal; n, nonisothermal.
[d]Type of WE–RE cell: s, simplest (cell without transference); t, cell with transference.

correct, it should be accepted with a certain reserve, in view of the practical experimental aspects of this problem. First of all, an EA cannot be determined in the simplest isothermal cell for any electrode reaction, in spite of the recommendations by IUPAC that $_{is}U^{\#}(\eta = 0)$ is the only correctly experimentally determined EA for electrode reactions.[250] It is not a simple task to satisfy the fundamental theoretical demands contained in the basic definition of EA [see Eqs. (7) and (8)] when an appropriate experimental procedure for EA determination is designed and implemented.

When data on EAs for electrode reactions in the literature are analyzed with respect to the diversity of definitions given in Table 5, some rather discouraging conclusions are reached. The majority of the data is categorized simply as $U^{\#}(\eta = 0)$, because explicit indications of the WE–RE cell type and the temperature regime used are mostly missing. The only information that is certain is that all these EAs are pertinent to WEs held in an equilibrium state. Hence, a classification of data on EAs from the literature with respect to the criteria contained in Table 5 cannot be made.

III. SIGNIFICANCE OF EA DETERMINED AT CONTROLLED OVERPOTENTIAL IN ISOTHERMAL CELLS WITHOUT TRANSFERENCE

It follows from Table 5 that EAs determined in the simplest cell in an isothermal regime, with overpotential as the experimental substitute for $\Delta\phi$ at WE, that is,

$$_{is}U^{\#}(\eta_c) = {}_{is}U^{\#}(\eta = 0) + \beta_c \cdot F \cdot \eta_c \tag{25a}$$

and

$$_{is}U^{\#}(\eta_a) = {}_{is}U^{\#}(\eta = 0) - \beta_a \cdot F \cdot \eta_a \tag{25b}$$

are the most convenient EAs for experimental determinations.

As stated earlier (see Section II.1), zero overpotential ($\eta = 0$) is assumed as the electrical reference point when Eqs. (17) and (25) are derived. For reversible electrode reactions, the equilibrium state ($\eta = 0$) is experimentally accessible, constant, and reproducible, and, in accordance with the definition of overpotential given by Eq. (13), it is formally temperature independent. An electrical reference point $\eta = 0$ is easily attained at any temperature within the WE–RE cell because temperature-induced changes of the electrical state of the WE are simultaneously compensated by identical changes at the RE, provided the equilibrium state of the reaction studied is maintained at both electrodes in the cell.

One of the principal requirements when the effect of temperature on the rate of electrode reactions is evaluated is to keep $\Delta\phi_{WE}$ = const at any temperature within the range studied [see Eq. (7)]. Is this requirement really satisfied when $\eta = 0$ is selected as the electrical reference point and the overpotential is the experimental substitute for $\Delta(\Delta\phi)$ in the WE–RE cell?

It follows from Eqs. (13) and (15)–(17) that

$$_{is}U^{\#}(\eta = 0) = \Delta H_{0,c}^{\#} + \beta_c \cdot F \cdot \Delta\phi_e = \Delta H_{0,a}^{\#} - \beta_a \cdot F \cdot \Delta\phi_e \qquad (26)$$

Hence, an is-EAZOP, as well as any is-EACOP, depends on $\Delta\phi_e$ of the reaction studied at the WE. However, the temperature independence of $\eta = 0$ does not imply a temperature-independent equilibrium state of the reaction considered, that is, $d\Delta\phi_e/dT = 0$. Any compilation of temperature coefficients of standard electrode potentials[253] clearly indicates that the equilibrium state of electrode reactions is generally affected by temperature, which means that electrode reactions are characterized by $\Delta_r S^{\ominus} \neq 0$.

If the electrical state of electrodes in the WE–RE cell is referred to any arbitrarily selected temperature, T_0, the equilibrium state of the electrode reaction occurring simultaneously at both electrodes will be affected by temperature as follows:

$$\Delta\phi_{WE}(T) = \Delta\phi_{WE}(T_0) + \int_{T_0}^{T} \left(\frac{\partial\Delta\phi_e}{\partial T}\right)_{c_i,p,\ldots} \cdot dT \qquad (27)$$

and

$$\Delta\phi_{RE}(T) = \Delta\phi_{RE}(T_0) + \int_{T_0}^{T} \left(\frac{\partial\Delta\phi_e}{\partial T}\right)_{c_i,p,\ldots} \cdot dT \qquad (28)$$

When the WE and the RE are held in equilibrium in the simplest isothermal cell, the electrical reference state will be always maintained, irrespective of the temperature of the cell, because at each temperature the equality $\Delta\phi_{WE} = \Delta\phi_{RE}$ is satisfied. However, if the electrical state of the WE is evaluated on the $\Delta\phi$ scale of potential under the same conditions,

$\Delta\phi_{WE}(T) \neq \Delta\phi_{WE}(T_0)$, which means that the principal requirement $\Delta\phi_{WE} = \text{const}$ is not satisfied when an EA is determined in the simplest isothermal cell, with the overpotential as the experimental substitute for $\Delta(\Delta\phi)$.

Selection of the equilibrium state of electrode reactions as the electrical reference point in determinations of EA, in accordance with Eqs. (16), is consequently appropriate when this state is treated thermodynamically. A zero overpotential has the properties of a thermodynamic reference state also, which means that at $\eta = 0$, all pertinent thermodynamic functions of the electrode reaction at the WE should have zero values. Consequently, when the temperature effect on the rate of electrode reactions is studied in the simplest isothermal cells with overpotential as the experimental substitute for $\Delta(\Delta\phi)$, the equilibrium state of any electrode reaction is characterized thermodynamically by $\Delta_r S^\circ = 0$, $\Delta_r H^\circ = 0$, and hence $\Delta_r G^\circ = 0$.[†] Due to the fact that at $\eta = 0$ the same value of is-EAZOP characterizes the reaction in both directions, and following the Arrhenius concept of EA,[19] one obtains

$$\Delta_r H^\circ = {}_{is}U^{\#}(\eta_c = 0) - {}_{is}U^{\#}(\eta_a = 0) = 0 \tag{29}$$

This result is consistent with $\eta = 0$ as both the electrical and the thermodynamic reference state, but it is not consistent with the existing experimental evidence on $\Delta_r H^\ominus$, which indicates that electrode reactions are not thermoneutral reactions.

On the overpotential scale, with $\eta = 0$ as the reference point, the equilibrium state of all electrode reactions is characterized by $\Delta_r G^\circ = 0$. Thus, all electrode reactions are "leveled" in the equilibrium state. The same conclusion follows from the dependence of is-EACOPs on overpotential, in accordance with Eqs. (25), as presented in Fig. 2.

The electrochemical enthalpy of an electrode reaction, for instance, in the cathodic direction, can be defined as follows:

$$\overline{\Delta_{rc}H}^\circ(\eta_c) = {}_{is}U^{\#}(\eta_c) - {}_{is}U^{\#}(\eta = 0) = \beta_c \cdot F \cdot \eta_c \tag{30a}$$

and analogously for the anodic direction:

[†]It should be noted that the right superscript, signifying the standard state of thermodynamic functions, is pertinent to the reference state considered, that is, $\eta = 0$; it should not be confused with the usual standard state symbols used in chemical thermodynamics ($\Delta_r S^\ominus$, $\Delta_r H^\ominus$, etc.), based on the SHE as the electrical reference state.

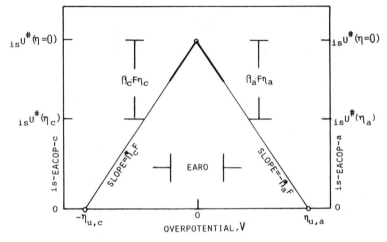

Figure 2. Schematic dependences of $_{is}U^{\#}(\eta)$ on overpotential for the cathodic and anodic directions of the reaction in Eq. (1). Temperature-independent symmetry factors and $\beta_c = \beta_a = 0.5$ are assumed. EARO, Experimentally accessible region of overpotential.

$$\overline{\Delta_{ra}H}°(\eta_a) = {}_{is}U^{\#}(\eta_a) - {}_{is}U^{\#}(\eta = 0) = -\beta_a \cdot F \cdot \eta_a \qquad (30b)$$

It may be concluded that in the case $\beta_c = \beta_a = 0.5$, the overpotential symmetrically affects the rate of all electrode reactions that may be represented by means of the stoichiometric equation (1), as presented in Fig. 2.

One of the characteristic parameters of any anodic or cathodic reaction is a particular value of the overpotential η_u, which is arbitrarily named the *ultimate overpotential*. As shown in Fig. 2, at their respective ultimate overpotentials both *is*-EACOP-c and *is*-EACOP-a reach their boundary value of zero.

The dependence of *is*-EACOP on overpotential may alternatively be presented as the effect of overpotential on the form of the energy profiles, which is schematically shown in Fig. 3 for the range of overpotential $0 < \eta < \eta_u$ for both the cathodic and anodic directions of the reaction in Eq. (1). All energy profiles in Fig. 3 are referred to the equilibrium state, characterized by zero change of standard enthalpy of reaction.

As follows from Figs. 2 and 3, when either direction of the reaction in Eq. (1) occurs at the corresponding ultimate overpotential, it is converted to a barrierless process.[97] Hence, at the ultimate overpotential for

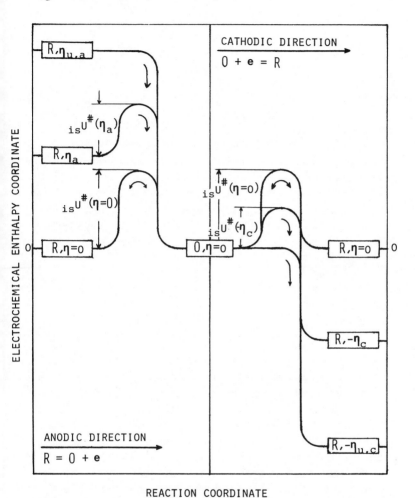

Figure 3. Energy profiles of the reaction in Eq. (1) at few characteristic values of the overpotential. It is assumed that the reaction occurs in an isothermal WE–RE cell without transference and is spontaneous in the cathodic and anodic directions presented. Energies on the ordinate are relative, referred to $\overline{\Delta_f H^\circ}$ $(0x, \eta = 0) = 0$, for presentation purposes only.

a given direction of the reaction, the WE/solution interface behaves as completely permeable for charge transfer in that direction, but perfectly irreversible in the reverse direction.

A definition of the ultimate overpotential follows from Eqs. (25):

$$\eta_{u,c} = -\frac{_{is}U^{\#}(\eta = 0)}{\beta_c \cdot F} \qquad (31a)$$

and

$$\eta_{u,a} = \frac{_{is}U^{\#}(\eta = 0)}{\beta_a \cdot F} \qquad (31b)$$

When the reaction in Eq. (1) occurs in the simplest isothermal cell, with $\beta_c = \beta_a = 0.5$, equal absolute values for the ultimate overpotentials are obtained, in accordance with perfect symmetry in the $_{is}U^{\#}(\eta)-\eta$ relationships for the cathodic and anodic directions. When Eqs. (31) are introduced into Eqs. (15), the corresponding ultimate current densities j_u are obtained, representing the highest possible rates of the reaction considered. In the case $\beta_c = \beta_a = 0.5$, $j_{u,c}$ and $j_{u,a}$ are symmetric with respect to the exchange current density j_0:

$$j_{u,c} = j_0 \cdot \exp(-\beta_c F\eta_{u,c}/RT) = j_0 \cdot \exp(_{is}U^{\#}(\eta = 0)/RT) \qquad (32a)$$

and

$$j_{u,a} = j_0 \cdot \exp(\beta_a F\eta_{u,a}/RT) = j_0 \cdot \exp(_{is}U^{\#}(\eta = 0)/RT) \qquad (32b)$$

It should be emphasized that for the reaction in Eq. (1), under the restrictions already mentioned, a pair of ultimate parameters (j_u,η_u), for both directions, can alternatively be obtained from the intersection points of the extrapolated Tafel lines at various temperatures, as first observed by Post and Hiskey[34] and schematically presented in Fig. 4. It may be concluded from Fig. 4 that the ultimate overpotentials possess the singular property of being temperature-independent.

Looking for experimental data to support a $_{is}U^{\#}(\eta)-\eta$ dependence, as given by Eqs. (25) and presented schematically in Fig. 2, led to the realization that such data are missing in the literature. Most of the available data on is-EACOP-c and is-EACOP-a for a single electrode reaction are referred to overpotentials where diffusion significantly affects the rate of the reaction in both directions (Refs. 48, 50, 51, 54, 69–71, 122, 211–213, 215, 220–222, 224, 228, 231, and 241). There are data on is-EACOP at a few activation overpotentials for certain reactions, but only in one direction (Refs. 75, 197, 217, 218, 223, 227, 229, and 232–236). Closer analysis of this evidence indicated that the majority of these data have been determined either in nonisothermal WE–RE cells or within a rather narrow

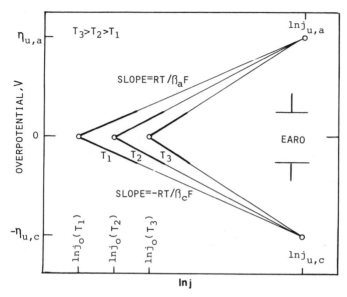

Figure 4. Schematic representation of Tafel lines for the cathodic and anodic directions of the reaction in Eq. (1), occurring in an isothermal WE–RE cell without transference, at various temperatures. It is assumed that $\beta_c = \beta_a = 0.5$ and that both symmetry factors are independent of temperature. EARO, Experimentally accessible region of overpotential.

region of overpotential. However, the kinetics of the hydrogen evolution reaction at mercury in acid solutions has been studied in a relatively wide region of overpotentials, but in the cathodic direction only. A number of papers reporting the effect of temperature on the kinetics of this reaction, studied in isothermal cells without transference,[16,31,34,110,138,229] contain experimental evidence that can be used to calculate values of $_{is}U^{\#}(\eta_c)$ at the various overpotentials at which these measurements were performed. The results obtained are presented in Fig. 5. A straight line with the theoretically predicted slope (i.e., $\beta_c \cdot F$, with $\beta_c = 0.5$) provides a good fit to the set of experimental data, in accordance with Eq. (25a). A representative $_{is}U^{\#}(\eta = 0)$ value of 90.5 kJ/mol is obtained from Fig. 5, which is close to the average value of the available data. It follows from Fig. 5 also that the cathodic ultimate overpotential for the hydrogen evolution reaction at mercury in acid solutions is −1.88 V; the same value was reported by Post and Hiskey[34] but was determined from extrapolated

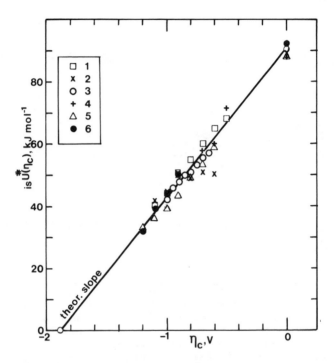

Figure 5. Experimental dependence of $_{is}U^{\#}(\eta_c)$ on η for the hydrogen evolution reaction at mercury in various acid solutions. □, 1, Iofa and Stepanova[16] (1.0M HCl), recalculated; x, 2, Bockris and Parsons[31] (0.1M HCl), recalculated; ○, 3, Post and Hiskey[34] (0.1M HCl), recalculated; +, 4, Tsionskii and Kriksunov[110] (0.3M HCl), recalculated; △, 5, Dekker et al.[138] (1.0M HCl), recalculated; ●, 6, Potapova et al.[229] (0.1M HClO$_4$ + 0.9M NaClO$_4$), original data.

Tafel lines at various temperatures. It is well known that the hydrogen electrode reaction in the anodic direction cannot be studied over a wide range of overpotentials because of certain limitations that have not been elucidated completely as yet. Therefore, the anodic branch of the $_{is}U^{\#}(\eta)-\eta$ dependence in Fig. 5 is missing.

It follows from Eq. (26) that is-EAZOPs are composed of two terms, and each of them is experimentally inaccessible. Both ΔH_0^* and $\Delta\phi_e$ are evaluated absolutely on the $\Delta\phi$ scale of potential, with $\Delta\phi = 0$ as the electrical reference point, which is also experimentally indeterminate. The term ΔH_0^* in Eq. (26) characterizes the kinetics of the reaction in Eq. (1) in the direction indicated by the right subscript, provided it occurs at a nonelectrified WE/solution interface, that is, at $\Delta\phi = 0$. The other term, $\Delta\phi_e$ (which is equal to $\Delta\phi_{RE}$ in the simplest isothermal cell), represents the experimentally inaccessible absolute value of the electrical state of the WE/solution interface when the reaction in Eq. (1) is held at thermodynamic equilibrium.

The single value of $_{is}U^{\#}(\eta = 0)$ for both directions of all electrode reactions indicates that distinctions characterizing the kinetics in the cathodic or the anodic direction are canceled by the equilibrium term $\beta \cdot F \cdot \Delta\phi_e$. Hence, $_{is}U^{\#}(\eta = 0)$ represents an experimentally accessible parameter but does not provide any diagnostic information on the kinetics in either direction of the reaction studied. This indicates why $_{is}U^{\#}(\eta = 0)$ is not used as a kinetic parameter in the elucidation of reaction mechanisms in electrode kinetics.

IV. POTENTIAL OF THE SHE AS THE ELECTRICAL REFERENCE POINT

Shortcomings of the choice of the equilibrium state as the electrical reference point in the evaluation of the temperature effect on the rate of electrode reactions, and consequently of the overpotential as an experimental substitute for $\Delta(\Delta\phi)$ in the WE–RE cell at various temperatures, have been discussed in the previous section. Hence, another reference point should be sought. From a theoretical point of view, the choice is unambiguous—it is the zero point on the relative electrode potential scale, defined by the SHE convention. Basically, this is also an equilibrium state, but of a single reaction selected by convention, namely, the reduction of two hydrogen ions to molecular hydrogen. The value of $\Delta\phi$ at the interface when this reaction is held at equilibrium, assuming all species involved are in standard thermodynamic states, is fixed by the SHE convention as zero. The same convention associates additional properties with this reference state: temperature, solvent, and solute independence. Formally, the properties of the SHE satisfy the principal theoretical requirements for the electrical reference point in the evaluation of the effect of temperature on the rate of electrode reactions.

However, the SHE is far from the ideal reference point for practical, experimental purposes. First of all, it is not an operative electrode, because it cannot be applied as a device in measurements. It is an electrode with a theoretically defined potential. Therefore, a few other reference electrodes with accurately determined electrode potentials on the SHE scale of potential are introduced into experimental electrochemistry.[251] Each of these operable reference electrodes, however, is characterized by the temperature coefficient of its standard electrode potential. Thus, when temperature effects on the rate of electrode reactions are evaluated, these reference electrodes have to be held at constant temperature, which means that isothermal cells should be eliminated in the usual experimental procedure.

Another drawback, which follows from the fact that the SHE must be substituted by an operable reference electrode, concerns the composition of solutions in the WE and RE compartments of the cell. Generally, these solutions are different, which means that some transference must be present in the WE–RE cell. Furthermore, the materials at the WE and RE are not identical, which means that a contact potential difference must also be present in the WE–RE cell. Thus, the choice of the SHE as the electrical reference point when EAs are determined assumes an experimental procedure in which the WE–RE cell is a nonisothermal cell with transference (see Fig. 1d). The measured potential of the WE–RE cell in this case is composed of terms representing all the $\Delta\phi$s within the cell [see Eq. (24)], giving the following expression for $\Delta\phi_{WE}(T)$:

$$\Delta\phi_{WE}(T) = \text{emf} + \Delta\phi_{RE}(T_0) - \Delta\phi_D(T_0) - \Delta\phi_C(T_0) - \Delta\phi_T \qquad (33)$$

Hence, the electrical state of the WE, which must be held constant irrespective of the temperature, is described by the experimentally accessible potential WE–RE cell, the constant contribution $\Delta\phi_{RE}(T_0)$, and three experimentally inaccessible terms pertaining to other phenomena besides the reaction occurring in the WE–RE cell. Principally, the requirement $\Delta\phi_{WE}(T) = \text{const}$ cannot be satisfied if the terms $\Delta\phi_D(T_0)$, $\Delta\phi_C(T_0)$, and $\Delta\phi_T$ are not known. On these grounds, Frumkin and Krishtalik[252] disqualified all procedures for the evaluation of temperature effects that are not based on isothermal cells without transference. However, the simple theoretical requirement $\Delta\phi_{WE}(T) = \text{const}$ is difficult to fulfill experimentally. Therefore, the use of nonisothermal cells with transference should be considered with less rigor, provided measurements of emf are performed under conditions such that Eq. (24) can be simplified to a more acceptable form.

The terms $\Delta\phi_D(T_0)$ and $\Delta\phi_C(T_0)$ are pertinent to the part of the WE–RE cell that is held at an arbitrarily selected constant temperature T_0 throughout the measurements (see Fig. 1d). Therefore, the sum of the terms $\Delta\phi_D(T_0) + \Delta\phi_C(T_0)$ represents an unknown but constant contribution to the measured emf of the cell. Theoretical calculations of $\Delta\phi_D$ in cells with transference, with properly selected solutions in contact, indicate that this term can be reduced to the 10^{-3} V range. For the majority of combinations of common electrode materials used in WE–RE cells, $\Delta\phi_C$ is also of the order of 10^{-3} V. Hence, when care in the choice of experimental conditions for the EA determinations is properly taken, the contribution of the two terms may be reduced to the range of 10^{-3} V. Consequently, in routine isothermal electrochemistry, the values of these two terms are not determined separately, but are merged into the true value of the relative electrode potential of the WE:

$$E_{WE}(T) = E_{WE}(\text{true},T) + \Delta\phi_D(T_0) + \Delta\phi_C(T_0) \tag{34}$$

Hence, there is an uncertainty in the experimentally determined potential of the WE at T = const [when T = const, $\Delta\phi_T = 0$ in Eq. (33)] due to the unknown, typically small, constant contribution of the terms $\Delta\phi_D(T_0)$ and $\Delta\phi_C(T_0)$, which is accepted as the error of the measurements.

When EA is determined for electrode reactions, it is important to keep the electrical state of the WE constant at various temperatures. The absolute value of $\Delta\phi_{WE}(T)$ is unknown anyway, and EA is not reported relative to $\Delta\phi_{WE}(T)$, but to an experimental substitute, such as $E_{WE}(T)$, which is evaluated on the SHE scale of potential. Any uncertainty in $E_{WE}(T)$ caused by the contributions of the terms $\Delta\phi_D(T_0) + \Delta\phi_C(T_0)$ is acceptable because it is usually within the experimental error of the measurements.

When the condition $\Delta\phi_{WE}(T)$ = const is satisfied in nonisothermal cells with transference, only the term $\Delta\phi_T$ in Eq. (33) changes with temperature in a way that is difficult for the experimental procedure to control. Therefore, the potential of any nonisothermal cell will be uncertain in terms of the contribution of the $\Delta\phi_T$ term, which is variable with temperature. This is a basic drawback of the choice of the SHE as the electrical reference point in determinations of EA for electrode reactions.

Evaluations of temperature effects on the rate of electrode reactions occurring in aqueous solutions, however, are performed mostly within narrow and almost constant temperature intervals (typically 300 ± 30 K) and mostly in solutions of constant ionic strength (typically, 10^0 mol/dm^3). Under these conditions, the maximum contribution of the $\Delta\phi_T$ term is of the order of 10^{-2} V. Hence, the uncertainty in $E_{WE}(T)$ will be 10^{-2} V, at most, which means that the corresponding EAs may be finally determined with an uncertainty of about 1 kJ/mol. In the case of electrode reactions with $_{nt}U^{\#}(E) > 50$ kJ/mol, the uncertainty mentioned will be less than 2%. If this level of experimental error can be accepted, nonisothermal cells with transference can be used for determinations of EAs. After $\Delta\phi_{WE}(T)$ is substituted with $E_{WE}(T)$, evaluated on the SHE scale of potential, Eq. (33) can be approximated to:

$$E_{WE}(T) = E_{WE}(\text{true},T) + \Delta\phi_D(T_0) + \Delta\phi_C(T_0) + \Delta\phi_T$$

$$= \text{emf} + E_{RE}(T_0) \tag{35}$$

1. Electrode Reactions with Temperature-Independent Symmetry Factors

By the choice of proper experimental conditions, uncertainties in the substitution of $\Delta\phi_{WE}(T) = $ const by $E_{WE}(T) = $ const can be ignored and experimental procedures for EA determinations in a nonisothermal WE–RE cell with transference accepted as suitable. In this case, it is fully valid to substitute $\Delta\phi_{WE}$ in Eq. (14) with E_{WE}, and Eqs. (18)–(20) are an adequate theoretical description of the phenomena considered in nonisothermal cells with transference.

One of the basic parameters in electrode kinetics, the exchange current density, is defined in accordance with Eqs. (18) as follows:

$$j_0 = j_{r,c} \cdot \exp(-\beta_c F E_e / RT) = j_{r,a} \cdot \exp(\beta_a F E_e / RT) \tag{36}$$

where E_e is the reversible potential of the reaction in Eq. (1), occurring at the WE.

When Eqs. (18) and (19) are combined, the physical meanings of the corresponding nt-EAZEPs are obtained:

$$_{nt}U^{\#}(E_c = 0) = \Delta H_{0,c}^{\#} + \beta_c \cdot F \cdot \Delta\phi_{SHE} \tag{37a}$$

and

$$_{nt}U^{\#}(E_a = 0) = \Delta H_{0,a}^{\#} - \beta_a \cdot F \cdot \Delta\phi_{SHE} \tag{37b}$$

where $\Delta H_{0,c}^{\#}$ and $\Delta H_{0,a}^{\#}$ are ideal or chemical enthalpies of activation of the reaction in Eq. (1) in the cathodic and the anodic direction, respectively. It should be noted that, in accordance with the SHE convention, the term $F \cdot \Delta\phi_{SHE} = 0$, because the SHE is one of the basic reference zero points in chemical thermodynamics. Therefore, when the SHE is selected as the electrical reference point in studies of temperature effects in electrode kinetics, data concerning the equilibrium state of electrode reactions obtained from chemical thermodynamics and from electrode kinetics are compatible. For instance, it follows from Eqs. (37) that

$$_{nt}U^{\#}(E_c = E^{\ominus}) - {}_{nt}U^{\#}(E_a = E^{\ominus}) = T \cdot \Delta_r S^{\circ} + (\beta_c + \beta_a) \cdot F \cdot \Delta\phi_{SHE}$$

$$= T \cdot \Delta_r S^{\ominus} \tag{38}$$

and, analogously,

$$_{nt}U^{\#}(E_c = 0) - {}_{nt}U^{\#}(E_a = 0) = \Delta H^{\#}_{0,c} - \Delta H^{\#}_{0,a} + (\beta_c + \beta_a) \cdot F \cdot \Delta\phi_{SHE}$$

$$= \Delta_r H^{\ominus} \tag{39}$$

Values of nt-EAZEPs, defined by Eqs. (37), provide evidence on the kinetics in each direction of the reaction in Eq. (1) separately. The kinetics of various electrode reactions occurring in the cathodic direction, for instance, can now be characterized by means of the corresponding nt-EAZEP-c values. These nt-EAZEP-c values can then be compared and analyzed in discussions of possible mechanisms, provided the electrode material and the composition of the solution are the same.

Dependences of nt-EACEPs on E, in accordance with Eqs. (20), together with various relationships between nt-EACEP-c and nt-EACEP-a for some characteristic values of the potential, are presented in Fig. 6. As follows from the corresponding nt-EACEP–E dependences in Fig. 6, there are particular values of potential, $E_{u,c}$ and $E_{u,a}$, arbitrarily named *ultimate potentials*, where the corresponding nt-EACEPs are equal to zero:

$$E_{u,c} = -\frac{_{nt}U^{\#}(E_c = 0)}{\beta_c \cdot F} = -\frac{\Delta H^{\#}_{0,c}}{\beta_c \cdot F} - \Delta\phi_{SHE} \tag{40a}$$

and

$$E_{u,a} = \frac{_{nt}U^{\#}(E_a = 0)}{\beta_a \cdot F} = \frac{\Delta H^{\#}_{0,a}}{\beta_a F} - \Delta\phi_{SHE} \tag{40b}$$

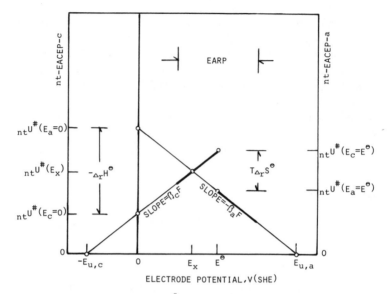

Figure 6. Schematic dependence of $_{nt}U^{\#}(E)$ on E for the cathodic and anodic directions of the reaction in Eq. (1). Temperature-independent symmetry factors, $\beta_c = \beta_a = 0.5$, and $dE^{\ominus}/dT > 0$ are assumed. EARP, Experimentally accessible region of overpotential.

In Fig. 7, nt-EACEP–E dependences are presented for electrode reactions with various thermodynamic properties, all characterized by temperature-independent symmetry factors, with $\beta_c + \beta_a = 1$, and $\beta_c = \beta_a = 0.5$. The case considered in Fig. 1a is presented in Fig. 6 in more detail.

At ultimate potentials, electrode reactions occurring at the WE in the direction indicated by the subscript are converted to barrierless processes.[97] When the condition $E = E_u$ is introduced into Eqs. (18), the corresponding maximum rates of the electrode reaction in Eq. (1), presented as ultimate current densities, are obtained:

$$j_{u,c} = F \cdot k_c \cdot c_O \cdot \exp(\Delta S^{\#}_{0,c}/R) \qquad (41a)$$

and

$$j_{u,a} = F \cdot k_a \cdot c_R \cdot \exp(\Delta S^{\#}_{0,a}/R) \qquad (41b)$$

A pair of ultimate parameters (E_u, j_u), signifying the electrical state of the WE when the reaction in Eq. (1) in the direction considered is

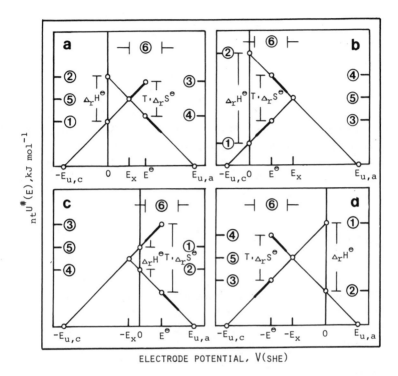

Figure 7. Schematic dependence of $_{nt}U^{\#}(E)$ on E, for the cathodic and anodic directions of the reaction in Eq. (1), characterized by temperature-independent symmetry factors and $\beta_c = \beta_a = 0.5$. Various thermodynamic properties for the reaction in Eq. (1) are postulated: (a) $\Delta_r H^{\ominus} < 0$, $dE^{\ominus}/dT > 0$, $E^{\ominus} > E_x$, and spontaneous in cathodic direction ($E^{\ominus} > 0$); (b) $\Delta_r H^{\ominus} < 0$, $dE^{\ominus}/dT < 0$, $E^{\ominus} < E_x$, and spontaneous in cathodic direction if $E^{\ominus} > 0$ or in anodic direction if $E^{\ominus} < 0$; (c) $\Delta_r H^{\ominus} > 0$, $dE^{\ominus}/dT > 0$, $E^{\ominus} > E_x$, and spontaneous in cathodic direction if $E^{\ominus} > 0$, or in anodic direction if $E^{\ominus} < 0$; (d) $\Delta_r H^{\ominus} > 0$, $dE^{\ominus}/dT < 0$, $E^{\ominus} < E_x$, and spontaneous in anodic direction ($E^{\ominus} < 0$). Legend to the circled numbers on the diagrams: 1, $_{nt}U^{\#}(E_c = 0)$; 2, $_{nt}U^{\#}(E_a = 0)$; 3, $_{nt}U^{\#}(E_c = E^{\ominus})$; 4, $_{nt}U^{\#}(E_a = E^{\ominus})$; 5, $_{nt}U^{\#}(E_x)$; 6, experimentally accessible region of electrode potential, where nt-EACOPs are determined.

converted to a barrierless process, represents the temperature-independent electrical state of the WE/solution interface, provided the symmetry factors are temperature independent.

When Eqs. (18), (20), and (41) are combined, rate laws similar to the Arrhenius equation are obtained:

$$j_c = j_{u,c} \cdot \exp\left(- \frac{{}_{nt}U^{\#}(E_c)}{RT}\right) \tag{42a}$$

and

$$j_a = j_{u,a} \cdot \exp\left(- \frac{{}_{nt}U^{\#}(E_a)}{RT}\right) \tag{42b}$$

As follows from Figs. 6 and 7, there is a particular potential E_x such that ${}_{nt}U^{\#}(E_c = E_x) = {}_{nt}U^{\#}(E_a = E_x)$ is satisfied for the reaction in Eq. (1). At this potential, the rates of the cathodic and anodic reactions, in accordance with Eqs. (42), are not identical, but satisfy the following relationship:

$$j_{x,c}/j_{x,a} = j_{u,c}/j_{u,a} \tag{43}$$

Assuming that the reaction in Eq. (1) occurs at $c_O = c_R = c^{\ominus}$, it follows from Eqs. (41) and (43) that

$$j_{x,c}/j_{x,a} = (k_c/k_a) \cdot \exp(\Delta_r S^{\ominus}/R) \tag{44}$$

Hence, the ratio of the cathodic and anodic rates of the reaction in Eq. (1) at the potential E_x is determined by the standard change of entropy of this reaction or, alternatively, by the temperature coefficient of the standard electrode potential of the reaction.

The potential E_x can be defined as follows from the basic condition ${}_{nt}U^{\#}(E_c = E_x) = {}_{nt}U^{\#}(E_a = E_x)$:

$$E_x = - \frac{\Delta_r H^{\ominus}}{(\beta_c + \beta_a) \cdot F} - \Delta\phi_{SHE} \tag{45}$$

From the simple geometry in Fig. 6, it follows that the position of E_x between $E_{u,c}$ and $E_{u,a}$ is determined by the reciprocal of the ratio of the symmetry factors for the cathodic and anodic reactions, provided the condition $\beta_c + \beta_a = 1$ is satisfied:

$$\frac{E_x - E_{u,c}}{E_x - E_{u,a}} = \frac{\beta_a}{\beta_c} \tag{46}$$

In the case of absolutely symmetric barriers for the cathodic and anodic reactions, for example, $\beta_c = \beta_a = 0.5$, the potential E_x is always the midpoint between $E_{u,c}$ and $E_{u,a}$ (see Fig. 6). The potential E_x can also be related to $E^{\ominus}$. When E_x and $E^{\ominus}$ are introduced into Eqs. (20), after the substraction ${}_{nt}U^{\#}(E^{\ominus}) - {}_{nt}U^{\#}(E_x)$, introduction of the relationship

$(\beta_c + \beta_a) \cdot F \cdot E^{\ominus} = \Delta_r G^{\ominus}$, and use of the Gibbs–Helmholtz equation, one obtains

$$(\beta_c + \beta_a) \cdot F \cdot (E^{\ominus} - E_x) = T \cdot \Delta_r S^{\ominus} \qquad (47)$$

giving finally

$$E_x = E^{\ominus} - \frac{T \cdot \Delta_r S^{\ominus}}{(\beta_c + \beta_a)F} = E^{\ominus} - \frac{T \cdot (dE^{\ominus}/dT)}{(\beta_c + \beta_a) \cdot F} \qquad (48)$$

Hence, it follows that $E_x < E^{\ominus}$ for electrode reactions with $(dE^{\ominus}/dT) > 0$, and $E_x > E^{\ominus}$ for electrode reactions with $(dE^{\ominus}/dT) < 0$.

If it is taken that $\Delta_r S^{\ominus} = \Delta S_{0,c}^{\#} - \Delta S_{0,a}^{\#}$, and taking into account Eqs. (38) and (47), it is reasonable to assume that the entropies of activation of the electrode reactions, in the direction indicated, may be expressed as follows:

$$\Delta S_{0,c}^{\#} = (\beta_c F/T) \cdot (E^{\ominus} - E_x) \qquad (49a)$$

and

$$\Delta S_{0,a}^{\#} = (\beta_a F/T) \cdot (E_x - E^{\ominus}) \qquad (49b)$$

Finally, when Eqs. (42) are considered for $E = E^{\ominus}$ and $E = E_u$, it follows that the position of the standard electrode potential of the reaction in Eq. (1), relative to the ultimate potential for the direction indicated, is determined by the ratio of the corresponding ultimate current density to the exchange current density:

$$E^{\ominus} = E_{u,c} + \frac{RT}{\beta_c \cdot F} \cdot \ln \frac{j_{u,c}}{j_0} \qquad (50a)$$

and

$$E^{\ominus} = E_{u,a} - \frac{RT}{\beta_a \cdot F} \cdot \ln \frac{j_{u,a}}{j_0} \qquad (50b)$$

2. Electrode Reactions with Symmetry Factors Linearly Dependent on Temperature

All the theoretical considerations regarding the effect of temperature on the rate of electrode reactions presented thus far were based on the assumption of temperature-independent symmetry factors. The treatment of the temperature dependence of the symmetry factor has an interesting history in electrode kinetics. At an early stage, entirely on the basis of experimental evidence,[26] the symmetry factor was considered as a temperature-dependent parameter. However, the development of modern electrode kinetics, essentially beginning with the keystone paper by Bockris in *Modern Aspects of Electrochemistry*, No. 1,[45] was based primarily on a temperature-independent symmetry factor, an assumption derived theoretically on the basis of analogies to homogeneous chemical kinetics. As a result, most of modern electrochemistry, including the treatment of the effect of temperature on the rate of electrode reactions, is based on this assumption. Recently, new experimental evidence was obtained that unambiguously supported the temperature dependence of the symmetry factor in certain reactions. Conway has reviewed the problem and classified the types of dependences observed.[95] In order to keep the existing model of electrode reactions, and elaborating on an idea initially suggested by Agar,[26] Conway put forward a formal generalization for the linear dependence of the symmetry factor on temperature:

$$\beta = \beta_H + \beta_S \cdot T \tag{51}$$

with β_H and β_S as temperature-independent parameters that are characteristic for each electrode reaction/working electrode/solution combination. Classical electrode kinetics corresponds to a limiting behavior of Eq. (51), that is, to the case of $\beta_S = 0$, with

$$\beta = \beta_H \tag{52}$$

Another limiting behavior, that is, the case of $\beta_H = 0$, with

$$\beta = \beta_S \cdot T \tag{53}$$

is not just a theoretical curiosity, as it has been reported for some electrode reactions.[95,151,196,254,255]

In the following discussion, it is assumed that the effect of temperature on the rate of electrode reactions is studied in nonisothermal cells with transference, with the SHE as the electrical reference point, employing all assumptions and simplifications mentioned earlier (see Section II.1).

(i) Case: $\beta = \beta_S \cdot T$

When Eq. (53) is introduced into Eqs. (18), one obtains

$$j_c = F \cdot k_c \cdot c_O \cdot \exp(-\Delta G_{0,c}^{\#}/RT)$$

$$\cdot \exp(-\beta_{S,c}F\Delta\phi_{SHE}/R) \cdot \exp(-\beta_{S,c}FE_c/R)$$

$$= j_{r,c}' \cdot \exp(-\beta_{S,c}FE_c/R) \tag{54a}$$

and

$$j_a = F \cdot k_a \cdot c_R \cdot \exp(-\Delta G_{0,a}^{\#}/RT)$$

$$\cdot \exp(\beta_{S,a}F\Delta\phi_{SHE}/R) \cdot \exp(\beta_{S,a}FE_a/R)$$

$$= j_{r,a}' \cdot \exp(\beta_{S,a}FE_a/R) \tag{54b}$$

where $j_{r,c}'$ and $j_{r,a}'$ are the characteristic current densities at the electrical reference point, that is, at $E_c = 0$ and $E_a = 0$, respectively.

When the rate laws given by Eqs. (54) are now introduced into Eqs. (7), assuming that $\Delta S_0^{\#}$ and $\Delta H_0^{\#}$ are not significantly affected by temperature, it follows that

$$\left(\frac{\partial \ln j_c}{\partial T^{-1}}\right)_{E_c, c_i, p, \ldots} = \left(\frac{\partial \ln j_{r,c}'}{\partial T^{-1}}\right)_{E_c=0, c_i, p, \ldots} = -(1/R) \cdot \Delta H_{0,c}^{\#} \tag{55a}$$

and

$$\left(\frac{\partial \ln j_a}{\partial T^{-1}}\right)_{E_a, c_i, p, \ldots} = \left(\frac{\partial \ln j_{r,a}'}{\partial T^{-1}}\right)_{E_a=0, c_i, p, \ldots} = -(1/R) \cdot \Delta H_{0,a}^{\#} \tag{55b}$$

Arrhenius plots of Eqs. (55), as presented schematically in Fig. 8, are parallel because experimental $_{nt}U^{\#}(E)$ values are not dependent on the electrode potential, but represent genuine ideal enthalpies of activation:

$$_{nt}U^{\#}(E_c) = {_{nt}U^{\#}}(E_c = 0) = \Delta H_{0,c}^{\#} \tag{56a}$$

and

$$_{nt}U^{\#}(E_a) = {_{nt}U^{\#}}(E_a = 0) = \Delta H_{0,a}^{\#} \tag{56b}$$

The Arrhenius plots in Fig. 8 are also characterized by the preexponential current densities, $j_{A,c}$ and $j_{A,a}$, parameters analogous to the preexponential factor in the Arrhenius equation.[19] These preexponential current densities are defined from Eqs. (54), when the limiting condition $T^{-1} = 0$ is introduced:

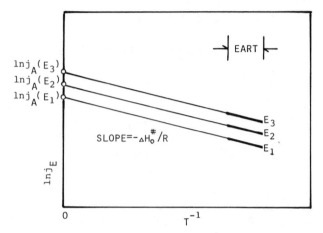

Figure 8. Arrhenius type plots of $\ln j_E$ versus T^{-1} for the reaction in Eq. (1), occurring in a nonisothermal WE–RE cell with transference, characterized by $\beta = \beta_S \cdot T$. EART, Experimentally accessible region of temperature.

$$j_{A,c} = F \cdot k_c \cdot c_O \cdot \exp(\Delta S_{0,c}^{\#}/R)$$

$$\cdot \exp(-\beta_{S,c} F \Delta \phi_{\mathrm{SHE}}/R) \cdot \exp(-\beta_{S,c} F E_c/R) \quad (57a)$$

and

$$j_{A,a} = F \cdot k_a \cdot c_R \cdot \exp(\Delta S_{0,a}^{\#}/R)$$

$$\cdot \exp(\beta_{S,a} F \Delta \phi_{\mathrm{SHE}}/R) \cdot \exp(\beta_{S,a} F E_a/R) \quad (57b)$$

As follows from Eqs. (57), the preexponential current densities are dependent on the electrode potential, as schematically presented in Fig. 9. Combination of Eqs. (54) and (57) gives:

$$j_c = j_{A,c} \cdot \exp\left(-\frac{\Delta H_{0,c}^{\#}}{RT}\right) \quad (58a)$$

and

$$j_a = j_{A,a} \cdot \exp\left(-\frac{\Delta H_{0,a}^{\#}}{RT}\right) \quad (58b)$$

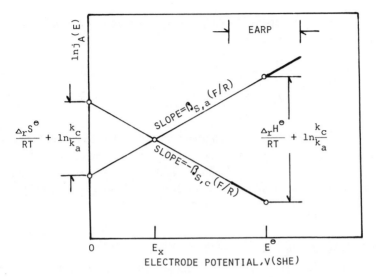

Figure 9. Schematic dependence of preexponential current densities on electrode potential for the cathodic and anodic directions of the reaction in Eq. (1), characterized by $\beta = \beta_S \cdot T$. EARP, Experimentally accessible region of potential.

Once the SHE is taken as the electrical reference potential in evaluations of the effect of temperature in electrode kinetics, combinations of kinetic parameters pertinent to the standard equilibrium state of electrode reactions may be related to the corresponding equilibrium data. For instance, taking the difference of the natural logarithms of the preexponential current densities from Eqs. (57) for $E = E^\ominus$, assuming $c_O = c_R = c^\ominus$ and $\beta_c + \beta_a = 1$, one obtains

$$\ln j_A(E_c = E^\ominus) - \ln j_A(E_a = E^\ominus) = \frac{1}{RT}[\Delta_r H^\circ + (\beta_c + \beta_a)F\Delta\phi_{SHE}] + \ln\frac{k_c}{k_a}$$

$$= \frac{1}{RT}\Delta_r H^\ominus + \ln\frac{k_c}{k_a} \tag{59}$$

Analogously, from the difference of the $\ln j_A$ values referred to $E = 0$, it follows that

$$\ln j_A(E_c = 0) - \ln j_A(E_a = 0) = \frac{1}{R} \cdot [\Delta_r S^\circ + (\beta_{S,c} + \beta_{S,a})F\Delta\phi_{SHE}] + \ln\frac{k_c}{k_a}$$

$$= \frac{1}{R} \cdot \Delta_r S^\ominus + \ln\frac{k_c}{k_a} \tag{60}$$

As follows from Fig. 9, extrapolated $\ln j_A$–E dependences intersect at the characteristic potential E_x, which is related to the ratio of rate constants for the cathodic and the anodic direction of the reaction in Eq. (1) and to the standard entropy change of the reaction occurring at the WE:

$$\ln \frac{k_c}{k_a} = \frac{F \cdot E_x - T \cdot \Delta_r S^{\ominus}}{RT} = \frac{F \cdot E_x - F \cdot T \cdot (dE^{\ominus}/dT)}{RT} \tag{61}$$

In accordance with transition-state theory applied to electrode reactions, the ratio k_c/k_a for the reaction in Éq. (1) represents essentially the ratio of electron-transfer probabilities for the reaction in the cathodic and the anodic direction. This parameter can be determined experimentally, as follows from Eq. (61), because both $dE^{\ominus}/dT$ and E_x are experimentally accessible parameters.

(ii) Case: $\beta = \beta_H + \beta_S \times T$

When Eq. (51) is introduced into Eqs. (18), it follows that

$$j_c = F \cdot k_c \cdot c_O \cdot \exp(-\beta_{S,c} F \Delta\phi_{SHE}/R) \cdot \exp(-\beta_{S,c} FE_c/R)$$

$$\cdot \exp(-\Delta G_{0,c}^{\#}/RT) \cdot \exp(-\beta_{H,c} F \Delta\phi_{SHE}/RT) \cdot \exp(-\beta_{H,c} FE_c/RT)$$

$$= j_{r,c}'' \cdot \exp(-\beta_{H,c} FE_c/RT) \tag{62a}$$

and

$$j_a = F \cdot k_a \cdot c_R \cdot \exp(\beta_{S,a} F \Delta\phi_{SHE}/R) \cdot \exp(\beta_{S,a} FE_a/R)$$

$$\cdot \exp(-\Delta G_{0,a}^{\#}/RT) \cdot \exp(\beta_{H,a} F \Delta\phi_{SHE}/RT) \cdot \exp(\beta_{H,a} FE_a/RT)$$

$$= j_{r,a}'' \cdot \exp(\beta_{H,a} FE_a/RT) \tag{62b}$$

where $j_{r,c}''$ and $j_{r,a}''$ are characteristic current densities at the electrochemical reference state, that is, at $E_c = 0$ and $E_a = 0$, respectively.

When the rate laws given by Eqs. (62) are introduced into Eqs. (7), assuming that $\Delta S_0^{\#}$ and $\Delta H_0^{\#}$ are not significantly affected by temperature, one obtains

$$\left(\frac{\partial \ln j_c}{\partial T^{-1}}\right)_{E_c, c_i, p, \ldots} = \left(\frac{\partial \ln j_{r,c}''}{\partial T^{-1}}\right)_{E_c=0, c_i, p, \ldots} - \frac{1}{R} \cdot \beta_{H,c} \cdot F \cdot E_c \tag{63a}$$

and

$$\left(\frac{\partial \ln j_a}{\partial T^{-1}}\right)_{E_a,c_i,p,\,\ldots} = \left(\frac{\partial \ln j''_{r,a}}{\partial T^{-1}}\right)_{E_a=0,c_i,p,\,\ldots} + \frac{1}{R} \cdot \beta_{H,a} \cdot F \cdot E_a \quad (63b)$$

with

$$\left(\frac{\partial \ln j''_{r,c}}{\partial T^{-1}}\right)_{E_c=0,c_i,p,\,\ldots} = -\frac{1}{R} \cdot (\Delta H^{\#}_{0,c} + \beta_{H,c}F\Delta\phi_{\text{SHE}}) \quad (64a)$$

and

$$\left(\frac{\partial \ln j''_{r,a}}{\partial T^{-1}}\right)_{E_a=0,c_i,p,\,\ldots} = -\frac{1}{R} \cdot (\Delta H^{\#}_{0,a} - \beta_{H,a} \cdot F \cdot \Delta\phi_{\text{SHE}}) \quad (64b)$$

The potential-dependent preexponential current densities are defined from Eqs. (62), for the limiting condition $T^{-1} = 0$:

$$j''_{A,c} = F \cdot k_c \cdot c_O \cdot \exp(\Delta S^{\#}_{0,c}/R)$$
$$\cdot \exp(-\beta_{S,c}F\Delta\phi_{\text{SHE}}/R) \cdot \exp(-\beta_{S,c}FE_c/R) \quad (65a)$$

and

$$j''_{A,a} = F \cdot k_a \cdot c_R \cdot \exp(\Delta S^{\#}_{0,a}/R)$$
$$\cdot \exp(\beta_{S,a}F\Delta\phi_{\text{SHE}}/R) \cdot \exp(\beta_{S,a}FE_a/R) \quad (65b)$$

The dependences of $\ln j''_{A,c}$ and $\ln j''_{A,a}$ on E, defined by Eqs. (65), are schematically presented in Fig. 10, where some relationships for characteristic values of the potential of the WE are also indicated. Thus,

$$\ln j''_A (E_c = 0) - \ln j''_A(E_a = 0) = \frac{1}{R} \cdot [\Delta_r S^{\circ} - (\beta_{S,c} + \beta_{S,a})F\Delta\phi_{\text{SHE}}] + \ln\frac{k_c}{k_a}$$
$$= \frac{1}{R} \cdot \Delta_r S^{\ominus} + \ln\frac{k_c}{k_a} \quad (66a)$$

and

$$\ln j''_A(E = E^{\ominus}) - \ln j''_A(E = E^{\ominus}) = \frac{1}{RT} \cdot [\Delta_r H^{\circ} + (\beta_c + \beta_a)F\Delta\phi_{\text{SHE}}] + \ln\frac{k_c}{k_a}$$
$$= \frac{1}{RT} \cdot \Delta_r H^{\ominus} + \ln\frac{k_c}{k_a} \quad (66b)$$

For the plots of Eqs. (66) in Fig. 10, the SHE convention (i.e., $\Delta\phi_{\text{SHE}} = 0$) is employed, and it is assumed that $c_O = c_R = c^{\ominus}$ and $\beta_c + \beta_a = 1$.

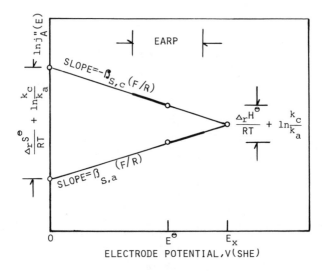

Figure 10. Schematic dependences of preexponential current densities on electrode potential for the cathodic and anodic directions of the reaction in Eq. (1), occurring in a nonisothermal WE–RE cell with transference, characterized by $\beta = \beta_H + \beta_S \cdot T$. EARP, Experimentally accessible region of potential.

When Eqs. (62) and (66) are combined, the following rate laws are obtained:

$$j_c = j_{A,c}'' \cdot \exp\left(-\frac{{}_{nt}U^{\#}(E_c)}{RT}\right) \tag{67a}$$

and

$$j_a = j_{A,a}'' \cdot \exp\left(-\frac{{}_{nt}U^{\#}(E_a)}{RT}\right) \tag{67b}$$

Arrhenius plots of Eqs. (63) are schematically presented in Fig. 11. Both ${}_{nt}U^{\#}(E)$ and $\ln j_A''$ are dependent on the electrode potential. The basic sources of information for Fig. 11 are the cathodic and anodic Tafel lines, obtained at different temperatures. When these Tafel lines are extrapolated beyond diffusion limiting currents, as schematically presented in Fig. 12, they do not intersect at a single point, as was the case when the symmetry factors were not dependent on temperature (see Fig. 4). Such behavior is

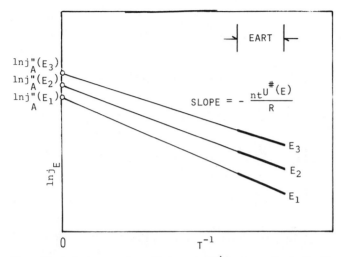

Figure 11. Arrhenius type plots of $\ln j_E$ versus T^{-1} for the reaction in Eq. (1), occurring in a nonisothermal WE–RE cell with transference, characterized by $\beta = \beta_H + \beta_S \cdot T$. EART, Experimentally accessible region of temperature.

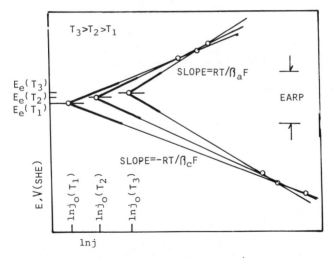

Figure 12. Schematic representation of Tafel lines for the cathodic and anodic directions of the reaction in Eq. (1), occurring in a nonisothermal WE–RE cell with transference, at various temperatures. The reaction in Eq. (1) is characterized by $\beta = \beta_H + \beta_S \cdot T$. EARP, Experimentally accessible region of potential.

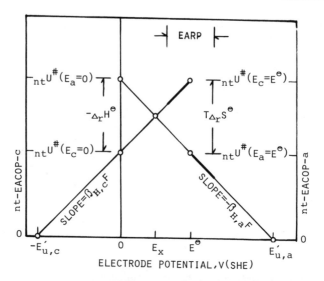

Figure 13. Schematic dependence of $_{nt}U^{\#}(E)$ on E for the cathodic and anodic directions of the reaction in Eq. (1), characterized by $\beta = \beta_H + \beta_S \cdot T$ and $dE^{\ominus}/dT > 0$. EARP, Experimentally accessible region of potential.

the result of simultaneous changes of both $_{nt}U^{\#}(E)$ and j_A'' with E, which are effected by the temperature dependence of the symmetry factors given by Eq. (51).

In Fig. 13, the $_{nt}U^{\#}(E)$–E relations described by Eqs. (63) and (64) are schematically represented. Formally, when the $_{nt}U^{\#}(E)$–E relations are extrapolated to zero, the values of $_{nt}U^{\#}(E)$, the corresponding ultimate potentials, are obtained:

$$E_{u,c}' = -\frac{\Delta H_{0,c}^{\#}}{\beta_{H,c}F} - \Delta\phi_{\text{SHE}} \tag{68a}$$

and

$$E_{u,a}' = \frac{\Delta H_{0,a}^{\#}}{\beta_{H,a}F} - \Delta\phi_{\text{SHE}} \tag{68b}$$

At the ultimate potentials $E_{u,c}'$ and $E_{u,a}'$, the electrode reaction in Eq. (1), when occurring in the cathodic and the anodic direction, respectively, is converted to a barrierless process.

When Fig. 13 is compared to Fig. 6, the basic difference is in the values of the slopes of the lines presented. In Fig. 13 the slopes of the lines are defined by $\beta_H F$, where β_H is the temperature-independent part of the symmetry factor. Thus, $\beta_H < \beta$. It should be noted that the values of the symmetry factor obtained from the slopes of the Tafel lines (Fig. 12) are different from those obtained from the slopes of the $_{nt}U^{\#}(E)-E$ relations (Fig. 13). This fact has been pointed out in the literature, although on other grounds.[110]

When Eqs. (68) are introduced into Eqs. (62), the corresponding ultimate current densities are obtained:

$$j'_{u,c} = F \cdot k_c \cdot c_O \cdot \exp(\Delta S^{\#}_{0,c}/R)$$

$$\cdot \exp(-\beta_{S,c}F\Delta\phi_{SHE}/R) \cdot \exp(-\beta_{S,c}FE_c/R) \qquad (69a)$$

and

$$j'_{u,a} = F \cdot k_a \cdot c_R \cdot \exp(\Delta S^{\#}_{0,a}/R)$$

$$\cdot \exp(\beta_{S,a}F\Delta\phi_{SHE}/R) \cdot \exp(\beta_{S,a}FE_a/R) \qquad (69b)$$

Following the simple geometry in Fig. 13, one obtains, analogously to Eq. (46),

$$\frac{E_x - E_{u,c}}{E_x - E_{u,a}} = \frac{\beta_{H,a}}{\beta_{H,c}} \qquad (70)$$

where E_x is a characteristic electrode potential defined through the requirement $_{nt}U^{\#}(E_c = E_x) = {}_{nt}U^{\#}(E_a = E_x)$. This characteristic potential E_x is related to the standard electrode potential of the reaction in Eq. (1) occurring at the WE, as follows:

$$E_x = E^{\ominus} - \frac{T \cdot \Delta_r S^{\ominus}}{(\beta_{H,c} + \beta_{H,a})F} = E^{\ominus} - \frac{T \cdot (dE^{\ominus}/dT)}{(\beta_{H,c} + \beta_{H,a})} \qquad (71)$$

The position of E_x with respect to $E^{\ominus}$ of the reaction occurring at the WE is dependent on the sign and the temperature coefficient of the standard electrode potential.

V. CONCLUSION

One hundred years after the discovery of the effect of temperature on the rate of electrode reactions, the understanding of the phenomenon may be summarized as follows. Formally, the description of the phenomenon has been theoretically based on the ideas of Bowden,[20] promoted later by Agar[26] and Temkin.[27] It postulates the equilibrium state of the electrode reaction studied as the temperature-independent electrical reference state, using the overpotential as an experimental substitute for $\Delta(\Delta\phi)$ in the WE–RE cell. Transition-state theory and classical electrode kinetics are used. The experimental behavior of isothermal cells without transference fits this theoretical description of the phenomenon. The theoretically defined energy of activation at zero overpotential determined in isothermal cells without transference, is-EAZOP, is generally accepted as the parameter to be used in quantitative evaluation of the phenomenon.[250] The majority of the experimental evidence accumulated in the literature may be classified as is-EAZOPs for various electrode reactions. The pertinent terminology and symbols used in the literature on this subject are confusing. This is perhaps one of the reasons why the impressive collection of experimental evidence is little used in electrode kinetics.

The real meaning of is-EAZOP for the equilibrium state of the reaction occurring at the WE has been discussed in this chapter. It has been shown that an evaluation of the phenomenon through the use of is-EAZOP values gives little information, because is-EAZOP does not discriminate between the energies of activation for the cathodic and anodic directions of the reaction occurring at the WE. Further, the selection of zero overpotential as the reference potential in evaluation of the phenomenon imposes certain attributes on the equilibrium state (for instance, $\Delta_r S^\circ = 0$, $\Delta_r H^\circ = 0$, etc.) which are not consistent with real properties of electrochemical reactions. Therefore, is-EAZOPs are not used in the elucidation of the mechanism of electrode reactions. Thus, the Agar–Temkin approach misses information on the rate of electrode reactions that are contained in temperature effect data. Although these defects of is-EAZOP have not yet been explicitly stated in the literature, some electrochemists are aware of them. The decreased interest in research of the phenomenon during the past 20 years illustrates this.

A slightly different approach to the phenomenon is proposed, and this respects the experimental conditions defining the determination procedure. The potential of the SHE is selected as the temperature-independent electrochemical reference point, and the electrode potential on this relative scale is used as a substitute for $\Delta(\Delta\phi)$ in the WE–RE cell. The transition-state theory and classical electrode kinetics are used to describe the phenomenon. Energies of activation are evaluated at a zero value of the electrode potential, determined in a nonisothermal cell with transference, for each direction of the reaction occurring at the WE, separately. These parameters, namely, nt-EAZEP-c and nt-EAZEP-a, are determined from the nt-EACEP–E relations extrapolated to $E = 0$, for each direction of the reaction. However, the theoretical equations defining nt-EACEPs are not fully compatible with the corresponding experimentally determined parameters, because the measured emf of the WE–RE cell contains unknown elements. Therefore, the experimentally determined nt-EACEPs contain contributions from parameters not related to the reaction occurring at the WE, but to some other phenomena. Nevertheless, under optimal conditions, the contributions of these parameters can be reduced to the level of typical measurement errors. In spite of the imperfect experimental procedure for determination of nt-EACEPs, they are parameters that offer the closest information on kinetics of the electrode reaction in either direction. When particular values of nt-EACEPs are combined for the cathodic and anodic directions (for instance, $E = E^{\ominus}$), some equilibrium parameters of the reaction occurring at the WE are obtained which are compatible with corresponding data from chemical thermodynamics. Values of nt-EAZEPs for various reactions can be used in the elucidation of reaction mechanisms, in discussions of electrocatalysis, and so on.[233,234,255]

A full treatment of the problems of the determination of energies of activation, which takes into account the complication of temperature-dependent symmetry factors, has been presented in this chapter for the simplest case of linear dependence of the symmetry factor on temperature. It is interesting to note that, among the experimental evidence available in the literature, it was not possible to find an adequate set of data to support the theoretical equations derived here. First of all, most studies are related to only one direction of the electrode reaction (for instance, hydrogen evolution, oxygen reduction, etc.), because of the rarity of electrode reactions in aqueous solutions that occur over a reasonably wide region of potential in both the cathodic and the anodic direction at rates that are not affected by diffusion or some other limitation. There are data on studies of the phenomenon for several reactions in both the cathodic and the anodic direction, but they pertain either to an isothermal regime in the WE–RE cell or to cases in which the rate of reaction is strongly affected by diffusion. New experimental evidence is needed, from carefully planned studies in accordance with all the requirements of nonisothermal cells with transference.

Finally, it may be concluded that the Agar–Temkin approach to the effect of temperature on the rate of electrode reactions is inadequate, in spite of the fact that the *is*-EAZOPs are determined by a correct procedure. Additional efforts are needed to find other theoretical descriptions or experimental innovations that would help to elucidate the phenomena of temperature dependence in accordance with its fundamental importance in electrode kinetics.

ACKNOWLEDGMENTS

Without help of my friends A. Damjanović and M. V. Vojnović, this chapter would never have been written. I want to thank them for years of discussions, helpful criticism, and encouragement.

Grateful acknowledgment is made to the Science Fund of Serbia for support of this work.

REFERENCES

[1] J. Roszkowsky, *Z. Phys. Chem.* **15** (1894) 267.
[2] E. Cohen and M. Newman, *Z. Phys. Chem.* **39** (1903) 353.
[3] J. Tafel, *Z. Phys. Chem.* **50** (1905) 641.
[4] G. N. Lewis and R. F. Jackson, *Z. Phys. Chem.* **56** (1906) 193.
[5] F. Kaufler, *Z. Elektrochem.* **13** (1907) 633.
[6] R. Sacerdoti, *Z. Elektrochem.* **17** (1911) 473.
[7] J. N. Pring and J. R. Curzon, *Trans. Faraday Soc.* **7** (1912) 237.
[8] E. K. Rideal, *J. Am. Chem. Soc.* **42** (1920) 94.
[9] S. J. Bircher and W. D. Harkins, *J. Am. Chem. Soc.* **45** (1923) 2890.
[10] M. Knobel and D. B. Joy, *Trans. Electrochem. Soc.* **44** (1924) 443.
[11] G. M. Westrip, *J. Chem. Soc.* **125** (1924) 1112.
[12] S. Glasstone, *J. Chem. Soc.* **125** (1924) 2646.
[13] W. D. Harkins and H. S. Adams, *J. Phys. Chem.* **29** (1925) 205.
[14] V. A. Roiter and R. B. Yampolskaya, *Zh. Fiz. Khim.* **9** (1937) 763; the same article was published in *Acta Physicochim. URSS* **7** (1937) 247.
[15] A. Hickling and F. W. Salt, *Trans. Faraday Soc.* **37** (1941) 33.
[16] Z. A. Iofa and V. Stepanova, *Zh. Fiz. Khim.* **19** (1945) 125.
[17] Z. A. Iofa and K. P. Mikulin, *Zh. Fiz. Khim.* **18** (1944) 137.
[18] J. O'M. Bockris, *Chem. Rev.* **43** (1948) 525.
[19] S. Arrhenius, *Z. Phys. Chem.* **4** (1889) 226.
[20] F. P. Bowden, *Proc. R. Soc. (London), Ser A.* **126** (1929) 107.
[21] T. Erdey-Gruz and M. Volmer, *Z. Phys. Chem.* **A150** (1930) 203.
[22] A. N. Frumkin, *Z. Phys. Chem.* **A164** (1933) 121.
[23] H. Eyring, S. Glasstone, and K. Laidler, *J. Chem. Phys.* **7** (1939) 1053.
[24] J. Horiuti and M. Polanyi, *Acta Physicochim. URSS* **16** (1942) 169.
[25] F. P. Bowden and J. A. N. Agar, *Annu. Rev. Chem. Soc.* **35** (1938) 90.
[26] J. N. Agar, *Disc. Faraday Soc.* **1** (1947) 81.
[27] M. Temkin, *Zh. Fiz. Khim.* **22** (1948) 1081.

[28]R. Agudubert, *J. Phys. Radium* **3** (1942) 81.

[29]H. P. Stout, *Trans. Faraday Soc.* **41** (1945) 64.

[30]H. P. Stout, *Disc. Faraday Soc.* **1** (1947) 246.

[31]J. O'M. Bockris and R. Parsons, *Trans. Faraday Soc.* **45** (1949) 916.

[32]S. V. Gorbachev, *Zh. Fiz. Khim.* **24** (1950) 888.

[33]V. A. Roiter, *Ukr. Khim. Zh.* **16** (1950) 225.

[34]B. Post and C. F. Hiskey, *J. Am. Chem. Soc.* **72** (1950) 4203.

[35]M. I. Temkin, *Tr. Soveshch. Elektrokhim.*, Akad. Nauk SSSR, Otdel Khim. Nauk, 1950, p. 181 (publ. 1953).

[36]J. O'M. Bockris, R. Parsons, and H. Rosenberg, *Trans. Faraday Soc.* **47** (1951) 766.

[37]R. Parsons and J. O'M. Bockris, *Trans. Faraday Soc.* **47** (1951) 914.

[38]R. Parsons, *Z. Elektrochem.* **55** (1951) 111.

[39]J. E. B. Randles, *Trans. Faraday Soc.* **48** (1952) 828.

[40]J. O'M. Bockris and N. Pentland, *Trans. Faraday Soc.* **48** (1952) 833.

[41]J. E. B. Randles and K. W. Somerton, *Trans. Faraday Soc.* **48** (1952) 937.

[42]J. O'M. Bockris and E. C. Potter, *J. Electrochem. Soc.* **99** (1952) 169.

[43]H. Gerischer, *Z. Phys. Chem.* **202** (1953) 302.

[44]A. L. Rotinyan, V. L. Kheifets, E. S. Kozich, and O. P. Kalganova, *Dokl. Akad. Nauk SSSR* **88** (1953) 301.

[45]J. O'M. Bockris, in *Modern Aspects of Electrochemistry*, No. 1, Ed. by J. O'M. Bockris and B. E. Conway, Butterworths, London, 1954, p. 180.

[46]P. Delahay, *New Instrumental Methods in Electrochemistry*, Interscience, New York, 1954.

[47]P. Delahay and C. Mattax, *J. Am. Chem. Soc.* **76** (1954) 5314.

[48]S. V. Gorbachev and Yu. N. Yurkevich, *Zh. Fiz. Khim.* **28** (1954) 1120.

[49]V. I. Arklarov, *Zh. Tekh. Fiz.* **24** (1954) 375.

[50]R. M. Vasenin and S. V. Gorbachev, *Zh. Fiz. Khim.* **28** (1954) 1922.

[51]S. V. Gorbachev and R. M. Vasenin, *Zh. Fiz. Khim.* **28** (1954) 135, 1795, 1928.

[52]H. J. Reiser and H. Fisher, *Z. Elektrochem.* **58** (1954) 668.

[53]O. M. Poltorak, *Zh. Fiz. Khim.* **28** (1954) 1845.

[54]S. V. Gorbachev and Ya. I. Vabel', *Zh. Fiz. Khim.* **29** (1955) 15, 23.

[55]M. I. Temkin and A. N. Frumkin, *Zh. Fiz. Khim.* **29** (1955) 1513.

[56]O. M. Poltorak, *Zh. Fiz. Khim.* **29** (1955) 2249.

[57]F. Gutmann and L. M. Simons, *Proceedings of the 6th International Committee on Electrochemical Thermodynamics and Kinetics*, Butterworths, London, 1955, p. 201.

[58]Yu. Yu. Matulis and O. K. Gal'dikene, *Liet. TSR Mokslu Akad. Darb.* **B3** (1955) 23; *Chem. Abstr.* **53** (1959) 11057c.

[59]H. Gerischer and K.-E. Staubach, *Z. Phys. Chem. (N. F.)* **6** (1956) 118.

[60]M. I. Temkin and A. N. Frumkin, *Zh. Fiz. Khim.* **30** (1956) 1162.

[61]A. V. Izmailov, *Zh. Fiz. Khim.* **30** (1956) 2813.

[62]I. A. Ammar and S. A. Awad, *J. Phys. Chem.* **60** (1956) 1290.

[63]D. P. Semchenko and I. I. Appenin, *Nauchn. Tr. Novocherkask. Politekh. Inst.* **34** (1956) 51; *Ref. Zh. Khim.* **1957**, Abstr. No. 26316.

[64]J. Horiuti, *J. Res. Inst. Catal. Hokkaido Univ.* **4** (1956) 55.

[65]A. V. Izmailov, *Trudy Chetvertogo Soveshchaniya Elektrokhim. Moscow,* 1956, p. 453 (publ. 1959); *Chem. Abstr.* **54** (1956) 13895a.

[66]M. Balkanski, *Z. Elektrochem.* **61** (1957) 141.

[67]B. E. Conway and J. O'M. Bockris, *Can. J. Chem.* **35** (1957) 1124.

[68]J. Horiuti, T. Keii, M. En'yo, and M. Fukuda, *J. Res. Inst. Catal. Hokkaido Univ.* **5** (1957) 40.

[69]S. V. Gorbachev and O. B. Khachaturyan, *Zh. Fiz. Khim.* **31** (1957) 2526.

[70]A. V. Pamfilova and A. I. Tsinman, *Ukr. Khim. Zh.* **23** (1957) 579.

[71]L. U. Sok, A. L. Rotinyan, and N. P. Fedot'ev, *Zh. Fiz. Khim.* **32** (1958) 952.

[72]B. E. Conway and J. O'M. Bockris, *Proc. R. Soc. (London), Ser. A* **248** (1958) 394.

[73]I. A. Ammar and S. Riad, *J. Phys. Chem.* **62** (1958) 660.

[74]H. Gerischer and M. Krause, *Z. Phys. Chem. (N. F.)* **14** (1958) 184.

[75]A. A. Vlcek, *Collect. Czech. Chem. Commun.* **24** (1959) 3538.

[76]J. Yeager, J. P. Cels, E. Yeager, and F. Hovorka, *J. Electrochem. Soc.* **106** (1959) 328.

[77]S. Minc and J. Sobkowski, *Bull. Acad. Pol. Sci., Ser. Sci. Chim. Geol. Geogr.* **7** (1959) 29.

[78]K. M. Ioshi, W. Mehl, and R. Parsons, *Transactions of the Symposium on Electrode Processes*, Philadelphia, 1959, Ed. by E. Yeager, John Wiley & Sons, New York, 1961, p. 249.

[79]B. E. Conway, *Theory and Principles of Electrode Processes*, Ronald Press, New York, 1965.

[80]N. A. Izgarishev and S. V. Gorbachev, *Kurs Teoreticheskoi Elektrokhimii*, Goskhimizdat, Moscow, 1971.

[81]V. V. Skorchelletti, *Teoreticheskaya Elektrokhimiya*, Khimiya, Moscow, 1974.

[82]E. Gileadi, E. Kirowa-Eisner, and J. Penciner, *Interfacial Electrochemistry*, Addison-Wesley, Reading, Massachusetts, 1975.

[83]J. O'M. Bockris and S. U. M. Khan, *Quantum Electrochemistry*, Plenum Press, New York, 1979.

[84]B. B. Damaskin and O. A. Petrii, *Vvedenie v Elektrokhimicheskuyu Kinetiku*, Vysshaya Shkola, Moscow, 1983.

[85]R. Tamamushi, *Rev. Polarog. (Kyoto)* **10** (1962) 1; cited in *Ref. Zh. Khim.* **24** (1962) B906.

[86]S. Barnartt, *Electrochim. Acta* **13** (1968) 901.

[87]M. J. Weaver, *J. Phys. Chem.* **80** (1976) 2645.

[88]B. E. Conway, *Trans. R. Soc. Can., Sect. III* **54** (1960) 19.

[89]M. Salomon, *J. Electrochem. Soc.* **113** (1966) 940.

[90]S. Barnart, *J. Phys. Chem.* **70** (1966) 940.

[91]L. I. Krishtalik, in *Advances in Electrochemistry and Electrochemical Engineering*, Vol 7, Ed. by P. Delahay, Interscience, New York, 1970, p. 283.

[92]A. J. Appleby, *Catal. Rev.* **4** (1970) 221.

[93]L. I. Krishtalik, in *Comprehensive Treatise of Electrochemistry*, Vol. 7, Ed. by B. E. Conway, J. O'M. Bockris, E. Yeager, S. U. M. Khan, and R. E. White, Plenum Press, New York, 1983, p. 87.

[94]L. I. Krishtalik, in *Dvoinoi Sloi i Elektrodnaya Kinetika*, Nauka, Moscow, 1981, p. 198.

[95]B. E. Conway, in *Modern Aspects of Electrochemistry*, No. 16, Ed. by B. E. Conway, R. E. White, and J. O'M. Bockris, Plenum Press, New York, 1985, p. 103.

[96]A. R. Despić and J. O'M. Bockris, *J. Chem. Phys.* **32** (1960) 389.

[97]L. I. Krishtalik, *Usp. Khim.* **34** (1965) 1891.

[98]P. P. Schmidt and H. B. Mark, *J. Chem. Phys.* **43** (1965) 3291.

[99]P. P. Schmidt and H. B. Mark, *J. Chem. Phys.* **45** (1966) 761.

[100]J. M. Hale, *J. Electroanal. Chem.* **11** (1966) 154.

[101]L. I. Krishtalik, *Electrochem. Acta* **13** (1968) 1045.

[102]B. E. Conway and D. J. MacKinnon, *J. Electrochem. Soc.* **116** (1969) 1665.

[103]G. Hills and S. Hsieh, *Chem.-Ing.-Tech.* **44** (1972) 216.

[104]B. G. Chauhan, W. R. Fawcett, and T. A. McCarrick, *J. Electroanal. Chem.* **58** (1975) 275.

[105]G. E. Titova and L. I. Krishtalik, *Elektrokhimiya* **13** (1977) 897.

[106]I. G. Medvedev, *Elektrokhimiya* **15** (1979) 713.

[107]L. I. Krishtalik, *J. Electroanal. Chem.* **100** (1979) 547.

[108]J. T. Hupp and M. J. Weaver, *J. Electroanal. Chem.* **145** (1983) 43.

[109]P. G. Dzhavalkhidze, A. A. Kornyshev, and L. I. Krishtalik, *J. Electroanal. Chem.* **228** (1987) 329.

[110] V. M. Tsionskii and L. B. Kriksunov, *Elektrokhimiya* **24** (1988) 311.

[111] N. S. Hush, *J. Chem. Phys.* **28** (1958) 962.

[112] K. J. Laidler, *Can. J. Chem.* **37** (1959) 138.

[113] R. A. Marcus, *Annu. Rev. Phys. Chem.* **15** (1964) 155.

[114] V. G. Levich, in *Advances in Electrochemistry and Electrochemical Engineering*, Vol. 4, Ed. by P. Delahay, Interscience, New York, 1966, p. 249.

[115] R. R. Dogonadze, in *Reactions of Molecules at Electrodes*, Ed. by N. S. Hush, Wiley-Interscience, London, 1971, p. 135.

[116] B. E. Conway and J. O'M. Bockris, *Electrochim. Acta* **3** (1961) 340.

[117] M. Salomon, C. G. Enke, and B. E. Conway, *J. Chem. Phys.* **43** (1965) 3989.

[118] D. B. Matthews and J. O'M. Bockris, in *Modern Aspects of Electrochemistry*, No. 6, Ed. by J. O'M. Bockris and B. E. Conway, Plenum Press, New York, 1971, p. 242.

[119] C. M. Marschoff and P. J. Aragon, *J. Electrochem. Soc.* **123** (1976) 213.

[120] A. J. Appleby, *Surf. Sci.* **27** (1971) 225.

[121] K. D. Allard and R. Parsons, *Chem.-Ing.-Tech.* **44** (1972) 201.

[122] S. V. Gorbachev, A. G. Atanasyants, and Yu. M. Senatorov, *Zh. Fiz. Khim.* **48** (1974) 3056.

[123] B. M. Grafov and A. N. Frumkin, *Elektrokhimiya* **11** (1975) 1833.

[124] P. Sideswaran, *Indian J. Chem.* **13** (1975) 1081.

[125] B. E. Conway, E. M. Beatty, and P. A. DeMaine, *Electrochim. Acta* **7** (1962) 39.

[126] R. Piontelli, L. Peraldo-Bicelli, M. Graziano, and A. LaVecchia, *Atti. Accad. Naz. Lincei, Rend. Cl. Sci. Fis. Mat. Nat.* **32** (1962) 445; *Chem. Abstr.* **58** (1963) 5263g.

[127] A. L. Rotinyan and N. M. Kozhevnikova, *Zh. Fiz. Khim.* **37** (1963) 1818.

[128] B. E. Conway and M. Salomon, *J. Chem. Phys.* **41** (1964) 3169.

[129] V. V. Batrakov, A. P. P'yankova, and Z. A. Iofa, *Zh. Fiz. Khim.* **38** (1964) 1340.

[130] H. Kita and O. Nomura, *J. Res. Inst. Catal. Hokkaido Univ.* **12** (1965) 107.

[131] M. J. Joncich, L. S. Stewart, and F. A. Posey, *J. Electrochem. Soc.* **112** (1965) 717.

[132] J. N. Butler and M. Dienst, *J. Electrochem. Soc.* **112** (1965) 226.

[133] J. N. Butler and M. L. Meehan, *Trans. Faraday Soc.* **62** (1965) 3524.

[134] J. O'M. Bockris and D. B. Matthews, *J. Chem. Phys.* **44** (1966) 298.

[135] J. O'M. Bockris and D. B. Matthews, *Electrochim. Acta* **11** (1966) 143.

[136] A. J. Vijh, *J. Phys. Chem.* **72** (1968) 1148.

[137] V. I. Bystrov and L. I. Krishtalik, *Elektrokhimiya* **4** (1968) 233.

[138] B. G. Dekker, M. Sluyters-Rehbach, and J. M. Sluyters, *J. Electroanal. Chem.* **21** (1969) 17.

[139] B. G. Dekker, M. Sluyters-Rehbach, and J. M. Sluyters, *J. Electroanal. Chem.* **23** (1969) 137.

[140] I. I. Pyshnograeva, A. M. Skundin, Yu. B. Vasil'ev, and V. S. Bagotskii, *Elektrokhimiya* **6** (1970) 142.

[141] T. S. Lee, *J. Electrochem. Soc.* **118** (1971) 1278.

[142] A. T. Kuhn and M. Byrne, *Electrochim. Acta* **16** (1971) 391.

[143] I. I. Kudryashov and E. S. Burmistrov, *Zh. Fiz. Khim.* **45** (1971) 838.

[144] U. V. Pal'm and T. T. Teno, *Elektrokhimiya* **10** (1974) 826.

[145] E. N. Baibatyrov, V. Sh. Palanker, and D. V. Sokol'skii, *Elektrokhimiya* **10** (1974) 158.

[146] D. V. Sokol'skii, V. Sh. Palanker, and E. N. Baibatyrov, *Electrochim. Acta* **20** (1975) 71.

[147] R. Notoya and A. Matsuda, *J. Res. Inst. Catal. Hokkaido Univ.* **27** (1979) 1.

[148] R. Notoya and A. Matsuda, *J. Res. Inst. Catal. Hokkaido Univ.* **27** (1979) 95.

[149] A. Hamelin, G. Picq, and P. Vennereau, *J. Electroanal. Chem.* **148** (1983) 61.

[150] A. Belanger and A. K. Vijh, *Surf. Coat. Technol.* **28** (1986) 93.

[151] U. Frese and U. Stimming, *J. Electroanal. Chem.* **198** (1986) 409.

[152] U. Frese and W. Schmickler, *Ber. Bunsenges. Phys. Chem.* **92** (1988) 1412.

[153] T. Hurlen, *Electrochim. Acta* **7** (1962) 653.

[154] D. L. Manning, *Talanta* **10** (1963) 225.

[155]Y. Ueno, M. Tsmiki, F. Hine, and S. Yoshizawa, *Denki Kagaku* **32** (1964) 211; *Chem. Abstr.* **62** (1965) 4885a.

[156]S. Gastev, *Zh. Prikl. Khim.* **40** (1967) 820.

[157]R. Yu. Bek, E. A. Nechaev, and N. T. Kudryavtsev, *Elektrokhimiya* **3** (1967) 1121.

[158]T. G. Clarke, N. A. Hampson, J. B. Lee, J. R. Morley, and B. Scanlon, *Can. J. Chem.* **46** (1968) 3437.

[159]M. C. Petit, *Electrochim. Acta* **13** (1968) 557.

[160] E. M. Solov'ev and B. P. Yur'ev, *Zh. Prikl. Khim.* **49** (1976) 2521.

[161]T. Hurlen, *Acta Chem. Scand.* **14** (1960) 1564.

[162]P. N. Kovalenko and K. N. Bagdasarov, *Ukr. Khim. Zh.* **26** (1960) 573.

[163]S. Inouye and H. Imai, *Bull. Chem. Soc. Jpn.* **33** (1960) 149.

[164]O. A. Khan, *Zh. Prikl. Khim.* **33** (1960) 1347.

[165]A. Rius, I. M. Tordesillas, and A. Sacristan, *Electrochim. Acta* **4** (1961) 62.

[166]T. Hurlen, *Acta Chem. Scand.* **15** (1961) 621.

[167]H. H. Bauer and R. B. Goodwin, *Aust. J. Chem.* **15** (1962) 391.

[168]G. Gavioli and P. Papoff, *Ric. Ser. II, Sez.*, *A1*, **1961**, 193; *Chem. Abstr.* **57** (1962) 9579f.

[169]K. S. Udupa, G. S. Subramanian, and H. V. K. Udupa, *Plating* **49** (1962) 1274.

[170]T. Hurlen, *Acta Chem. Scand.* **16** (1962) 1353.

[171]Yu. M. Loshkarev, *Ukr. Khim. Zh.* **29** (1963) 918.

[172]R. Tamamushi and N. Tanaka, *Z. Phys. Chem. (N. F.)* **39** (1963) 117.

[173]M. Ya. Poperka and O. I. Avramenko, *Zh. Prikl. Khim.* **38** (1965) 1783.

[174]A. I. Alekperov and F. S. Novruzova, *Elektrokhimiya* **5** (1969) 97.

[175]A. I. Krasil'shchikov, *Elektrokhimiya* **6** (1970) 341.

[176]S. G. Meibuhr, *J. Electrochem. Soc.* **117** (1970) 56.

[177]S. A. Awad and A. Kassab, *J. Electroanal. Chem.* **26** (1970) 127.

[178]N. A. Hampson and R. J. Latham, *J. Electroanal. Chem.* **32** (1971) 175.

[179]N. S. Ageenko, *Zashch. Met.* **10** (1974) 460.

[180]B. I. Tomilov and M. A. Loshkarev, *Dokl. Akad. Nauk SSSR* **151** (1963) 894.

[181]H. P. Agarwal, *J. Electroanal. Chem.* **5** (1963) 236.

[182]A. C. Makrides, *J. Electrochem. Soc.* **111** (1964) 392, 400.

[183]R. Parsons and E. Passeron, *J. Electroanal. Chem.* **44** (1973) 367.

[184]J. O'M. Bockris, R. J. Mannan, and A. Damjanovic, *J. Chem. Phys.* **48** (1968) 1898.

[185]D. Galizzioli and S. Trasatti, *J. Electroanal. Chem.* **44** (1973) 367.

[186]P. Bindra, H. Gerischer, and L. M. Peter, *J. Electroanal. Chem.* **57** (1974) 435.

[187]L. M. Peter, W. Dür, P. Bindra, and H. Gerischer, *J. Electroanal. Chem.* **71** (1976) 31.

[188]A. M. T. Olmedo, R. Pereiro, and D. J. Schiffrin, *J. Electroanal. Chem.* **74** (1976) 19.

[189]A. J. Appleby, *J. Electroanal. Chem.* **27** (1970) 325, 335, 347.

[190]A. J. Appleby, *J. Electrochem. Soc.* **117** (1970) 328, 1157, 1373.

[191]S. M. Park, S. Ho, S. Aruliach, M. F. Weber, C. A. Ward, and R. D. Venter, *J. Electrochem. Soc.* **133** (1986) 1641.

[192]G. Coturier, D. W. Kirk, P. J. Hyde, and S. Srinivasan, *Electrochim. Acta* **32** (1987) 995.

[193]A. J. Appleby and B. S. Baker, *J. Electrochem. Soc.* **125** (1978) 404.

[194]J. C. Huang, R. K. Sen, and E. Yeager, *J. Electrochem. Soc.* **126** (1979) 786.

[195]M. Inai, C. Iwakura, and H. Tamura, *Denki Kagaku* **48** (1980) 229.

[196]A. Damjanovic, A. T. Walsh, and D. B. Shepa, *J. Phys. Chem.* **94** (1990) 1967.

[197]H. Wroblova, B. J. Piersma, and J. O'M. Bockris, *J. Electroanal. Chem.* **6** (1963) 401.

[198]S. Toshima and H. Okaniwa, *Denki Kagaku* **34** (1966) 958.

[199]R. V. Marvet and O. A. Petrii, *Elektrokhimiya* **3** (1967) 518.

[200]M. Aubar, *J. Am. Chem. Soc.* **89** (1967) 1263.

[201]J. W. Johnson and L. D. Gilmartin, *J. Electroanal. Chem.* **15** (1967) 231.

[202]L. M. Dané, L. J. J. Jansen, and J. G. Hoogland, *Electrochim. Acta* **13** (1968) 507.

[203]S. F. Belewski and L. A. Leonova, *Elektrokhimiya* **6** (1970) 440.

[204]N. R. DeTaconi, A. J. Calandra, and A. J. Arvia, *J. Electroanal. Chem.* **57** (1974) 325.

[205]R. D. Grypa and J. T. Maloy, *J. Electrochem. Soc.* **122** (1975) 509.

[206]M. Inai, C. Iwakura, and H. Tamura, *Electrochim. Acta* **24** (1979) 993.

[207]R. W. Tsang, D. C. Johnson, and G. R. Luecke, *J. Electrochem. Soc.* **131** (1984) 2369.

[208]B. E. Conway, *Electrochemical Data*, Elsevier, Amsterdam, 1952.

[209]N. Tanaka and R. Tamamushi, *Electrochim. Acta* **9** (1964) 963.

[210]A. V. Izmailov, *Nauchn. Dokl. Vyssh. Shk. Khim. Khim. Tekhnol.* **1959**, 23, 28, 435; *Chem. Abstr.* **53** (1959) 911h, 912e, 914j.

[211]Yu. A. Korostelin, S. V. Gorbachev, and Z. N. Ryantseva, *Zh. Fiz. Khim.* **40** (1960) 1909.

[212]S. V. Gorbachev and V. P. Kondrat′ev, *Zh. Fiz. Khim.* **35** (1961) 2400.

[213]E. M. Schmidt and S. V. Gorbachev, *Zh. Fiz. Khim.* **36** (1962) 2795.

[214]G. A. Emel′yanenko and V. I. Atanasenko, *Zh. Fiz. Khim.* **37** (1963) 1854.

[215]S. V. Gorbachev and A. M. Aboimov, *Zh. Fiz. Khim.* **40** (1966) 1406.

[216]I. Madi, *J. Inorg. Nucl. Chem.* **26** (1964) 2135, 2149.

[217]J. W. Johnson, H. Wroblova, and J. O′M. Bockris, *J. Electrochem. Soc.* **111** (1964) 863.

[218]J. O′M. Bockris, H. Wroblova, E. Gileadi, and B. J. Piersma, *Trans. Faraday Soc.* **61** (1965) 2531.

[219]L. I. Krishtalik, *Elektrokhimiya* **2** (1966) 1176.

[220]S. V. Gorbachev and A. M. Aboimov, *Zh. Fiz. Khim.* **40** (1966) 2757.

[221]Yu. Bochkov, A. M. Aboimov, and S. V. Gorbachev, *Zh. Fiz. Khim.* **40** (1966) 1148.

[222]Yu. A. Korostelin and S. V. Gorbachev, *Zh. Fiz. Khim.* **40** (1966) 2324.

[223]A. Damjanovic, A. Dey, and J. O′M. Bockris, *Electrochim. Acta* **11** (1966) 791.

[224]S. V. Gorbachev and N. D. Kalugina, *Zh. Fiz. Khim.* **41** (1967) 884, 1521.

[225]N. I. Gusev and P. V. Rakchev, *Zh. Fiz. Khim.* **42** (1968) 1983.

[226]M. S. Shapnik, K. A. Zinkicheva, and N. V. Gudin, *Zashch. Met.* **5** (1969) 647.

[227]R. Mraz, V. Srb, and S. Tichy, *Electrochim. Acta* **18** (1973) 551.

[228]S. V. Gorbachev and A. M. Aboimov, *Zh. Fiz. Khim.* **47** (1973) 2039.

[229]E. N. Potapova, L. I. Krishtalik, and I. A. Bagotskaya, *Elektrokhimiya* **10** (1974) 49, 53.

[230]S. D. Kamyshchenko, I. V. Kudryashov, M. A. Nikulenkova, and E. M. Makaryan, *Zh. Fiz. Khim.* **48** (1974) 1755, 1761.

[231]A. V. Babanin, A. M. Aboimov, and S. V. Gorbachev, *Zh. Fiz. Khim.* **49** (1975) 1290.

[232]D. W. DeBerry and N. A. Hackerman, *J. Electrochem. Soc.* **123** (1976) 1174.

[233]D. B. Šepa, M. V. Vojnović, Lj. M. Vračar, and A. Damjanovic, *Electrochim. Acta* **29** (1984) 1169.

[234]D. B. Šepa, M. V. Vojnović, Lj. M. Vračar, and A. Damjanovic, *Electrochim. Acta* **31** (1986) 91, 97.

[235]A. Damjanovic, D. B. Šepa, Lj. M. Vračar and M. V. Vojnović, *Ber. Bunsenges. Phys. Chem.* **90** (1986) 1231.

[236]V. M. Tsionskii and I. B. Kriksunov, *J. Electroanal. Chem.* **204** (1986) 131.

[237]V. I. Shimulis and P. Oiola, *Zh. Fiz. Khim.* **45** (1971) 2105.

[238]N. S. Fedorova and O. B. Khachaturyan, *Zh. Fiz. Khim.* **46** (1972) 68.

[239]B. B. Graves, in *Proceedings of the Symposium on Electrocatalysis*, Ed. by M. Breiter, The Electrochemical Society, Princeton, New Jersey, 1974, p. 365.

[240]G. C. Baker and A. W. Gardner, *J. Electroanal. Chem.* **65** (1975) 95.

[241]A. M. Aboimov, S. V. Gorbachev, A. V. Babanin, and V. N. Gryzlov, *Zh. Fiz. Khim.* **49** (1975) 1333.

[242]M. L. Brown and G. N. Walton, *J. Appl. Electrochem.* **6** (1976) 551.

[243]R. P. VanDyne and C. N. Reilley, *Anal. Chem.* **44** (1972) 141, 153, 159.

[244]M. J. Weaver, *J. Phys. Chem.* **83** (1979) 1748.

[245]S. O. Bernhardsson, *Corros. Sci.* **14** (1974) 611.

[246]D. R. Crow and S. L. Ling, *Talanta* **19** (1972) 915.

[247]E. A. Guggenheim, *J. Chem. Phys.* **33** (1929) 842.

[248] R. Parsons, in *Modern Aspects of Electrochemistry*, No. 1, Ed. by J. O'M. Bockris and B. E. Conway, Butterworths, London, 1954, p. 103.

[249] S. Trasatti, *J. Electroanal. Chem.* **52** (1974) 313.

[250] International Union for Pure and Applied Chemistry, *Pure Appl. Chem.* **37** (1974) 499.

[251] D. J. Ives and G. I. Janz, *Reference Electrodes*, Academic Press, New York, 1961.

[252] A. N. Frumkin and L. I. Krishtalik, *Electrokhimiya* **11** (1975) 1793.

[253] G. Milazzo and S. Caroli, *Tables of Standard Electrode Potentials*, John Wiley & Sons, New York, 1978.

[254] D. B. Šepa, M. V. Vojnović, M. Stojanović, and A. Damjanovic, *J. Electrochem. Soc.* **134** (1987) 845.

[255] D. B. Šepa, Lj. M. Vračar, M. V. Vojnović, and A. Damjanovic, *Electrochim. Acta* **31** (1986) 1401.

2

The Electrochemical Activation of Catalytic Reactions

Constantinos G. Vayenas, Milan M. Jaksic,
Symeon I. Bebelis, and Stylianos G. Neophytides

Department of Chemical Engineering, University of Patras, Patras GR-26500, Greece

I. INTRODUCTION

The use of electrochemistry to activate and precisely tune heterogeneous catalytic processes is a new development[1–7] which originally emerged due to the existence of solid electrolytes. Depending on their composition, these specific anionic or cationic conductor materials exhibit substantial electrical conductivity at temperatures between 25 and 1000°C. Within this broad temperature range, which covers practically all heterogeneous catalytic reactions, solid electrolytes can be used as reversible *in situ* promoter donors or poison acceptors to affect the catalytic activity and product selectivity of metals deposited on solid electrolytes in a very pronounced, reversible, and, to some extent, predictable manner. This is accomplished by cofeeding the reactants (e.g., C_2H_4 plus O_2 or H_2 plus O_2) on the working electrode of a solid electrolyte cell:

gaseous reactants	catalyst working electrode	|solid electrolyte|	counter electrode	auxiliary gas
(e.g., $C_2H_4 + O_2$)	(e.g., Pt, Rh, Ag)	(e.g., $ZrO_2 - Y_2O_3$)	(e.g., Pt)	(e.g., O_2)

and using the working electrode both as an electrode and as a catalyst for the heterogeneous catalytic reaction under study, for example, C_2H_4 or H_2 oxidation. Upon varying the potential of the working electrode/catalyst,

Modern Aspects of Electrochemistry, Number 29, edited by John O'M. Bockris *et al.*
Plenum Press, New York, 1996.

it is found that not only the electrocatalytic (net charge-transfer) reaction rate is affected, but also the catalytic (no net charge-transfer) reaction rate changes in a very pronounced, controlled, and reversible manner. The rate of the catalytic reaction can be up to 100 times larger than the open-circuit catalytic rate and up to 3×10^5 times larger than the rate of ion supply to the catalyst through the solid electrolyte. This is why this novel effect has been termed nonfaradaic electrochemical modification of catalytic activity (NEMCA effect[1–6]). The terms electrochemical promotion[5] (EP) and *in situ* controlled promotion[7] (ICP) have been also proposed for the NEMCA effect. This effect has been studied already for more than 30 catalytic reactions and does not seem to be limited to any specific type of metal or solid electrolyte, particularly in view of the recent demonstration of NEMCA using aqueous electrolyte solutions.[8]

Wagner was the first to propose the use of solid electrolytes to measure *in situ* the thermodynamic activity of oxygen on metal catalysts.[9] This led to the technique of solid electrolyte potentiometry.[10] Pancharatnam *et al.* were the first to use solid electrolyte cells to carry out electrocatalytic reactions such as NO decomposition.[11] The use of solid electrolyte cells for "chemical cogeneration," that is, for the simultaneous production of electrical power and industrial chemicals, was first demonstrated in 1980.[12] The first "nonfaradaic" enhancement in heterogeneous catalysis was reported in 1981 for the case of ethylene epoxidation on Ag electrodes,[13] but it was only in 1988 that it was realized that electrochemical promotion is a general phenomenon.[1–5,14] In addition to the group which first reported the electrochemical promotion effect,[1–4,8,13,14] the groups of Sobyanin,[15,16] Haller,[17,18] Lambert,[7,19] Stoukides,[20] and Comninellis[21] have also made significant contributions in this area.

The importance of NEMCA in electrochemistry and in heterogeneous catalysis and surface science has been discussed by Bockris and Minevski[6] and Pritchard,[5] respectively.

Detailed[4] and shorter[22–27] reviews of the electrochemical promotion literature have been published recently, mainly addressed to the catalytic community. Earlier applications of solid electrolytes in catalysis, including solid electrolyte potentiometry and electrocatalysis, have been reviewed previously.[28–31] The present chapter is the first detailed review of the electrochemical activation of catalytic reactions addressed to the electrochemical community. We stress the electrochemical aspects of electrochemical promotion and hope that the text will be found useful and easy to follow by all readers including those not frequently using the catalytic and surface science methodology and terminology. Throughout the text, we use the term "catalytic" to denote reactions involving no net charge transfer, such as, for example, the reaction of $H_2(g)$ and $O_2(g)$ to produce $H_2O(g$ or $l)$ on a Pt surface, and reserve the term "electrocatalytic" for reactions involving a net charge transfer, such as, for example, the anodic oxidation of H_2 or the cathodic reduction of O_2.

As will be shown in this chapter, the effect of electrochemical promotion (EP), or NEMCA, or *in situ* controlled promotion (ICP), is due to an electrochemically induced and controlled migration (backspillover) of ions from the solid electrolyte onto the gas-exposed, that is, catalytically active, surface of metal electrodes. It is these ions that, accompanied by their compensating (screening) charge in the metal, form an effective

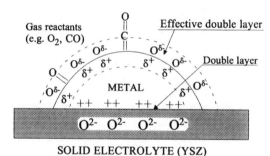

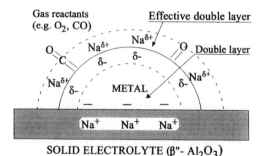

Figure 1. Schematic representations of a metal electrode deposited on an O^{2-}-conducting and on a Na^+-conducting solid electrolyte, showing the location of the metal–electrolyte double layer and of the effective double layer created at the metal/gas interface due to potential-controlled ion migration (backspillover).

electrochemical double layer on the gas-exposed catalyst surface (Fig. 1) and affect the catalytic phenomena taking place there in a very pronounced, reversible, and controlled manner.

II. SOLID ELECTROLYTES AND THEIR PREVIOUS CATALYTIC AND ELECTROCATALYTIC APPLICATIONS

1. Solid Electrolytes

Michael Faraday was the first to observe, in 1834, that solid PbF_2 when heated at 500°C becomes an electrical conductor. It took almost a century to explain this observation and establish that PbF_2 is a F^--ion conductor. In the meantime, other solid electrolytes such as AgI, a Ag^+ conductor, had been discovered by Tubandt and Strock, and it soon became apparent that ions can diffuse as rapidly in some solids as in aqueous salt solutions. The atomistic interpretation of ionic conduction in solids was largely established by the pioneering work of Joffé, Frenkel, Wagner, and Schottky in the twenties and early thirties.[32] These works established that ion conduction can take place either by hopping of ions through a series of interstitial sites (Frenkel disorder) or by hopping of vacancies among lattice positions (Schottky disorder).

Today, the term solid electrolyte or fast ionic conductor or, sometimes, superionic conductor is used to describe solid materials whose conductivity is wholly due to ionic displacement. Mixed conductors exhibit both ionic and electronic conductivity. Solid electrolytes range from hard, refractory materials, such as 8 mol % Y_2O_3-stabilized ZrO_2 (YSZ) or sodium β-Al_2O_3 ($Na_{1+x}Al_{11}O_{17+x/2}$), to soft proton-exchange polymeric membranes, such as Du Pont's Nafion, and include compounds that are stoichiometric (AgI), nonstoichiometric (sodium β-Al_2O_3), or doped (YSZ). The preparation, properties, and some applications of solid electrolytes have been discussed in a number of books[33,34] and reviews.[35,36] The main commercial application of solid electrolytes is in gas sensors.[37] Another emerging application is in solid oxide fuel cells.[38]

The classification of solid electrolytes is usually based on the ion mainly responsible for the conductivity. There exist:

(i) Oxygen ion conductors. They are solid solutions of divalent and trivalent metal oxides (e.g., Y_2O_3, Yb_2O_3, CaO) in quadrivalent metal oxides (e.g., ZrO_2, ThO_2, CeO_2). Calcia- or yttria-stabilized zirconia (YSZ), containing 5–15 mol % CaO or 6–10 mol % Y_2O_3 in ZrO_2, is widely used in oxygen sensors, normally in the temperature range 400–1200°C.

(ii) H^+ and Li^+ conductors. Several polymeric solid electrolytes belong to this category. Of particular importance are the proton exchange membranes (PEM), such as Nafion 117, which is a copolymer of poly(tetrafluoroethylene) and poly(sulfonylfluoride) containing pendant sulfonic acid groups, which exhibit substantial conductivity at room temperature. High cationic conductivity is also exhibited by several alkali salt solutions in poly(ethylene oxide). Proton conduction is also exhibited by $CsHSO_4$,[15] by H^+-substituted β''-Al_2O_3,[34,36] and by $SrCeO_3$-based compounds.[39]

(iii) Na^+ conductors. These are β- and β''-aluminas, which are non-stoichiometric compounds corresponding to $Na_{1+x}Al_{11}O_{17+x/2}$ ($0.15 \leq x \leq 0.3$) and $Na_{1+x}M_xAl_{11-x}O_{17}$, respectively, where M is a divalent metal (e.g., Mg^{2+}, Ni^{2+}, Zn^{2+}). They exhibit high conductivity in the range 150–300°C.

(iv) K^+, Cs^+, Rb^+, Tl^+ conductors. They are substituted β- and β''-Al_2O_3 and are conductive in the range 200–400°C.

(v) Ag^+ conductors, e.g., α-AgI, $RbAg_4I_5$, and Ag_2HgI_4, which are conductive in the range 150–350°C.

(vi) Cu^+ conductors, e.g., Cu_2Se and KCu_4I_5, which are conductive in the range 250–400°C.

(vii) F^- conductors, e.g., PbF_2 and CaF_2, which are conductive above 500 and 600°C, respectively.

Detailed information about the conductivity of solid electrolytes can be found elsewhere.[4,33–36] As shown in Fig. 2, the temperature dependence of the ionic conductivity σ can, in general, be described by the semiempirical equation

$$\sigma = (\sigma_o/T) \exp(-E_A/k_bT) \tag{1}$$

where σ_o is a function of the ionic charge, the concentration of the mobile ions, and the frequency with which these ions attempt to move to a neighboring site (attempt frequency); E_A is the activation energy for defect motion, and k_b is the Boltzmann constant. The activation energy E_A is usually on the order of 0.5–2 eV. The minimum ionic conductivity value of a solid electrolyte for practical fuel cell applications[38] is 0.1– 1 Ω^{-1} cm^{-1}. This places very stringent restrictions on the choice of material and operating temperature. For catalytic (promotional) and sensor applications, however, much lower conductivity values (~10^{-4} Ω^{-1} cm^{-1}) are

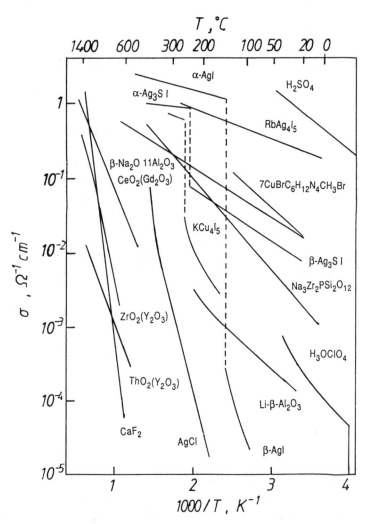

Figure 2. Temperature dependence of the ionic conductivity of some solid electrolytes.[4,33] The conductivity of concentrated H_2SO_4 (37 wt %) is included for comparison. (Reprinted, with permission from Elsevier Science Publishers B.V.,
Amsterdam, from Ref. 4.)

usually sufficient. This permits the use of a large variety of solid electrolytes over a very wide temperature range. Electrochemical promotion (EP) studies have so far utilized:

(i) yttria-stabilized-zirconia (YSZ), an O^{2-} conductor, at temperatures between 280 and 650°C.[1,2,4,17,40,41,43–54]

(ii) β''-Al_2O_3, a Na^+ conductor at temperatures between 180 and 400°C.[3,4,7,18,19,42]

(iii) $CsHSO_4$[15] and Nafion,[55] which are proton conductors, at 150 and 25°C, respectively.

(iv) CaF_2, a F^- conductor, at temperatures between 550 and 700°C.[56]

(v) Aqueous KOH solutions (0.01–0.2 M) at temperatures between 25 and 60°C.[8,57]

2. Solid Electrolyte Potentiometry (SEP)

When a solid electrolyte component is interfaced with two electronically conducting (e.g., metal) films (electrodes), a solid electrolyte galvanic cell is formed (Fig. 3). Cells of this type with YSZ solid electrolyte are used as oxygen sensors.[37] The potential difference V_{WR}^o that develops spontaneously between the two electrodes (W and R designate the working and the reference electrode, respectively) is given by

$$V_{WR}^o = (RT/4F)\ln(P_{O_2,W}/P_{O_2,R}) \tag{2}$$

where F is Faraday's constant (96,487 C/mol), and $P_{O_2,W}$ and $P_{O_2,R}$ are the oxygen partial pressures over the two electrodes. The superscript "o" designates hereafter open-circuit conditions; that is, there is no current ($I = 0$) flowing between the two electrodes. Equation (2), the Nernst equation, is valid provided there is equilibrium between gaseous oxygen and oxygen, O(tpb), adsorbed at the metal–solid electrolyte–gas three-phase boundaries (tpb). It is also necessary that the net-charge-transfer (electrocatalytic) reaction at the tpb is

$$O(a) + 2e^- \quad O^{2-}(YSZ) \tag{3}$$

that is, that there is no interference from other gases, such as H_2 and CO, which may react with O^{2-} (YSZ) at the tpb, establishing mixed potentials.[4]

Wagner first proposed the use of such galvanic cells in heterogeneous catalysis, to measure *in situ* the thermodynamic activity of oxygen O(a) adsorbed on metal electrodes during catalytic reactions.[9] This led to the technique of solid electrolyte potentiometry (SEP).[4,10,28–31,58] In this tech-

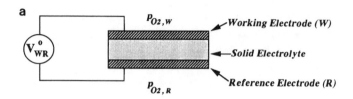

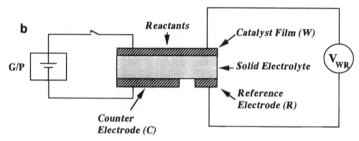

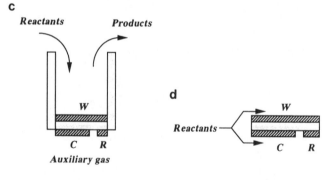

Figure 3. Electrode configuration for SEP (a) and for PPR or NEMCA studies (b). The latter can be carried out using the fuel-cell type configuration (c) or the single-pellet type configuration (d). G/P, Galvanostat/potentiostat.

nique the working electrode W (e.g., Pt) is exposed to the reactive gas mixture (e.g., C_2H_4 plus O_2) and also serves as the catalyst for a catalytic reaction, for example:

$$C_2H_4 + 3O_2 \rightarrow 2CO_2 + 2H_2O \tag{4}$$

The measured potential difference V_{WR}^o is related to the oxygen activity, a_O, on the catalyst surface via[4,10,28–31,58]:

$$V_{WR}^o = (RT/4F) \ln(a_O^2/P_{O_2,R}) \tag{5}$$

which is again derived on the basis of the equilibrium given by Eq. (3). The SEP technique, when used in conjunction with kinetic studies, is a useful tool for mechanistic investigations and is particularly suitable for the study of oscillatory reactions.[4,28–31,58] The limitations of Eq. (5) together with detailed reviews of the SEP literature can be found elsewhere.[4,28–31] Today, it is well established both theoretically[4] and experimentally[3,4,59] that SEP with metal catalyst electrodes is a work-function $(e\Phi)$ measuring technique:

$$eV_{WR}^o = e\Phi_W - e\Phi_R \tag{6}$$

where $e\Phi_W$ and $e\Phi_R$ are the (average[4]) work functions $e\Phi$ of the gas-exposed surfaces of the working (W) and reference (R) electrodes. Equation (6) is more general than Eq. (5) as it does not depend on the nature of the solid electrolyte and does not require the establishment of any specific charge-transfer equilibrium [e.g., Eq. (3)] at the tpb.[4] It shows that solid electrolyte galvanic cells are work-function probes for their gas-exposed, that is, catalytically active, electrode surfaces. This important point is analyzed in detail in Section III, in relation to the effect of electrochemical promotion.

In addition to solid electrolyte potentiometry, the techniques of cyclic voltammetry[46,60] and linear potential sweep[60] have also been used recently in solid electrolyte cells to investigate catalytic phenomena occurring on the gas-exposed electrode surfaces. The latter technique, in particular, is known in catalysis under the term potential-programmed reduction (PPR).[60] With appropriate choice of the sweep rate and other operating parameters, both techniques can provide valuable kinetic[46] and thermo-dynamic[60] information about catalytically active chemisorbed species and also about the NEMCA effect,[46,60] as analyzed in detail in Section III.

3. Electrocatalytic Operation of Solid Electrolyte Cells

Solid electrolyte fuel cells have been investigated intensively during the last three decades.[38,61–63] Their operating principle is shown schematically in Fig. 4. The positive electrode (cathode) acts as an electrocatalyst to promote the electrocatalytic reduction of $O_2(g)$ to O^{2-}:

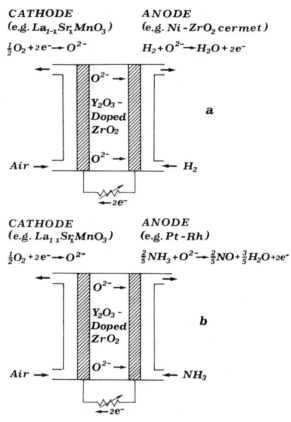

Figure 4. Operating principle of a solid oxide fuel cell (a) and of a chemical cogenerator (b).

$$\frac{1}{2}O_2(g) + 2e^- \rightarrow O^{2-} \tag{7}$$

Although several metals, such as Pt and Ag, can also act as electrocatalysts for the reaction in Eq. (7), the most efficient electrocatalysts known so far are perovskites such as $La_{1-x}Sr_xMnO_3$. These materials are mixed conductors; that is, they exhibit both anionic (O^{2-}) and electronic conductivity. This, in principle, can extend the electrocatalytically active zone to include not only the tpb but also the entire gas-exposed electrode surface.

The negative electrode (anode) acts as an electrocatalyst for the reaction of O^{2-} with the fuel, for example, H_2:

$$H_2 + O^{2-} \rightarrow H_2O + 2e^- \tag{8}$$

A Ni-stabilized ZrO_2 cermet is used as the anodic electrocatalyst in state-of-the-art solid oxide fuel cells. Nickel is a good electrocatalyst for the reaction in Eq. (8), and the Ni-stabilized ZrO_2 cermet is used to create a large metal–solid electrolyte–gas three-phase-boundary length, since it is exactly at these three phase boundaries that electrocatalytic reactions usually take place in solid electrolyte cells. Thus, the use of Ni in a Ni–ZrO_2 cermet to carry out electrocatalytic reactions is analogous to the use of Ni, or other transition metals, in a highly dispersed form on SiO_2, Al_2O_3, or TiO_2 supports to carry out catalytic reactions. Throughout this chapter, the term electrocatalytic reaction is used to denote reactions such as those in Eqs. (7) and (8) involving a net charge transfer.

Fuel cells such as the one shown in Fig. 4a convert H_2 to H_2O and produce electrical power with no intermediate combustion cycle. Thus, their thermodynamic efficiency compares favorably with thermal power generation which is limited by Carnot-type constraints. One important advantage of solid electrolyte fuel cells is that, owing to their high operating temperature (typically 700°C to 1100°C), they offer the possibility of "internal reforming," which permits the use of fuels such as methane without a separate external reformer.[63]

In recent years, it was shown that solid electrolyte fuel cells with appropriate electrocatalytic anodes can be used for "chemical cogeneration," that is, for the simultaneous production of electrical power and useful chemicals. This mode of operation, first demonstrated for the case of NH_3 conversion to NO,[12,64,65] combines the concepts of a fuel cell and of a chemical reactor (Fig. 4b). The economics of chemical cogeneration have been discussed and modeled.[66] Several other exothermic reactions have been investigated recently, including the oxidation of H_2S to SO_2,[67] of CH_3OH to H_2CO,[68] and of methane to ethylene.[69] In the latter case, it was found that ethylene yields up to 85% can be obtained in a gas-recycle solid electrolyte cell reactor–separator using an Ag–Sm_2O_3 anode and a molecular sieve adsorbent.[69] It is possible that if solid electrolyte fuel cells operating on H_2 or natural gas become commercially available, then they can also be used, with appropriate anodic electrocatalysts, by several chemical industries.

Table 1
Electrocatalytic Reactions Investigated in Doped ZrO$_2$ Solid Electrolyte
Fuel Cells for Chemical Cogeneration

Reaction	Electrocatalyst	Reference(s)
$2NH_3 + 5O^{2-} \rightarrow 2NO + 3H_2O + 10e^-$	Pt, Pt–Rh	12, 64, 65
$CH_4 + NH_3 + 3O^{2-} \rightarrow HCN + 3H_2O + 6e^-$	Pt, Pt–Rh	70
$CH_3OH + O^{2-} \rightarrow H_2CO + H_2O + 2e^-$	Ag	68
$C_6H_5-CH_2CH_3 + O^{2-} \rightarrow C_6H_5-CH{=}CH_2 + H_2O + 2e^-$	Pt, Fe$_2$O$_3$	71, 72
$H_2S + 3O^{2-} \rightarrow SO_2 + H_2O + 6e^-$	Pt	67
$C_3H_6 + O^{2-} \rightarrow C_3 \text{ dimers} + 2e^-$	Bi$_2$O$_3$–La$_2$O$_3$	73
$2CH_4 + 2O^{2-} \rightarrow C_2H_4 + 2H_2O + 4e^-$	Ag, Ag–Sm$_2$O$_3$	69

In Table 1 the anodic reactions that have been studied so far in small cogenerative solid oxide fuel cells are listed. One simple and interesting rule which has emerged from these studies is that the selection of the anodic electrocatalyst for a selective *electrocatalytic* oxidation can be based on the heterogeneous catalytic literature for the corresponding selective *catalytic* oxidation. Thus, the selectivity of Pt and Pt–Rh alloy electrocatalysts for the anodic NH$_3$ oxidation to NO turns out to be comparable (>95%) to the selectivity of Pt and Pt–Rh alloy catalysts for the corresponding commercial catalytic oxidation.[12,64,65] The same applies for Ag, which turns out to be equally selective as an electrocatalyst for the anodic partial oxidation of methanol to formaldehyde,[68]

$$CH_3OH + O^{2-} \rightarrow H_2CO + H_2O + 2e^- \tag{9}$$

as it is a catalyst for the corresponding catalytic reaction:

$$CH_3OH + \frac{1}{2}O_2 \rightarrow H_2CO + H_2O \tag{10}$$

Aside from chemical cogeneration studies, where the electrocatalytic anodic and cathodic reactions are driven by the voltage spontaneously generated by the solid electrolyte cell, several other electrocatalytic reactions have been investigated in solid electrolyte cells.[11,74–86] These reactions are listed in Table 2. Earlier studies by Kleitz and co-workers had focused mainly on the investigation of electrocatalysts for H$_2$O electrolysis.[74] Huggins and co-workers were the first to show that other electrocatalytic reactions, such as NO decomposition[11,75] and CO

Table 2

Electrocatalytic Reactions Investigated in Doped ZrO$_2$ Solid Electrolyte Fuel Cells with External Potential Application

Reaction	Electrocatalyst	Reference(s)
$H_2O + 2e^- \rightarrow H_2 + O^{2-}$	Ni	74, 87
$2NO + 4e^- \rightarrow N_2 + 2O^{2-}$	Pt, Au	11, 75
$CO + 2H_2 + 2e^- \rightarrow CH_4 + O^{2-}$	Pt, Ni	76, 77
$C_3H_6 + O^{2-} \rightarrow C_3H_6O + 2e^-$	Au	78
$CH_4 + yO^{2-} \rightarrow C_2H_6, C_2H_4, CO, CO_2, H_2O + 2ye^-$	Ag, Ag–MgO, Ag–Bi$_2$O$_3$, Ag–Sm$_2$O$_3$	69, 79–86

hydrogenation,[76,77] can be carried out in zirconia cells. More recently, the groups of Otsuka[79,81,82] and Stoukides,[80][84–86] among others, have concentrated on the study of the oxidative coupling of CH$_4$ utilizing a variety of metal and metal oxide electrodes. These systems are of great technological interest.

With the exception of H$_2$O electrolysis,[74,87] it is likely that, for all other electrocatalytic reactions listed in Table 2, catalytic phenomena taking place on the gas-exposed electrode surface or also on the solid electrolyte surface, had a certain role in the observed kinetic behavior. However, this role cannot be quantified, since the measured increase in reaction rate was, similarly to the case of the reactions listed in Table 1, limited by Faraday's law, that is,

$$\Delta r \approx I/2F \qquad (11)$$

where I is the cell current, and Δr (expressed in moles of O) is the measured change in global reaction rate. As already mentioned, the fact that no nonfaradaic behavior, that is, no NEMCA behavior, was observed during these interesting earlier studies can be safely attributed to the low polarizability of the metal/solid electrolyte interface. This dictated that electrocatalysis at the three-phase boundaries, rather than catalysis on the gas-exposed electrode surface, dominated the global kinetic picture. It is worth noting that in some more recent studies of the oxidative coupling of CH$_4$ in solid electrolyte cells involving more polarizable metal/solid electrolyte interfaces, the observed Δr was found to significantly exceed the rate $I/2F$ of O^{2-} transport to the anode.[45,88,89] Also, NEMCA behavior has been found recently for the hydrogenation of CO on Pd, as discussed in Section IV.

III. ELECTROCHEMICAL PROMOTION OF CATALYTIC REACTIONS: THE NEMCA EFFECT

1. Experimental Setup

(i) The Catalyst Film

A typical experimental setup for NEMCA studies is shown in Fig. 5. The catalyst film, typically 3–30 μm thick, is connected to a wire that acts as electron collector or supplier. The film must have enough porosity so that its gas-exposed, that is, catalytically active, surface area A_c can give a measurable catalytic reaction rate in the desired operating temperature range. Typically, A_c must be 30 to 3000 times larger than the electrolyte surface area A_E on which the catalyst film is supported. Scanning electron micrographs of Pt and Ag catalyst films used in some previous studies are shown in Figs. 6a, 6b, and 7. Both the top view of the films and cross sections of the catalyst film/solid electrolyte interface are shown. Figure 6c shows a scanning tunneling micrograph (STM) of the surface of a Pt catalyst film.

As discussed below, the porosity and surface area of the catalyst film are controllable to a large extent by the sintering temperature during catalyst preparation. This, however, affects not only the catalytically

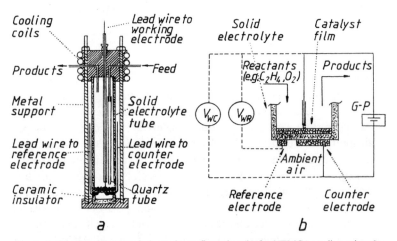

Figure 5. Reactor cell (a) and electrode configuration (b) for NEMCA studies using the fuel-cell type design. G-P, Galvanostat–potentiostat. (Reprinted, with permission from Elsevier Science Publishers B.V., Amsterdam, from Ref. 4.)

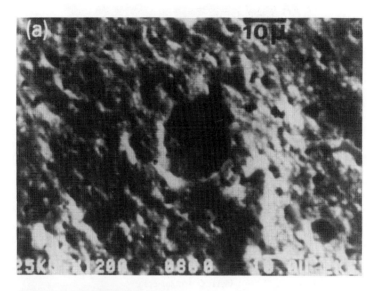

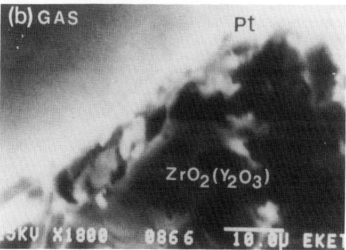

Figure 6. Scanning electron micrographs of the top side of a porous Pt catalyst film (a) and of a section perpendicular to the Pt catalyst/yttria-stabilized zirconia (YSZ) interface (b). (c) Scanning tunneling micrograph of the surface of a Pt catalyst used in NEMCA studies. Scan size, 62 Å; V_{bias} = 0.5 V, I_{tunnel} = 15 nA. (Reprinted, with permission from Elsevier Science Publishers B.V., Amsterdam, from Ref. 4.)

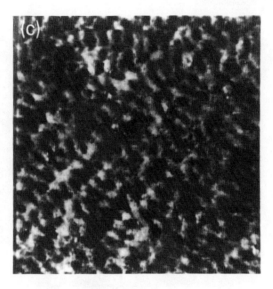

Figure 6. (continued)

active surface area A_c but also the length l of the three-phase boundaries between the solid electrolyte, the catalyst film, and the gas phase (Fig. 8). Electrocatalytic reactions, such as the transformation of O^{2-} from the zirconia lattice to oxygen adsorbed *on the film* at or near the three-phase boundaries, which we denote by O(a), have been found to take place primarily at these three-phase boundaries.[46,90–92] This electrocatalytic reaction will be denoted by

$$O^{2-} \leftrightharpoons O(a) + 2e^- \qquad (12)$$

or, in Kröger–Vink notation,

$$O_O \leftrightharpoons O(a) + V_O^{\bullet\bullet} + 2e^{|} \qquad (13)$$

where O_O denotes an oxygen anion O^{2-} in the yttria-stabilized zirconia (YSZ) lattice and $V_O^{\bullet\bullet}$ stands for an O^{2-} vacancy in the lattice with a double positive charge. There is some experimental evidence that the reaction in Eq. (12) can also take place, to some extent, at the two-phase boundaries between the zirconia and the metal, followed by diffusion of oxygen through the metal or through the two-phase boundaries to the three-phase boundaries.[92] However, this scheme becomes important ap-

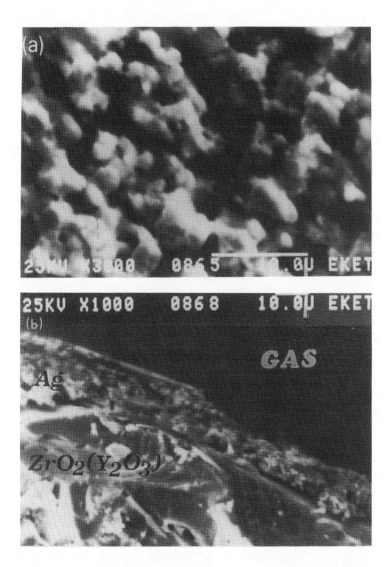

Figure 7. Scanning electron micrographs of a porous Ag catalyst film (a) and of a section perpendicular to the Ag catalyst/yttria-stabilized zirconia (YSZ) interface (b). (Reprinted, with permission from Elsevier Science Publishers B.V., Amsterdam, from Ref. 4.)

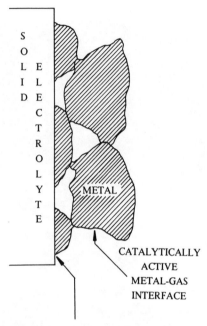

S
O
L   E
I   L
D   E
    C
    T
    R
    O
    L
    Y   METAL
    T
    E

CATALYTICALLY
ACTIVE
METAL-GAS
INTERFACE

ELECTROCATALYTICALLY ACTIVE
METAL-SOLID ELECTROLYTE-GAS
THREE-PHASE BOUNDARIES

Figure 8. Schematic representation of the location of
electrocatalytically and catalytically active sites in a
section perpendicular to the catalyst film/solid electrolyte interface. (Reprinted, with permission from
Elsevier Science Publishers B.V., Amsterdam, from
Ref. 4.)

parently only with metals having a high oxygen solubility, such as Ag,[43]
or when metal foils are used instead of porous metal films or with very
dense nonporous metal films acting as blocking electrodes. There is also
some evidence that under reducing gas-phase conditions and large applied
negative currents, that is, under conditions of O^{2-} *removal from* the metal
catalyst–zirconia–gas three-phase boundaries, so that the applied potential
is near the potential required for the decomposition of zirconia (~2.4 V at
800 K), then the electrocatalytic reaction can also take place directly at

the zirconia/gas interface.[93] The above conditions are usually referred to as "zirconia blackening" conditions, as the electrolyte surface indeed turns black. Such operation has not been employed in any of the NEMCA studies reviewed here. Also, the porosity of the catalyst films employed was usually high enough so that the *electrocatalytic* reaction in Eq. (12) can be safely assumed to occur at the three-phase boundaries only, except in the case of Ag electrodes, which act as oxygen "sponges"[43] and where the reaction in Eq. (12) takes place at the Ag/YSZ interface as well.[43] A method for estimating the three-phase boundary length *l* by the use of solid electrolyte cyclic voltammetry has been described recently.[46]

Metal catalyst films used in NEMCA studies[1-4,7,14-27,40-54] are usually prepared by using commercial unfluxed metal pastes. Thin coatings of the paste are applied on the solid electrolyte surface, followed by drying and calcining. The calcination temperature program obviously depends on the metal to be deposited and on paste composition and plays a key role in obtaining a well-adhered film, which is, of course, essential for NEMCA studies. Thus, for Pt catalyst preparation, a proven procedure[2,4,7,41,44,46] is to use a thin coating of Engelhard A1121 Pt paste followed by drying and then calcining in air, first for 2 h at 400°C and then for 20 min at 820°C. When a Demetron Pt paste was used, final calcination temperatures at least 100°C higher were found necessary to produce sufficiently polarizable Pt/zirconia interfaces. For Ag catalyst film preparation, one may use thin coatings of an Ag solution in butyl acetate[4,13,40,43] followed by drying at 80°C and calcining at 650°C. In general, it appears preferable to use a slow heating rate (e.g., 2°C/min) during calcination and to always maintain a high flow rate of air (e.g., 0.2 l/min) through the furnace. Increasing calcination temperatures even by only 20–30°C for a given metal paste leads to increased sintering and decreases in both catalyst surface area and length of the metal–zirconia–gas three-phase boundaries. The latter has a beneficial effect for increasing the polarizability of the metal/solid electrolyte interface and thus observing strongly nonfaradaic catalytic rate changes. The decrease in catalyst surface area may, however, be undesirable in many cases, if the intrinsic catalytic rate is very small. In general, the optimal calcination temperature for each catalyst and paste has to be found by trial and error.

A proven procedure for enhancing the adherence of metal or metal oxide films on flat YSZ surfaces is to first roughen the YSZ surface by adding a slurry containing fine (1–2 μm) YSZ powder, followed by drying and calcining at 1450–1500°C. Catalyst deposition on such a roughened surface may, of course, lead to a long three-phase-boundary line length and to a concomitant decreased polarizability of the zirconia/solid electrolyte interface, but this is definitely preferable to a not well-adhered film.

In recent years, the deposition of thin conductive oxide films on flat zirconia components has received considerable attention for fuel cell applications[38,61–63] and also for SEP studies.[94] The interested reader is referred to the original references for experimental details.

A final and obvious but important caution about catalyst film preparation: The thickness and surface area A_c of the catalyst film must be low enough so that the catalytic reaction under study is not subject to external or internal mass-transfer limitations within the desired operating temperature range. Direct impingement of the reactant stream on the catalyst surface[1–4,40–52] can be used to diminish the external mass-transfer resistance.

(ii) Counter and Reference Electrodes

As shown in Fig. 5, the counter and reference electrodes are deposited on the gas-impervious solid electrolyte component on the side opposite the catalyst film. The electrolyte thickness, typically $500\ \mu$m to 2 mm, is not crucial, but it is preferable to keep it low, so that the ohmic drop in it is small during operation, preferably below 100–300 mV.

Both the counter and the reference electrode are essential for fundamental NEMCA studies. They need not be of the same material as the catalyst, although for the reference electrode this is convenient in data interpretation, as then the measured potential difference V_{WR} (Fig. 5) can be directly related to the average work-function difference between the catalyst (W) and reference (R) electrode gas-exposed surfaces.[3,4,59] The counter electrode/solid electrolyte interface does not have to be polarizable. In fact, it is advantageous when it is not, because then most of the applied potential difference ends up as overpotential at the catalyst and not at the counter electrode.

The reference electrode/solid electrolyte interface must also be nonpolarizable, so that rapid equilibration is established for the electrocatalytic charge-transfer reaction. Thus, it is generally advisable to sinter the counter and reference electrodes at a temperature which is lower than that used for the catalyst film. Porous Pt and Ag films exposed to ambient air have been employed in most NEMCA studies to date.[1–4,25,40–46]

2. Electrocatalytic Reactions and the Measurement of the Catalyst–Solid Electrolyte Exchange Current, I_0

Although NEMCA is a catalytic effect taking place over the entire catalyst gas-exposed surface, it is important for its description to concentrate first on the electrocatalytic reactions taking place at the catalyst–solid electrolyte–gas three-phase boundaries (tpb). Throughout this chapter, the term electrocatalytic reaction denotes a reaction where there is a net charge transfer, such as the usual reaction taking place at the metal-stabilized zirconia–gas tpb:

$$O^{2-} \leftrightharpoons O(a) + 2e^- \qquad (12)$$

It should be mentioned that in the presence of oxidizable reactants over the catalyst surface, other electrocatalytic reactions may also take place in parallel with that in Eq. (12) at the tpb. Thus, in the presence of high CO concentrations, direct reaction of CO with O^{2-} can also take place:

$$CO(a) + O^{2-} \rightarrow CO_2(g) + 2e^- \qquad (14a)$$

The extent to which such reactions take place in parallel with the dominant reaction (Eq. 12) is, in general, difficult to quantify[95,96] as the overall reaction given by Eq. (14a) may consist of Eq. (12) as the elementary step followed by reaction between adsorbed CO and adsorbed oxygen on the metal surface:

$$CO(a) + O(a) \rightarrow CO_2(g) \qquad (14b)$$

When other types of solid electrolytes are used, such as the Na^+-conducting β''-Al_2O_3, then the dominant electrocatalytic reaction at the tpb is

$$Na^+ + e^- \leftrightharpoons Na(a) \qquad (15)$$

where Na(a) stands for Na adsorbed on the catalyst surface. However, in the presence of H_2O, other parallel reactions can also take place[42] such as:

$$H_2O + e^- \leftrightharpoons OH^- + H(a) \qquad (16)$$

Again, the extent to which such parallel reactions contribute to the measured current is not very easy to quantify. However, fortunately, such a quantification is not necessary for the description of NEMCA. What is needed is only a measure of the overall electrocatalytic activity of the metal/solid electrolyte interface or, equivalently, of the tpb, and this can be obtained by determining the value of a single electrochemical parameter, the exchange current I_0, which is related to the exchange current density i_0 via

$$i_0 = I_0/A_E \qquad (17)$$

Strictly speaking, I_0 is a measure of the electrocatalytic activity of the tpb for a given electrocatalytic reaction. It expresses the rates of the forward (anodic) and reverse (cathodic) directions of the electrocatalytic reaction under consideration, for example, Eq. (12), when there is no *net* current crossing the metal/solid electrolyte interface or, equivalently, the tpb. In this case, the rates of the forward and the reverse reaction are obviously equal. It has been recently shown that, in most cases, as one would intuitively expect, I_0 is proportional to the length l of the tpb.[46]

The measurement of I_0, or i_0, is based on the classical Butler–Volmer[97–101] equation:

$$I = I_0[\exp(\alpha_a F\eta_{ac,j}/RT) - \exp(-\alpha_c F\eta_{ac,j}/RT)] \qquad (18)$$

where α_a and α_c are the anodic and cathodic transfer coefficients, and $\eta_{ac,j}$ is the *activation overpotential* of the electrode j under consideration. Before discussing the use of the Butler–Volmer equation (18) to extract the values of I_0 and of α_a and α_c, it is important to first discuss some issues regarding the activation overpotential.

When a current I flows in a galvanic cell, such as the one shown in Fig. 5, between the catalyst, or working (W), electrode and the counter (C) electrode, then the potential difference V_{WC} deviates from its open-circuit value V^o_{WC}. The galvanic cell overpotential η_{WC} is then defined by

$$\eta_{WC} = V_{WC} - V^o_{WC} \qquad (19)$$

The cell overpotential η_{WC} is the sum of three terms:

$$\eta_{WC} = \eta_W + \eta_C + \eta_{ohmic,WC} \qquad (20)$$

where η_W and η_C are the overpotentials of the catalyst (W) and counter (C) electrodes, respectively, and $\eta_{ohmic,WC}$ is the ohmic overpotential due to the resistance of the electrolyte between the working and counter electrodes. The overpotentials of the catalyst and counter electrodes are defined as the deviation of the inner (or Galvani) potentials of these electrodes from their open-circuit values:

$$\eta_W = \varphi_W - \varphi_W^0 \qquad (21)$$

$$\eta_C = \varphi_C - \varphi_C^0 \qquad (22)$$

It is worth emphasizing that although overpotentials are usually associated with electrode/electrolyte interfaces, in reality they refer to, and are measured as, deviations of the potential of the electrodes only. Thus, the concept of overpotential must be associated with *an electrode* and not with an electrode/electrolyte interface, although the nature of this interface will, in general, dictate the magnitude of the measured overpotential.

The overpotential η of an electrode, for example, W, can be considered to be the sum of three terms:

$$\eta_W = \eta_{ac,W} + \eta_{con,W} + \eta_{ohmic,W} \qquad (23)$$

The activation overpotential $\eta_{ac,W}$ is due to slow charge-transfer reactions at the electrode/electrolyte interface and is related to current via the Butler–Volmer equation (18). A slow chemical reaction (e.g., adsorption or desorption) preceding or following the charge-transfer step can also contribute to the development of activation overpotential.

The concentration overpotential $\eta_{con,W}$ is due to slow mass transfer of reactants and/or products involved in the charge-transfer reaction. There exist simple equations for computing its magnitude in terms of mass-transfer coefficients or, more frequently, in terms of the limiting current I_L, which is the maximum current obtained when the charge-transfer reaction is completely mass-transfer-controlled.[100,102] Contrary to aqueous electrochemistry, where concentration overpotential is frequently important due to low reactant and/or product diffusivities in the aqueous phase, in solid electrolyte cells mass transfer in the gas phase is fast, and, consequently, concentration overpotential is negligible, particularly in NEMCA applications, where the currents involved are usually very small.

The ohmic overpotential $\eta_{ohmic,W}$ is also negligible, provided the catalyst electrode is sufficiently conductive.

Thus, to a good approximation one can rewrite Eq. (23) for the working (W) and counter (C) electrodes as:

$$\eta_W = \eta_{ac,W} \qquad (24)$$

$$\eta_C = \eta_{ac,C} \qquad (25)$$

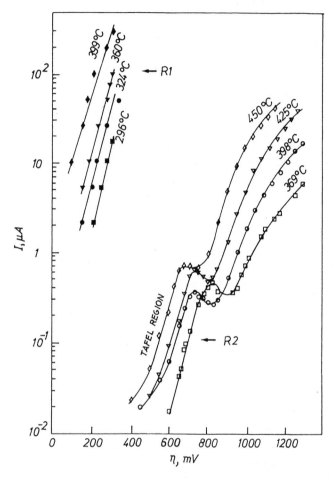

Figure 9. Typical Tafel plots for Pt catalyst/YSZ interfaces. The large differences in I_0 values are due to the higher calcination temperature of the Pt film R2 ($P_{O_2} = 4.8 \times 10^{-2}$ bar, $P_{ET} = 0.4 \times 10^{-2}$ bar) as compared to that of the Pt film R1 ($P_{O_2} = 6.4 \times 10^{-2}$ bar, $P_{ET} = 0.4 \times 10^{-2}$ bar). (Reprinted, with permission from Academic Press, from Ref. 2.)

The usefulness of the reference electrode can now be appreciated if the aim is to measure η_W instead of the sum $\eta_W + \eta_c$ (Eq. 20). Similarly to Eq. (20), one has

$$\eta_{WR} = \eta_W + \eta_R + \eta_{\text{ohmic},WR} \tag{26}$$

Ideally, no current flows through the reference electrode, and therefore η_R and $\eta_{ohmic,WR}$ should both be zero. In practice, the assumption that $\eta_R = 0$ is usually good for reasonably nonpolarizable reference electrodes, since the parasitic uncompensated current flowing via the reference electrode is usually very small.[4] The ohmic drop, however, between the working and reference electrodes, $\eta_{ohmic,WR}$, may, in general, be not negligible and must be determined using the current interruption technique in conjunction with a recording oscilloscope.[2,4,90] The ohmic component decays to zero within less than 1 μs, and the remaining part of η_{WR} is η_W. As in aqueous electrochemistry, the reference electrode must be placed as near to the catalyst as possible to minimize $\eta_{ohmic,WR}$.

The usual procedure for extracting the exchange current I_0 is then to measure η_W as a function of I and to plot $\ln I$ versus η_W (Tafel plot). Such plots are shown in Figs. 9 and 10 for Pt and Ag catalyst electrodes. Throughout the remainder of this chapter, we omit the subscript "W" from η_W and simply write η, since the only overpotential of interest is that of the catalyst film. When $|\eta| > 100$ mV, then the Butler–Volmer equation (18) reduces to its "high-field approximation" form,[100] i.e.,

$$\ln(I/I_0) = \alpha_a F\eta/RT \qquad (27a)$$

for anodic ($I > 0$, $\eta > 0$) operation and to

$$\ln(-I/I_0) = -\alpha_c F\eta/RT \qquad (27b)$$

for cathodic ($I < 0$, $\eta < 0$) operation. Thus, by extrapolating the linear parts of the $\ln |I|$ versus η plot to $\eta = 0$, one obtains I_0. The slopes of the linear parts of the plot give the transfer coefficients α_a and α_c. One can then plot I versus η and use the "low-field" approximation of the Butler–Volmer equation, which is valid for $|\eta| < 10$ mV, i.e.,

$$I/I_0 = (\alpha_a + \alpha_c)F\eta/RT \qquad (28)$$

in order to check the accuracy of the extracted I_0, α_a, and α_c values.

It is worth noting that I_0 is, in general, strongly dependent both on temperature and on gaseous composition. It increases with temperature with an activation energy of typically 35–45 kcal/mol for Pt and 20–25 kcal/mol for Ag films deposited on stabilized zirconia.[90–92,103] The dependence of I_0 on gaseous composition is usually complex. Thus, Manton[103] has shown that I_0 goes through a maximum with increasing P_{O_2} at any fixed temperature for Pt/ZrO$_2$(Y$_2$O$_3$) catalyst films. These results can

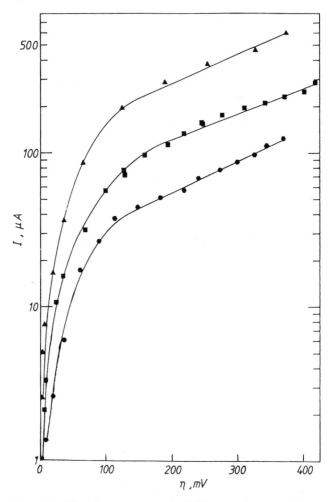

Figure 10. Effect of temperature on the Tafel plots and corresponding I_0 values of an Ag catalyst/YSZ interface, $P_{C_2H_4} = 2.30 \times 10^{-2}$ bar; $P_{O_2} = 3.15 \times 10^{-2}$ bar; ▲, $T = 436°C$; ■, $T = 398°C$; ●, $T = 368°C$. (Reprinted, with permission from Elsevier Science Publishers B.V., Amsterdam, from Ref. 4.)

be described adequately on the basis of Langmuir-type adsorption of oxygen at the tpb, i.e.,

$$\theta_O = K_O P_{O_2}^{1/2}/(1 + K_O P_{O_2}^{1/2}) \tag{29}$$

where θ_O is the oxygen coverage. It can be shown[90–92] that

$$I_0 \sim [\theta_O(1 - \theta_O)]^{1/2} \tag{30a}$$

or, equivalently,

$$I_0 \sim K_O P_{O_2}^{1/4}/(1 + K_O P_{O_2})^{1/2} \tag{30b}$$

which explains nicely the observed maxima and the fact that I_0 is proportional to $P_{O_2}^{1/4}$ for low P_{O_2} and to $P_{O_2}^{-1/4}$ for high P_{O_2}.[4,90–92,103] According to this successful model, the I_0 maxima correspond to $\theta_O = \frac{1}{2}$. It has been found, however, that for low T and high P_{O_2} the situation becomes more complicated owing to the formation of surface Pt oxide, PtO_2.[103–107]

When other gases are present in the gas phase in addition to O_2, then I_0 can be affected for two different reasons: first, because θ_O may be affected due to a catalytic reaction and/or due to competitive chemisorption and, second, because these gases may react with O^{2-} at the tpb. In general, it is difficult to determine experimentally which one of these two factors is more important.

The exchange current I_0 is an important parameter for the quantitative description of NEMCA. As subsequently analyzed in this chapter, it has been found both theoretically and experimentally[3,4] that the order of magnitude of the absolute value $|\Lambda|$ of the NEMCA enhancement factor Λ defined by

$$\Lambda = \Delta r/(I/2F) \tag{31}$$

where Δr is the NEMCA-induced change in catalytic rate and I is the applied current, can be estimated for any catalytic reaction from

$$|\Lambda| \approx 2F r_0/I_0 \tag{32}$$

where r_0 is the regular, that is, open-circuit, catalytic rate. When using Eq. (32) to estimate the order of magnitude of the enhancement factor Λ expected for a given catalytic reaction, one must use the I_0 value measured in the presence of the reacting gas mixture.

The fact that I_0 increases exponentially with temperature in conjunction with the fact that Λ is proportional to I_0^{-1} explains why most NEMCA studies have been limited to temperatures typically below 600°C.

3. The Work Function of Catalyst Films Deposited on Solid Electrolytes

It has been found recently both theoretically[4] and experimentally[3,59] that:

1. Solid electrolyte cells can be used to alter significantly the work function $e\Phi$ of the gas-exposed (i.e., catalytically active) catalyst-electrode surface by polarizing the catalyst/solid electrolyte interface.

2. Solid electrolyte cells are work-function probes for the gas-exposed, catalytically active catalyst-electrode surfaces; that is, the change $\Delta(e\Phi)$ in catalyst surface average work function $e\Phi$ is equal to $e\Delta V_{WR}$. The catalyst potential V_{WR} with respect to a reference electrode can be varied both by changing the gaseous composition and/or by polarizing the catalyst/solid electrolyte interface.

The above two observations play a key role in understanding and interpreting the NEMCA effect, and it is therefore important to discuss them in some detail. We start by considering a schematic representation of a porous metal film deposited on a solid electrolyte, for example, on Y_2O_3-stabilized ZrO_2 (Figs. 8 and 11). The catalyst surface is divided into two distinct parts. One part, having a surface area A_E, is in contact with the electrolyte. The other, having a surface area A_c, is not in contact with the electrolyte. It constitutes the gas-exposed (i.e., catalytically active) film surface area. *Catalytic reactions* take place on this surface only. In the subsequent discussion we will use the letters E (for electrochemical) and C (for catalytic), respectively, to denote these two distinct parts of the catalyst film surface. Regions E and C are separated by the three-phase boundaries (tpb), where *electrocatalytic reactions* take place. Since, as previously discussed, electrocatalytic reactions can also take place to, usually, a minor extent on region E, one may consider the tpb to be part of region E as well. It will become apparent below that the essence of NEMCA is the following. One uses electrochemistry (i.e., a slow electrocatalytic reaction) to alter the electronic properties of the metal/solid electrolyte interface E. This perturbation is then propagated via the spatial constancy of the Fermi level E_F throughout the metal film to the metal/gas interface C, altering its electronic properties and thus causing ion migra-

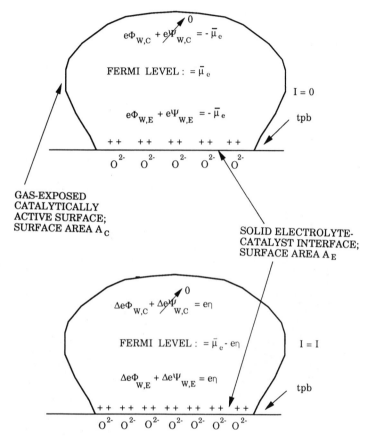

Figure 11. Schematic representation of a metal crystallite deposited on YSZ and of the changes induced in its electronic properties upon polarizing the catalyst/solid electrolyte interface and changing the Fermi level (or electrochemical potential of electrons) from an initial value $\bar{\mu}$ to a new value $\bar{\mu} - e\eta$. (Reprinted, with permission from Elsevier Science Publishers B.V., Amsterdam, from Ref. 4.)

tion and thus influencing catalysis, that is, catalytic reactions taking place on the metal/gas interface C.

We then concentrate on the meaning of V_{WR}, that is, of the (ohmic-drop-free) potential difference between the catalyst film (W, for working electrode) and the reference film (R). The measured (by a voltmeter)

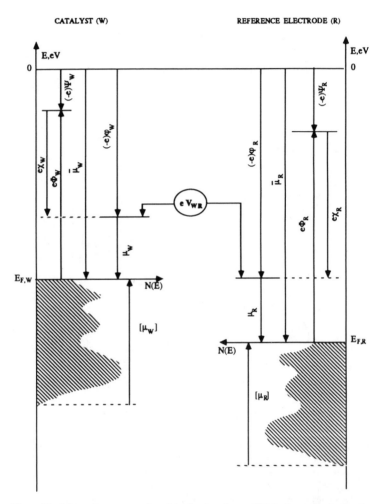

Figure 12. Schematic representation of the density of states $N(E)$ in the conduction band and of the definitions of work function $e\Phi$, chemical potential of electrons μ, electrochemical potential of electrons or Fermi level $\bar{\mu}$, surface potential χ, Galvani (or inner) potential φ, Volta (or outer) potential Ψ, and Fermi energy $[\mu]$ for the catalyst (W) and for the reference electrode (R). The measured potential difference V_{WR} is, by definition, the difference in Galvani potentials; φ, μ, and $\bar{\mu}$ are spatially uniform; $e\Phi$ and Ψ can vary locally on the metal sample surfaces; and the Ψ potentials vanish, on the average, for the gas-exposed catalyst and reference electrode surfaces. (Reprinted, with permission from Elsevier Science Publishers B.V., Amsterdam, from Ref. 4.)

potential difference V_{WR} is, by definition.[100] the difference between the inner (or Galvani) potentials φ of two electrodes:

$$V_{WR} = \varphi_W - \varphi_R \tag{33}$$

The Galvani potential is the electrostatic potential of electrons inside the metal film. The electrochemical potential of electrons in a metal $\bar{\mu}$ is related to φ via

$$\bar{\mu} = \mu + (-e)\varphi \tag{34}$$

where μ is the chemical potential of electrons in the metal, a purely bulk property. It should be recalled that $\bar{\mu}$ can be shown[108,109] to be identical with the Fermi level E_F in the metal (see Fig. 12 and Refs. 109–115, which provide an excellent introduction to the meaning of the various potentials discussed here). In view of Eq. (34), one can rewrite Eq. (33) as

$$eV_{WR} = \bar{\mu} - \bar{\mu}_W + (\mu_W - \mu_R) \tag{35}$$

Equation (34) represents the electrochemical way of counting the energy difference between zero (defined in the present discussion as the potential energy of an electron at its ground state at "infinite" distance from the metal[100]) and the Fermi level E_F. The latter must not be confused with the Fermi energy $[\mu]$, which is the energy difference between the Fermi level and the energy at the bottom of the conduction band and provides a measure of the average kinetic energy of electrons at the Fermi level (Fig. 12). The electrochemical way of splitting the energy difference from zero to E_F is a conceptual one, as the absolute values of φ and μ are not accessible to direct experimental measurement. The second way to split this energy difference is to consider it as the sum of the work function $e\Phi$ (Φ is the electron extraction potential) and $e\Psi$, where Ψ is the outer (or Volta) potential (Figs. 12 and 13).

The work function $e\Phi$ is the work required to bring an electron from the Fermi level of the metal to a point outside the metal where the image forces are negligible, that is, typically 10^{-4} to 10^{-5} cm outside the metal surface.[100,110,111] The Volta potential Ψ at this point is defined so that the energy required to bring an electron from that point to an "infinite" distance from the metal surface is $e\Psi$:

$$\bar{\mu} = -e\Phi - e\Psi \tag{36}$$

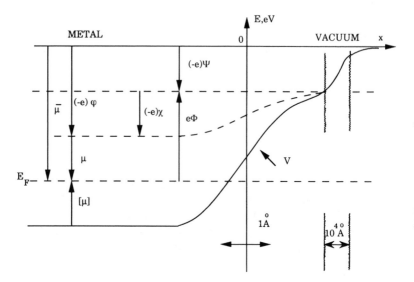

Figure 13. Schematic representations of the definitions of work function $e\Phi$, chemical potential of electrons μ, electrochemical potential of electrons or Fermi level $\overline{\mu} = E_F$, surface potential χ, Galvani (or inner) potential φ, Volta (or outer) potential Ψ, and Fermi energy $[\mu]$ and of the variation in the mean effective potential energy V of electrons in the vicinity of a metal/vacuum interface according to the jellium model (based on Refs. 4 and 115). (Reprinted, with permission from Elsevier Science Publishers B.V., Amsterdam, from Ref. 4.)

It is important to emphasize that $e\Phi$ and Ψ (which are both accessible to experimental measurement)[100] are not, in general, spatially uniform over the metal surface. Different crystallographic planes are well known to have different $e\Phi$ values, and, thus, nontrivial variations in $e\Phi$ and $e\Psi$ are to be expected on the surface of polycrystalline samples. It is important, however, to notice that their sum has to be spatially uniform (Eq. 36) since the electrochemical potential $\overline{\mu}$ or, equivalently, the Fermi level E_F is spatially uniform. This is true even when an electrical current is passing through the metal film under consideration, provided that the ohmic drop in the film is negligible (less than a few millivolts), which is always the case with the conductive metal films and low currents employed in NEMCA studies. It is also important to notice that, by definition, Ψ vanishes if there is no *net* charge on the metal surface under consideration.

One can then combine Eqs. (35) and (36) to obtain

$$eV_{WR} = e\Phi_W - e\Phi_R + e(\Psi_W - \Psi_R) + (\mu_W - \mu_R) \qquad (37)$$

It is worth emphasizing that Eq. (37) is valid under both open-circuit and closed-circuit conditions and that it holds for *any* part of the surfaces of the catalyst and the reference electrodes. Thus, referring to the metal electrode surfaces in contact with the electrolyte (region E), it is

$$eV_{WR} = e\Phi_{W,E} - e\Phi_{R,E} + e(\Psi_{W,E} - \Psi_{R,E}) + (\mu_W - \mu_R) \qquad (38)$$

while for the gas-exposed (i.e., catalytically active) electrode surfaces (region C), it is

$$eV_{WR} = e\Phi_{W,C} - e\Phi_{R,C} + e(\Psi_{W,C} - \Psi_{R,C}) + (\mu_W - \mu_R) \qquad (39)$$

In order to understand the origin of NEMCA, one needs to concentrate only on Eq. (39), which refers to the gas-exposed, catalytically active, film surface. As already stated, different crystallographic planes will, in general, have different $e\Phi$ values; thus, even over region C the work function $e\Phi$ need not be spatially uniform. These local spatial variations in $e\Phi$ and Ψ are not expected to be significant in polycrystalline films with large (~ 1 μm) crystallites such as the ones used in most NEMCA studies[4] since the surface must consist primarily of low Miller index planes, for example, of the (111) plane in the case of Pt films.[4] This is also supported by recent STM information (e.g., Fig. 6c). We will thus first assume that $e\Phi$ and Ψ are spatially uniform over region C and will treat the more general case below. Returning to Eq. (39), we note that when V_{WR} is changed by varying the gaseous composition over the catalyst or by polarizing the catalyst/solid electrolyte interface by means of a current, then the properties of the reference electrode remain unaffected, and thus μ_W, which is a pure bulk property, also remains constant; therefore,

$$e\Delta V_{WR} = \Delta e\Phi_{W,C} + e\Delta\Psi_{W,C} \qquad (40)$$

We then concentrate on the Volta potential terms $\Psi_{W,C}$ and $\Delta\Psi_{W,C}$. It follows from simple electrostatic considerations,[4] and taking into account that the solid electrolyte has an abundance of charge carriers while the gas phase does not, that all the net charge in the metal will be localized in region E facing an equal and opposite charge in the solid electrolyte. To further elucidate this point, consider the (initially uncharged) metal film brought into contact with the (also initially uncharged) solid electrolyte. When the film and the solid electrolyte are brought in contact, the system has to remain neutral. Thus, if a charge q develops in the film, a charge $-q$

will develop in the solid electrolyte facing the metal.[100] Thus, if the film charge q were to be split into a part q_E remaining in region E and a part q_C developing in region C, then a net charge $-(q - q_E) = -q_C$ would remain in region E, attracting the charge q_C back to region E. Thus, no *net* charge can be sustained in region C, and therefore $\Psi_{W,C} = 0$ under both open-circuit and closed-circuit conditions.

One can then rewrite Eqs. (39) and (40) as

$$eV_{WR} = e\Phi_{W,C} - e\Phi_{R,C} + (\mu_W - \mu_R) \tag{41}$$

$$e\Delta V_{WR} = \Delta e\Phi_{W,C} \tag{42}$$

It is worth noting that if the reference electrode is of the same bulk material as the catalyst and both are at the same temperature, then $\mu_W = \mu_R$. Using the superscript "o" to denote open-circuit conditions ($I = 0$), one can rewrite Eq. (41) as

$$eV_{WR}^{o} = e\Phi_{W,C} - e\Phi_{R,C} \tag{43}$$

Equation (43) shows that the emf eV_{WR}^{o} of solid electrolyte cells with electrodes made of the same bulk material provides a direct measure of the difference in work function of the gas-exposed (i.e., catalytically active) electrode surfaces. Thus, *solid electrolyte cells are work-function probes for their gas-exposed electrode surfaces.*

Equation (42) is equally important, as it shows that the work function of the catalytically active catalyst electrode surface can be varied at will by varying the (ohmic-drop-free) catalyst potential. This can be done either by varying the gaseous composition over the catalyst or by using a potentiostat. Catalytic chemists are familiar with the former mode: When the gaseous composition changes, then surface coverages will change with a concomitant change in work function. But what about the latter mode? For the work function to change, the coverages and/or dipole moments of species already adsorbed on the surface must change, or new species must get adsorbed. As discussed in this chapter, it is now firmly established that the induced work-function change on the surface is predominantly due to the migration (backspillover) of ions originating from the solid electrolyte. We use the term "backspillover" instead of "spillover" for the following reason. In catalysis, the term "spillover" denotes migration of adsorbates from a supported metal catalyst to an oxide carrier (support). The term "backspillover" denotes migration of species in the opposite direction, that

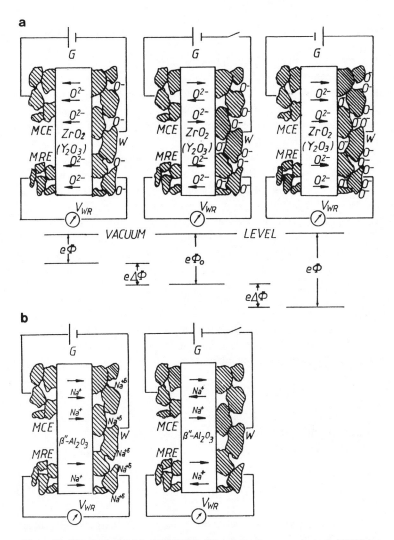

Figure 14. The physical origin of NEMCA. When a metal counter electrode (MCE) is used in conjunction with a galvanostat (G) to supply or remove ions [O^{2-} in the case of the doped ZrO_2 (a), Na^+ in the case of β''-Al_2O_3 (b)] to or from the polarizable solid electrolyte/catalyst (or working electrode, W) interface, backspillover ions [O^- in (a), $Na^{+\delta}$ in (b)] together with their compensating charge in the metal are produced or consumed at the solid electrolyte–catalyst–gas three-phase boundaries (tpb). This causes an increase (right) or decrease (left) in the work function $e\Phi$ of the gas-exposed catalyst surface. In all cases, $\Delta e\Phi = e\,\Delta V_{WR}$, where ΔV_{WR} is the overpotential measured between the catalyst and the reference electrode (MRE). (Reprinted, with permission from Elsevier Science Publishers B.V., Amsterdam, from Ref. 4.)

is, from the support to the metal catalyst, as is the case here. These backspillover ions (oxygen anions in the case of doped ZrO_2, partly ionized Na in the case of β''-Al_2O_3), accompanied by their compensating charge in the metal, thus forming spillover dipoles, spread over the catalytically active surface, altering its work function and catalytic properties (Fig. 14).

One can then address the question of the meaning of Eqs. (42) and (43) in the case of significant spatial variations in the work function $e\Phi$ of the polycrystalline catalyst surface. In this case, due to the constancy of the Fermi level, slightly different nonzero excess free charge densities will exist on different planes with different $e\Phi$, causing local variations in Ψ. Surface physicists would refer to this as a local variation in the "vacuum level." In this case the average surface work function $e\Phi$ is defined by[111]

$$e\Phi = \sum_j f_j e\Phi_j \tag{44}$$

where f_j is the fraction of the total catalyst surface corresponding to a crystallographic plane with a work function $e\Phi_j$. One can then apply Eq. (39) to each crystallographic plane j:

$$eV_{WR} = e\Phi_{W,Cj} - e\Phi_{R,C} + e(\Psi_{W,C,j} - \Psi_{R,C}) \tag{45}$$

where, for simplicity, it is assumed that the reference electrode is of the same material as the catalyst ($\mu_W = \mu_R$) By multiplying Eq. (45) by f_j, summing for all planes, and noting that $\Psi_{R,C} = 0$, one obtains

$$eV_{WR} = \sum_j f_j e\Phi_{W,Cj} - e\Phi_{R,C} + \sum_j f_j e\Psi_{W,C,j} \tag{46}$$

Since Ψ is proportional to the local excess free charge, it follows that the term $\Sigma_j f_j e\Psi_{W,C,j}$ is proportional to the net charge stored in the metal in region C. This net charge, however, was shown above to be zero, and thus $\Sigma_j f_j e\Psi_{W,C,j}$ must also vanish. Consequently, Eq. (46) takes the same form as Eq. (43), where $e\Phi$ now stands for the average surface work function. The same holds for Eq. (42).

As shown in Figs. 15 and 16a, Eqs. (42) and (43) have been verified experimentally by directly measuring the work function of catalyst electrodes using a Kelvin probe (vibrating condenser method) under reaction conditions.[3,59]

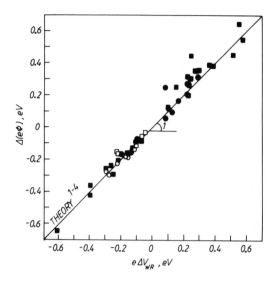

Figure 15. Steady-state effect of the change in the ohmic-drop-free catalyst potential V_{WR} on catalyst surface work function $e\Phi$. ■ and □, Y_2O_3-doped ZrO_2 electrolyte, $T = 300°C$; ● and ○, β''-Al_2O_3 electrolyte, $T = 240°C$; filled symbols: closed-circuit operation; open symbols: open-circuit operation. (Reprinted, with permission from *Nature*, Macmillan Magazines Ltd., from Ref. 3.)

Figure 15 shows that the change in the work function of the gas-exposed (i.e., catalytically active) surface of the catalyst is $e\Delta V_{WR}$, under both closed- and open-circuit conditions. In the former case, V_{WR} was varied by changing the polarizing current with the catalyst exposed to air or to $NH_3/O_2/He$ and $CO/O_2/He$ mixtures whereas in the latter case only the gaseous composition was varied. Both doped ZrO_2 and β''-Al_2O_3 solid electrolytes were used.[3,59]

As shown in Fig. 16, the equality $\Delta(e\Phi) = e\Delta V_{WR}$ also holds to a good approximation during transients. In this case, a constant current is applied at $t = 0$ between the catalyst and the counter electrode, and one follows the time evolution of V_{WR} by using a voltmeter and of $e\Phi$ by using a Kelvin probe.

Work-function transients of the type shown in Fig. 16 can be used to estimate initial dipole moments of the spillover dipoles on the catalyst

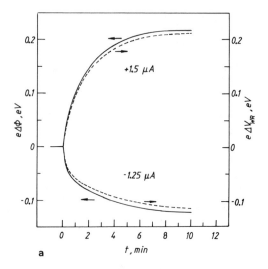

Figure 16. (a) Transient response of catalyst work function $e\Phi$ and potential V_{WR} upon imposition of constant currents I between the Pt catalyst (C2) and the Pt counter electrode. β''-Al$_2$O$_3$ solid electrolyte; $T = 240°C$, $P_{O_2} = 21$ kPa. Na ions are pumped to ($I < 0$) or from ($I > 0$) the catalyst surface at a rate I/F. (b) Effect of applied current on induced work-function change. ---, catalyst C1, $T = 291°C$, $P_{O_2} = 5$ kPa, $P_{C_2H_4} = 2.1 \times 10^{-2}$ kPa; —, catalyst C2, $T = 240°C$, $P_{O_2} = 21$ kPa. Inset: Effect of applied current on computed initial dipole moment of Na on Pt; •, $I > 0$; ▲, $I < 0$. (Reprinted, with permission from North-Holland Publishing Co., from Ref. 59.)

surface.[42,59] Thus, referring to Na supply onto a Pt catalyst surface with surface area A_C via a β''-Al$_2$O$_3$ solid electrolyte, one can use Faraday's law to obtain

$$\frac{d\theta_{Na}}{dt} = -\frac{N_{AV}(I/F)}{A_c \cdot N_{Pt}} \tag{47a}$$

where N_{AV} is Avogadro's number, and $N_{Pt} = 1.53 \times 10^{19}$ atoms/m^2 is the surface Pt atom concentration on the Pt(111) plane. One can then combine Eq. (47a) with the definition of the Na coverage $\theta_{Na}(= N_{Na}/N_{Pt})$ and with the differential form of the Helmholtz equation:

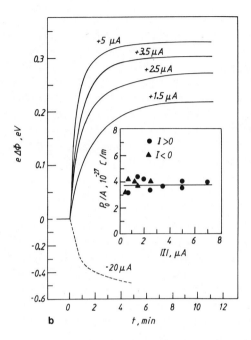

Figure 16. (continued)

$$\frac{d(e\Phi)}{dt} = -\frac{eP_o}{\varepsilon_o} \cdot \frac{dN_{Na}}{dt} \tag{47b}$$

where N_{Na} denotes adsorbed Na atoms/m^2, $\varepsilon_o = 8.85 \times 10^{-12}$ C^2/J·m, and P_o is the initial dipole moment of Na on Pt to obtain:

$$\frac{d(e\Phi)}{dt} = \frac{P_o I}{\varepsilon_o A_C} \tag{48}$$

Using Eq. (48) and the initial slope in Fig. 16, one computes the initial dipole moment P_o of Na on Pt to be 2.15×10^{-29} C·m or 6.5 D, that is, 22% higher than the literature value of 5.3 D for Na on a clean Pt(111) surface.[111,116] This is excellent agreement, in view of the fact that in the case of Fig. 16 the Pt surface is essentially saturated in oxygen,[59] which has been shown for systems like Cs/W(110) and Cs/Ni(100)[117] to give P_o values typically 20–30% higher than on the clean metal surface. As shown in Fig. 16, the computed P_0 value is independent of the magnitude and sign of the applied current, which confirms the validity of the approach. One additional conclusion which may be drawn is that Na introduced on metal surfaces via β''-Al_2O_3 to induce NEMCA is not different from Na introduced as a dopant using standard metal dispenser sources.[118,121] An important advantage, however, in using the electrochemical approach, that is, employing a solid electrolyte as the dopant donor, is that the doping is reversible; that is, the dopant can be removed electrochemically. Furthermore, the amount and coverage of the dopant on the surface can be accurately determined by integrating Eq. (47a), that is, by using Faraday's law.

When doped ZrO_2 is used as the ion donor, then the situation is slightly more complicated, as the backspillover oxygen anions may eventually form chemisorbed oxygen and desorb or react with the reactants, albeit at a slow rate.[4] Consequently, the coverage of backspillover oxygen anions is more difficult to measure. Aside from direct measurements by X-ray photoelectron spectroscopy (XPS) in ultrahigh vacuum (UHV) under NEMCA conditions,[122–125] a new method has been recently reported for measuring *in situ* the backspillover oxide ion coverage $\theta_{O^{\delta-}}$ via analysis of the catalytic rate response upon current interruption.[56,126]

It is worth noting that, in general,

$$e\Delta V_{WR} = \Delta(e\Phi) = -\frac{eN_M}{\varepsilon_0} \sum_j \Delta(P_{o,j} \cdot \theta_j) \tag{49}$$

where N_M is the surface metal atom concentration (atoms/m^2), and j stands for all adsorbed species on the catalyst surface, including the backspillover promoters but also the adsorbed reactants and intermediate species ($P_{o,j}$ is taken here positive for electropositive species and negative for electronegative ones). Consequently, upon varying V_{WR} and thus $e\Phi$, it follows that the coverages and/or dipole moments of adsorbed reactants and intermediates may also change, although the effect of promoting ions (Na^+, $O^{\delta-}$) must be dominant due to their large absolute dipole moments. This is also supported by the above-mentioned good agreement between $P_{o,Na}$ values measured under UHV conditions (no coadsorbates) and under NEMCA conditions. In the case, however, that no backspillover ions can be supplied to the catalyst surface (e.g., negative current application to metal/YSZ systems, which also leads frequently to pronounced NEMCA behavior[4,14]), the imposed $e\Delta V_{WR}$ and $\Delta(e\Phi)$ change is accommodated

(Eq. 49) by changes in the coverages and, more importantly, in the dipole moments of the adsorbates, for example, weakening in the C=O bond of chemisorbed CO and eventual dissociation caused by very negative V_{WR} values on Pt.[14]

4. Electrochemical Promotion: Galvanostatic Transients

A typical electrochemical promotion (EP), or *in situ* controlled promotion (ICP) or NEMCA, experiment utilizing YSZ, an O^{2-} conductor, as the promoter donor is shown in Fig. 17. The reaction under study is the oxidation of C_2H_4 on Pt. The Pt film, with a surface area corresponding to $N = 4.2 \times 10^{-9}$ mol Pt, measured via surface titration of oxygen with C_2H_4,[2,4] is exposed to $P_{O_2} = 4.6$ kPa and $P_{C_2H_4} = 0.36$ kPa in a continuous-flow gradientless (CSTR) reactor having a volume of 30 cm^3. The rate of CO_2 formation is monitored by use of an infrared analyzer.[2,4]

Figure 17 depicts a typical galvanostatic transient experiment. Initially ($t < 0$), the circuit is open ($I = 0$, $V^0_{WR} = -150$ mV), and the

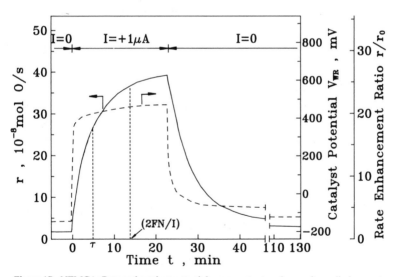

Figure 17. NEMCA: Rate and catalyst potential response to step changes in applied current during C_2H_4 oxidation on Pt. $T = 370°C$, $P_{O_2} = 4.6$ kPa, $P_{C_2H_4} = 0.36$ kPa. The experimental (τ) and computed ($2FN/I$) rate relaxation time constants are indicated on the figure. See text for discussion. $r_0 = 1.5 \times 10^{-8}$ mol O/s, $\Delta r = 38.5 \times 10^{-8}$ mol O/s, $I/2F = 5.2 \times 10^{-12}$ mol O/s, $\rho_{max} = 26$, $\Lambda_{max} = 74000$, $N = 4.2 \times 10^{-9}$ mol Pt. (Reprinted, with permission from Academic Press, from Ref. 2.)

unpromoted catalytic rate r_0 is 1.5×10^{-8} mol O/s. The corresponding turnover frequency, TOF,[127] based on the reactive oxygen uptake of 4.2×10^{-9} mol O, that is, the number of oxygen atoms reacting per surface site per second, is 3.57 s^{-1}.

At $t = 0$, the galvanostat is used to apply a constant current $I = 1 \ \mu$A between the catalyst and the counter electrode (Fig. 5). Consequently, O^{2-} is supplied to the catalyst at a rate $I/2F = 5.2 \times 10^{-12}$ mol O/s. The catalytic rate, r, increases gradually to a steady-state value of 4.0×10^{-7} mol O/s, which is 26 times larger than r_0. The new TOF is 95.2 s^{-1}. The increase in catalytic rate $\Delta r = r - r_0 = 3.85 \times 10^{-7}$ mol O/s is 74,000 times larger than the rate, $I/2F$, of supply of O^{2-} to the Pt catalyst. Thus, each O^{2-} supplied to the catalyst, creating a backspillover O$^{\delta-}$ species, causes at steady state 74,000 additional chemisorbed oxygen atoms to react with C$_2$H$_4$ and form CO$_2$ and H$_2$O. This is why this phenomenon was termed nonfaradaic electrochemical modification of catalytic activity (NEMCA).[1-4] It is worth noting that, at steady state, eV_{WR}, and thus $e\Phi$, has increased by 0.65 eV. Upon current interruption (Fig. 17), r and V_{WR} return to their initial unpromoted values owing to the gradual consumption of O$^{\delta-}$ by C$_2$H$_4$.

5. Definitions and Some Key Aspects of Electrochemical Promotion

(i) NEMCA Time Constant τ

The NEMCA time constant τ is defined[2,4] as the time required for the rate increase Δr to reach 63% of its steady-state value during a galvanostatic transient, such as the one shown in Fig. 17. Such rate transients can usually be approximated reasonably well by

$$\Delta r = \Delta r_{max}[1 - \exp(-t/\tau)] \qquad (50)$$

A general observation in NEMCA studies with O^{2-} conductors is that the magnitude of τ can be predicted from

$$\tau \approx 2FN/I \qquad (51)$$

where N (mol) is the reactive oxygen uptake of the metal catalyst,[4] which expresses, approximately, the moles of surface Pt. The parameter $2FN/I$ expresses the time required to form a monolayer of O$^{\delta-}$ on the Pt surface. Equation (51) is nicely rationalized on the basis of recent XPS investigations of Pt films during electrochemical promotion studies[125] which showed that the maximum coverages of $O^{\delta-}$ and normally chemisorbed, that is, reactive, atomic oxygen are comparable.

(ii) Enhancement Factor or Faradaic Efficiency Λ

The enhancement factor Λ is defined by

$$\Lambda = \Delta r/(I/2F) \tag{52}$$

where Δr is the promotion-induced change in catalytic rate expressed in moles of O per second. More generally, Λ is computed by expressing Δr in gram-equivalents and dividing by I/F.

In the experiment of Fig. 17, the Λ value at steady state is 74,000. A reaction is said to exhibit the NEMCA effect when $|\Lambda| > 1$.

An important step in the elucidation of the origin of NEMCA was the observation[2,4] that the magnitude of $|\Lambda|$ can be estimated for any reaction as

$$|\Lambda| \approx 2Fr_0/I_0 \tag{53}$$

where r_0 is the unpromoted catalytic rate, and I_0 is the exchange current[4,100] of the catalyst/solid electrolyte interface. As previously noted, this parameter can be measured easily by fitting η versus I data to the classical Butler–Volmer equation[4,100]:

$$I/I_0 = \exp(\alpha_a F\eta /RT) - \exp(-\alpha_c F\eta /RT) \tag{54}$$

where α_a and α_c, usually on the order of unity,[4,100] are the anodic and cathodic transfer coefficients.

As previously discussed, Eq. (53) has been also derived theoretically[4] and is in good qualitative agreement with experiment over five orders of magnitude (Fig. 18). This important relationship shows that:

(i) Highly polarizable, that is, low-I_0,[4,100,101] metal/solid electrolyte interfaces are required to obtain large $|\Lambda|$ values. The magnitude of I_0 is proportional to the tpb length[4,46] and can be controlled during the catalyst film preparation by appropriate choice of the sintering temperature,[4,46] which in turn determines the metal grain size and thus the tpb length.[4]

(ii) The measurement of Λ is important for determining whether a reaction exhibits the NEMCA effect, but its magnitude is not a fundamental characteristic of a catalytic reaction, since for the same reaction different $|\Lambda|$ values can be obtained by varying I_0 during catalyst prepara-

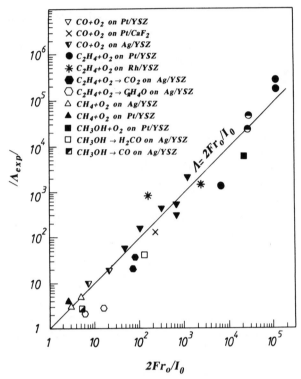

Figure 18. Comparison of predicted and measured values of the enhancement factor Λ for some of the catalytic reactions already found to exhibit the NEMCA effect (Table 3).

tion. Thus, the magnitude of $|\Lambda|$ depends both on catalytic (r_0) and on electrocatalytic (I_0) kinetics.

For $\Lambda \gg 1$, the faradaic efficiency Λ has an interesting physical meaning[48]: For oxidation reactions, it expresses the ratio of the reaction rates of normally chemisorbed atomic oxygen on the promoted surface and of backspillover oxide ions with the oxidizable species, for example, C_2H_4. It thus follows that the backspillover oxide ions are significantly less reactive than normally chemisorbed oxygen, in agreement with the recent XPS investigation of NEMCA on Pt/YSZ under UHV conditions.[125]

The maximum Λ value measured so far in NEMCA studies utilizing YSZ is 3×10^5 (Refs. 2 and 4), as shown in Table 3, which lists all catalytic reaction studied so far under electrochemical promotion conditions. When the promoting ion does not react at all with any of the reactants (e.g., Na$^+$), then, in principle, "infinite" Λ values are expected. In practice, Λ is always measurable due to a very slow consumption of Na$^+$ to form surface oxides and carbonates.[7,19,42] Nevertheless in all cases the catalytically important parameter is the promotion index[7] P_i defined below and also shown in Table 3.

6. Spectroscopic Studies

X-ray photoelectron spectroscopic (XPS) studies of Ag[122,123] and Pt[124,125] films deposited on YSZ under positive current application conditions have confirmed the proposition[2-4] that NEMCA with oxide-ion-conducting solid electrolytes is due to an electrochemically induced and controlled backspillover of oxide ions on the catalyst surface.

The early studies of Arakawa *et al.*[122,123] focused on Ag films. Upon positive current application, chemisorbed atomic oxygen is immediately formed (O $1s$ binding energy at 532.6 eV) followed by the gradual appearance of anionic oxygen (O $1s$ at 529.2 eV), which eventually causes a small decrease in the amount of chemisorbed atomic oxygen. This transient behavior is in good qualitative agreement with catalytic rate transients during ethylene epoxidation on Ag under similar temperature and imposed current conditions at atmospheric pressure.[43] More recently, Göpel and co-workers have used XPS, ultraviolet photoelectron spectroscopy (UPS), and electron energy loss spectroscopy (EELS) to study Ag/YSZ surfaces under NEMCA conditions.[124] Their XPS spectra are similar to those in Refs. 122 and 123.

A detailed XPS study of Pt films interfaced with YSZ[125] has shown that:

(i) Backspillover oxide ions (O $1s$ at 528.8 eV) are generated on the Pt surface with a time constant of $2FN/I$ (Eq. 51), where I is the applied current (peak δ in the top panel of Fig. 19).

(ii) Normally chemisorbed atomic oxygen (O $1s$ at 530.2 eV) is also formed with applied current (peak γ in the top panel of Fig. 19). The maximum coverages of the γ and δ states of oxygen (based on the number of surface Pt atoms) are comparable, each being of the order of 0.5.[125]

Table 3
In Situ Controlled Promotional NEMCA Studies

Reactants	Product(s)	Catalyst	Electrolyte (promoting ion)	T (°C)	Λ	ρ	P_i	Reference(s)
I. Electrophobic reactions $[\partial r/\partial(e\Phi) > 0; \partial r/\partial V_{WR} > 0; \partial r/\partial\Lambda > 0; \Lambda > 0]$								
C_2H_4, O_2	CO_2	Pt	YSZ (O^{2-})	260–450	3×10^5	55	55	2, 4
C_2H_4, O_2	CO_2	Rh	YSZ (O^{2-})	250–400	5×10^4	90	90	48
C_2H_4, O_2	CO_2	IrO_2	YSZ (O^{2-})	350–400	200	6	5	21
C_2H_4, O_2	C_2H_4O, CO_2	Ag	YSZ (O^{2-})	320–470	300	30^a	30	13, 43, 148
C_2H_6, O_2	CO_2	Pt	YSZ (O^{2-})	270–500	300	20	20	49
C_3H_6, O_2	C_3H_6O, CO_2	Ag	YSZ (O^{2-})	320–420	300	2^a	1	13
CH_4, O_2	CO_2	Pt	YSZ (O^{2-})	600–750	5	70	70	44^b
CH_4, O_2	CO_2	Pt	YSZ (O^{2-})	590	50	3	3	142
CH_4, O_2	CO_2, C_2H_4, C_2H_6	Ag	YSZ (O^{2-})	650–850	5	30^a	30	45
CO, O_2	CO_2	Pt	YSZ (O^{2-})	300–550	2×10^3	3	2	14^b
CO, O_2	CO_2	Pd	YSZ (O^{2-})	400–550	10^3	2	1	4
CO, O_2	CO_2	Ag	YSZ (O^{2-})	350–450	20	5	4	4, 147^b
CH_3OH, O_2	H_2CO, CO_2	Pt	YSZ (O^{2-})	300–500	10^4	4^a	3	17, 41^b
CO_2, H_2	CH_4, CO	Rh	YSZ (O^{2-})	300–450	200	3^a	2	51
CO, H_2	C_xH_y, $C_xH_yO_z$	Pd	YSZ (O^{2-})	300–370	10	3^a	2	51
CH_4, H_2O	CO, CO_2	Ni	YSZ (O^{2-})	600–900	12	2^a	1	52
H_2S	S_x, H_2	Pd	YSZ (O^{2-})	600–750	—	11	10	152
C_2H_4, O_2	CO_2	Pt	β''-Al_2O_3 (Na^+)	180–300	5×10^4	0.25	−30	4, 42
CO, O_2	CO_2	Pt	β''-Al_2O_3 (Na^+)	300–450	10^5	0.3	−30	7
C_6H_6, H_2	C_6H_{12}	Pt	β''-Al_2O_3 (Na^+)	100–150	—	$\sim$0	−10	18
CH_4	C_2H_6, C_2H_4	Ag	$SrCe_{0.95}Yb_{0.05}O_3$ (H^+)	750	—	11^a	10	20

C_2H_4, H_2	C_2H_6	Ni	$CsHSO_4$ (H^+)	150–170	6–300	0.16–2	12	15
H_2, O_2	H_2O	Pt	Nafion (H^+)	25	20	6	5	55
H_2, O_2	H_2O	Pt	$KOH–H_2O$ (OH^-)	25–50	20	6	5	8
CO, O_2	CO_2	Pt	CaF_2 (F^-)	500–700	200	2.5	1.5	56
C_2H_4, O_2	CO_2	Pt	TiO_2 (TiO_x^+, O^{2-})	450–600	5×10^3	20	20	154[b]
II. Electrophilic reactions [$\partial r/\partial(e\Phi) < 0;\ \partial r/\partial V_{WR} < 0;\ \partial r/\partial I < 0;\ \Lambda < 0$]								
C_2H_6, O_2	CO_2	Pt	YSZ (O^{2-})	270–500	−100	7	—	49
C_3H_6, O_2	CO_2	Pt	YSZ (O^{2-})	350–480	-3×10^3	6	—	49
CO, O_2	CO_2	Pt	YSZ (O^{2-})	300–550	−500	6	—	14
CO, O_2	CO_2	Au	YSZ (O^{2-})	450–600	−60	3	—	16, 54
CO_2, H_2	CO	Pd	YSZ (O^{2-})	500–590	−50	10	—	4, 51
CH_3OH, O_2	H_2CO, CO_2	Pt	YSZ (O^{2-})	300–550	-10^4	15[a]	—	17, 41
CH_3OH	H_2CO, CO, CH_4	Ag	YSZ (O^{2-})	550–750	−25	6[a]	—	4, 40
CH_3OH	H_2CO, CO, CH_4	Pt	YSZ (O^{2-})	400–500	−10	3[a]	—	4
CH_4, O_2	CO_2	Au	YSZ (O^{2-})	700–750	−3	3	—	16, 54
CH_4, O_2	C_2H_4, C_2H_6, CO_2	Ag	YSZ (O^{2-})	700–750	−1.2	8[a]	—	16, 54
CO, O_2	CO_2	Pt	β''-Al_2O_3 (Na^+)	300–450	-10^5	8	250	7
C_2H_4, NO	CO_2, N_2	Pt	β''-Al_2O_3 (Na^+)	400	—	∞	500	19

[a] Promotion-induced change in product selectivity.
[b] Electrophilic behavior also observed for negative currents.

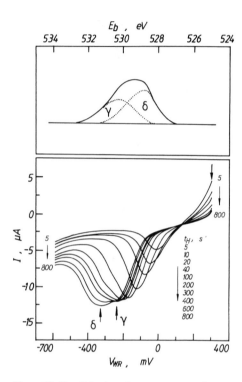

Figure 19. *Top*: O 1*s* photoelectron spectrum of oxygen adsorbed on a Pt electrode supported on YSZ
under UHV conditions after application of a constant
overpotential ΔV_{WR} = 1.2 V, corresponding to a
steady-state current $I = 40\,\mu$A, for 15 min at 673 K.[125]
The same O 1*s* spectrum was maintained after the
potentiostat was turned off and the electrode was
rapidly cooled to 400 K.[125] The γ-state is normally
chemisorbed atomic oxygen (E_b = 530.2 eV), and the
δ-state is backspillover oxidic oxygen (E_b = 528.8
eV). *Bottom*: Linear potential sweep voltammogram
obtained at T = 653 K and P_{O_2} = 0.1 kPa on a Pt
electrode supported on YSZ, showing the effect of
holding time t_H at V_{WR} = 300 mV on the reduction of
the γ- and δ-states of adsorbed oxygen; sweep rate: 30
mV/s.[27,60,107]

(iii) Oxidic backspillover oxygen (δ state) is much less reactive than normally chemisorbed oxygen (γ state) with the reducing (H_2 and CO) UHV background.[125]

The creation of two types of chemisorbed oxygen on Pt surfaces subject to NEMCA conditions has been also confirmed by the potential-programmed-reduction (PPR) technique[60,107] (Fig. 19, bottom panel). The high-V_{WR} oxygen reduction peak corresponds to normally chemisorbed atomic oxygen (γ state), and the low-V_{WR} peak, which develops only after prolonged positive current application,[60] corresponds to δ-oxygen, that is, backspillover oxide ions originating from the YSZ solid electrolyte.

The above observations provide a straightforward explanation for the physicochemical origin of NEMCA, or electrochemical promotion, when O^{2-}-conducting solid electrolytes, such as YSZ, are used. Backspillover oxide ions $O^{\delta-}$ (O $1s$ at 528.8 eV) generated at the tpb upon electrochemical O^{2-} pumping to the catalyst spread over the gas-exposed catalyst surface. They are accompanied by their compensating (screening) charge in the metal, thus forming surface dipoles. An "effective electrochemical double layer"[17] is thus established on the catalyst surface (Fig. 1) which increases the work function of the metal and affects the strength of chemisorptive bonds such as that of normally chemisorbed oxygen via through-the-metal or through-the-vacuum interactions. The change in chemisorptive bond strengths causes the observed dramatic changes in catalytic rates.

The observed phenomena are very similar to the effects on chemisorptive bond strength induced via electrical polarization (0.3 V/Å) and concomitant work-function change on well-characterized surfaces under UHV conditions.[128]

Sodium backspillover as the origin of electrochemical promotion when β''-Al$_2$O$_3$, a Na$^+$ conductor, is used as the solid electrolyte has been recently confirmed by Harkness and Lambert[19] by means of *ex situ* XPS. These authors found evidence that backspillover Na forms a surface carbonate during the C_2H_4 + NO reaction that acts as the promoting Na species.[19]

7. Catalytic Aspects of *in Situ* Controlled Promotion

Having surveyed the dual electrocatalytic/catalytic aspects of electrochemical promotion and established that NEMCA is due to the controlled introduction of promoters on the catalyst surface, we concentrate now on the purely catalytic aspects of *in situ* controlled promotion.

(i) Rate Enhancement Ratio ρ

The rate enhancement ratio ρ is defined as the ratio of the promoted, r, to the unpromoted, r_0, catalytic rate:

$$\rho = r/r_0 \tag{55}$$

In the galvanostatic transient experiment of Fig. 17, the steady-state ρ value is 26. The maximum ρ value measured for C_2H_4 oxidation on Pt/YSZ is $60^{2,4}$ versus 100 for the same reaction on Rh/YSZ (Fig. 20). Even higher ρ values have been obtained recently by Harkness and Lambert[19] for NO reduction by C_2H_4 on Pt/β''-Al_2O_3. The rate was practically nil on the unpromoted surface and quite significant on the Na-promoted surface owing to enhanced NO dissociation.[19] At the opposite extreme, Cavalca and Haller[18] have recently obtained ρ values approaching zero for the hydrogenation of benzene to cyclohexane on

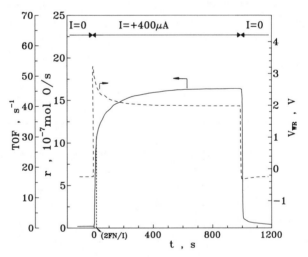

Figure 20. NEMCA: Rate and catalyst potential response to step changes in applied current during C_2H_4 oxidation on Rh. $T = 350°C$, $P_{O_2} = 2.6$ kPa, $P_{C_2H_4} = 5.9$ kPa, $r_0 = 1.8 \times 10^{-6}$ mol O/s, $\Delta r = 1.6 \times 10^{-6}$ mol O/s, $I/2F = 2.1 \times 10^{-9}$ mol O/s, $\Lambda = 770$, $\rho = 88$.

Pt/β''-Al_2O_3. They showed that Na coverages of less than 0.1 suffice to completely poison the rate of hydrogenation.[18]

(ii) Promotion Index P_i

From a catalytic viewpoint, the promotion index P_i is the most important phenomenological parameter for quantifying the promoting or poisoning effect of a given coadsorbed species i (e.g., O^{2-}, F^-, Na^+, H^+) on the rate of a catalytic reaction. It is defined by

$$P_i = \frac{\Delta r/r_0}{\Delta \theta_i} \tag{56}$$

where r_0 is the unpromoted catalytic reaction rate, and θ_i is the coverage of the promoting or poisoning species. Thus, for a coadsorbed species i that just blocks surface sites, $P_i = -1$. When $P_i > 0$, the species i is a promoter for the reaction. When $P_i < -1$, the species i is a poison for the reaction. Values of the promotion index P_{Na} up to 250 and down to -30 have been measured during CO oxidation on Pt/β''-Al_2O_3[7] at different gaseous compositions and θ_{Na} values, implying strong electronic interactions. Since P_i is often found to be strongly dependent on θ_i, it is also useful to define a differential promotion index p_i as

$$p_i = \frac{dr/r_0}{d\theta_i} \tag{57}$$

It follows then that

$$P_i = \int_0^{\theta_i} p_i \, d\theta_i' \tag{58}$$

The measurement of P_i and p_i, is quite straightforward when the promoting species (e.g., Na^+) does not react appreciably with any of the reactants or products. In this case, the coverage θ_{Na} of sodium on the metal surface can be easily measured coulometrically via[4,7]

$$\theta_{Na} = -\int_0^t \frac{I \, dt'}{FN} \tag{59}$$

where N is the catalyst surface area (moles of metal), and t is the time of current application. When I is constant, then Eq. (59) simplifies to

$$\theta_{Na} = -It/FN \tag{60}$$

When the promoting species is also partially consumed by one of the reactants, as, for example, in the case of C_2H_4 oxidation when YSZ is used as the solid electrolyte catalyst support (in which case C_2H_4 reacts with the promoting species $O^{\delta-}$ at a rate Λ times smaller than with normally chemisorbed oxygen), then the measurement of P_i is more complicated because of the difficulties in measuring the surface coverage of $O^{\delta-}$. In such cases, a conservative estimate of P_i can be obtained from $P_i = (\Delta r/r_o)_{max}$, that is, by assuming[125] that the maximum rate enhancement is obtained for $\Delta\theta_{O^{\delta-}} = 1$. With this assumption, it can be shown easily that

$$P_i = \rho_{max} - 1 \tag{61}$$

Consequently, the maximum $P_{O^{\delta-}}$ values measured so far are on the order of 60 and 100, respectively, for the oxidation of C_2H_4 on Pt[2,4] and on Rh.[48]

(iii) Electrophobic and Electrophilic Reactions

Depending on the rate behavior upon varying the catalyst potential V_{WR} and, equivalently, the work function $e\Phi$, catalytic reactions are divided into two large groups (Table 3), namely, electrophobic and electrophilic reactions.

A catalytic reaction is termed electrophobic when its rate increases with increasing catalyst work function $e\Phi$:

$$\partial r/\partial(e\Phi) > 0 \tag{62}$$

There are several equivalent definitions:

$$\partial r/\partial V_{WR} > 0 \tag{63a}$$

$$\Lambda > 0 \tag{63b}$$

$$P_{O^{2-}} > 0; \; P_{Na^+} < 0 \tag{63c}$$

A typical example of an electrophobic reaction is the oxidation of C_2H_4 on Pt[2,4] (Fig. 17), Rh,[48] and Ag.[43]

A catalytic reaction is termed electrophilic when its rate increases with decreasing catalyst work function $e\Phi$:

$$\partial r/\partial(e\Phi) < 0 \qquad (64)$$

or equivalently:

$$\partial r/\partial V_{WR} < 0 \qquad (65a)$$

$$\Lambda < 0 \qquad (65b)$$

$$P_{O^{2-}} < 0; \ P_{Na^+} > 0 \qquad (65c)$$

Typical examples of electrophilic reactions are the reduction of NO by ethylene on Pt[19] and CO oxidation on Pt under fuel-rich conditions.[14]

Some reactions exhibit both electrophobic and electrophilic behavior over different V_{WR} and $e\Phi$ ranges, leading to volcano-type[23] or inverted-volcano-type behavior.[44]

The electrophobicity or electrophilicity of a catalytic reaction appears to depend on the polarity of the metal–adsorbate or intra-adsorbate bond broken in the rate-limiting step (rls).[4] Thus, most catalytic oxidations on metals are found to be electrophobic reactions (Table 3). In these cases, cleavage of the metal–chemisorbed oxygen bond is involved in the rls. Increasing $e\Phi$ is known to weaken metal–electron acceptor bonds,[4] thus leading to a rate increase with increasing $e\Phi$, that is, to electrophobic behavior, as experimentally observed (Table 3). In the case of electrophilic reactions, such as NO reduction or CO hydrogenation, decreasing $e\Phi$ is known to strengthen the metal–adsorbate bond and thus to weaken the intra-adsorbate N=O or C=O bonds, the cleavage of which is usually rate-limiting. Consequently, decreasing $e\Phi$ enhances the catalytic rate, leading to electrophilic behavior, as experimentally observed.

8. Dependence of Catalytic Rates and Activation Energies on Catalyst Work Function $e\Phi$

A general observation that has emerged from *in situ* controlled promotion studies[1-4,7,8,22-27,40-54] is that over wide ranges of catalyst work function $e\Phi$ (0.2–1.0 eV), catalytic rates depend exponentially on catalyst work function $e\Phi$:

$$\ln(r/r_0) = \alpha(e\Phi - e\Phi^*)/k_b T \qquad (66)$$

where α and $e\Phi^*$ are catalyst- and reaction-specific constants. The "NEMCA coefficient" α is positive for electrophobic reactions and negative for electrophilic ones and typically takes values between -1 and 1.[4]

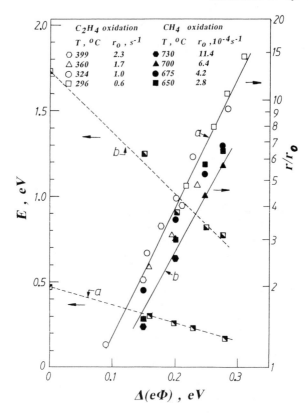

Figure 21. Effect of catalyst work function $e\Phi$ on the activation energy E and catalytic rate enhancement ratio r/r_0: (a) C_2H_4 oxidation on Pt, P_{O_2} = 4.8 kPa, $P_{C_2H_4}$ = 0.4 kPa; (b) CH$_4$ oxidation on Pt, P_{O_2} = 2.0 kPa, P_{CH_4} = 2.0 kPa. (Reprinted, with permission from Kluwer Academic Publishers, from Ref. 25.)

Typical examples of electrophobic reactions are shown in Fig. 21 for the catalytic oxidation of C_2H_4 and of CH$_4$ on Pt/YSZ. As shown in this figure, increasing $e\Phi$ also causes a linear variation in activation energy E:

$$E = E^{\circ} + \alpha_H \Delta(e\Phi) \tag{67}$$

where α_H is a constant which is usually negative for electrophobic reactions.

This linear decrease in E with $\Delta(e\Phi)$, which, for these systems, is due to the weakening of the Pt=O bond strength with increasing $e\Phi$,[4] is

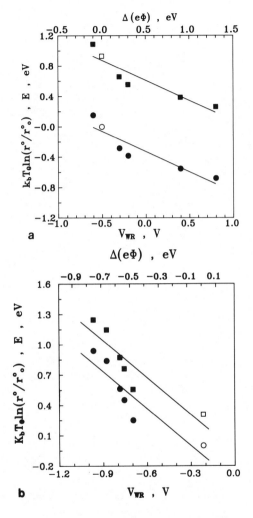

Figure 22. Effect of catalyst potential V_{WR} and work function $e\Phi$ on the activation energy E (squares) and preexponential factor r^0 (circles) of C_2H_4 oxidation on Rh[48] (a) and C_3H_6 oxidation on Pt (b); open symbols: open-circuit conditions.[49] T_Θ is the isokinetic temperature [372°C in (a) and 398°C in (b)], and r_0^0 is the open-circuit preexponential factor. Conditions: (a) $P_{O_2} = 1.3$ kPa, $P_{C_2H_4} = 7.4$ kPa; (b) $P_{O_2} = 3$ kPa, $P_{C_3H_6} = 0.4$ kPa.

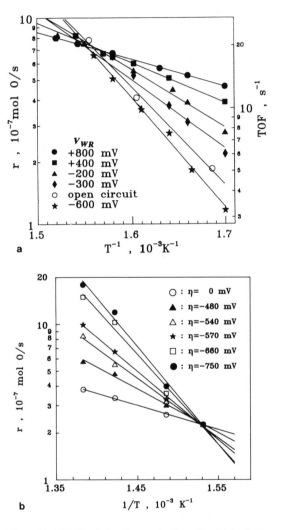

Figure 23. NEMCA-induced compensation effect in the isoki-netic point for C_2H_4 oxidation on Rh^{48} (a) and C_3H_6 oxidation on Pt^{49} (b). Conditions: (a) P_{O_2} = 1.3 kPa, $P_{C_2H_4}$ = 7.4 kPa; (b) P_{O_2} = 3 kPa, $P_{C_3H_6}$ = 0.4 kPa.

accompanied by a concomitant linear variation in the logarithm of the preexponential factor r^0, defined from

$$r = r^0 \exp(-E/k_b T) \tag{68}$$

as shown in Fig. 22 for the cases of C_2H_4 oxidation on Rh[48] and C_3H_6 oxidation on Pt.[49] This leads to the appearance of the well-known compensation effect (Fig. 23) with the isokinetic point lying within the temperature range of the kinetic investigation.

9. Selectivity Modification

One of the most promising applications of *in situ* controlled promotion is in product selectivity modification. Two examples regarding the epoxidation of ethylene on Ag are shown in Figs. 24 and 25. In the former case,[50] the Ag film is supported on YSZ. For $V_{WR} > 0$, ethylene oxide and CO_2 are the only products, and the selectivity to ethylene oxide is 55%. Decreasing the catalyst potential to $V_{WR} = -0.6$ V causes a dramatic shift

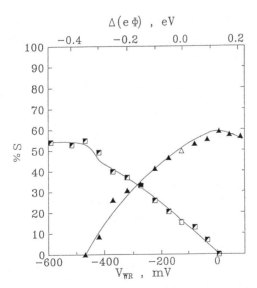

Figure 24. Effect of catalyst potential V_{WR} and catalyst work function $e\Phi$ on the selectivity (S) of ethylene oxidation to C_2H_4O (▲) and CH_3CHO (◨). $T = 270°C$, $P = 500$ kPa, 8.5% O_2, 7.8% C_2H_4.

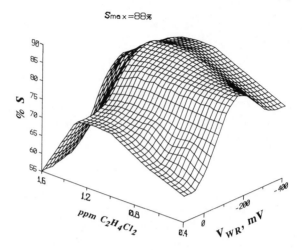

Figure 25. Effect of catalyst potential and of gas-phase 1,2-$C_2H_4Cl_2$ partial pressure on the selectivity (S) of ethylene epoxidation on Ag/β''-Al$_2$O$_3$.[50]

in selectivity. The selectivity to ethylene oxide vanishes, and acetaldehyde becomes the dominant product with a selectivity of 55%.

In the latter case (Fig. 25), the Ag catalyst is supported on β''-Al$_2$O$_3$ and traces of $C_2H_4Cl_2$ "moderator" are also added to the gas phase.[50] Ethylene oxide and CO_2 are the only products. The figure shows the combined effect of the partial pressure of $C_2H_4Cl_2$ and of the catalyst potential on the selectivity to ethylene oxide. For $V_{WR} = 0$ and –0.4 V, the Na coverage is nil and 0.04, respectively. As shown in the figure, there is an optimal combination of $V_{WR}(\theta_{Na})$ and $P_{C_2H_4Cl_2}$ leading to a selectivity to ethylene oxide of 88%. This is one of the highest values reported for the epoxidation of ethylene. Figure 25 exemplifies how *in situ* controlled promotion can be used for a systematic investigation of the role of promoters in technologically important systems.

10. Promotional Effects on Chemisorption

As previously noted, the strength of chemisorptive bonds can be varied *in situ* via controlled promotion. Figure 26 shows the effect of catalyst potential V_{WR} and work function $e\Phi$ on the Temperature Programmed Desorption (TPD) peak desorption temperature T_p and on the binding

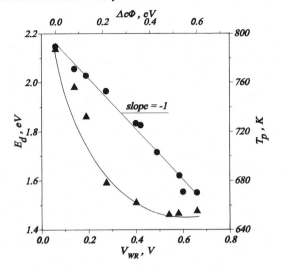

Figure 26. Effect of catalyst potential V_{WR} and work function $e\Phi$ on the TPD peak desorption temperature T_p (triangles) and on the binding strength E_d (circles) of oxygen chemisorbed on Pt/YSZ.[129]

strength E_d of oxygen dissociatively chemisorbed on Pt supported on YSZ.[129] Increasing eV_{WR} and $e\Phi$ by 0.6 eV causes a 150°C decrease in T_p and a 0.6-eV decrease in E_d. The latter is computed by varying the heating rate β via the modified Redhead equation of Falconer and Madix[130]:

$$\ln(\beta/T_p^2) = \ln(Rv_n Co^{n-1}/E_d) - (E_d/R)(1/T_p) \tag{69}$$

where β is the heating rate, v_n is the preexponential factor, and Co is the initial coverage.

It is important to notice that E_d decreases linearly with $e\Phi$ with a slope of -1, in excellent agreement with the observed decrease in activation energy E with $e\Phi$ in the Pt-catalyzed oxidation of C_2H_4 and CH_4 (Fig. 21).

The effect of V_{WR} and $e\Phi$ on the kinetics of oxygen adsorption and desorption on Ag deposited on YSZ has also been investigated recently.[43] It was found that decreasing $e\Phi$ causes a fivefold increase in the rate of atomic oxygen adsorption, a sixfold decrease in the rate of atomic oxygen desorption, and a twofold increase in the equilibrium atomic oxygen uptake.[43]

These results establish that increasing/decreasing $e\Phi$ causes a decrease/increase in the chemisorptive bond strength of electron acceptor adsorbates such as chemisorbed atomic oxygen and an increase/decrease in the chemisorptive bond strength of electron donor adsorbates such as benzene[18] or dissociatively chemisorbed hydrogen.[8] This is a key observation for rationalizing the *in situ* controlled promotional phenomena.

IV. ELECTROCHEMICAL PROMOTION WITH O^{2-}-CONDUCTING SOLID ELECTROLYTES

Most electrochemical promotion studies have been carried out so far using 8 mol % Y_2O_3 in ZrO_2 (YSZ), an O^{2-} conductor, as the solid electrolyte. The nature of the promoting species, $O^{\delta-}$, is reasonably well established via *in situ* XPS and cyclic voltammetric studies as analyzed in Section III. This type of promotion is unique in that the promoting species, $O^{\delta-}$, cannot form via gas-phase adsorption of oxygen. The promotional action of $O^{\delta-}$ is not limited to oxidation reactions but is also important in hydrogenation and dehydrogenation reactions.

1. Complete Oxidation Reactions

(i) Ethylene Oxidation on Pt

The kinetics and mechanism of ethylene oxidation on Pt have been studied for years on Pt films deposited on doped ZrO_2.[2,58,131] It has been found that at temperatures above 280°C the open-circuit catalytic kinetics can be described quantitatively by the rate expression

$$r_0 = k k_{ad} P_{C_2H_4} P_{O_2} / (k P_{C_2H_4} + k_{ad} P_{O_2}) \qquad (70)$$

and that the activity a_O of atomic oxygen on the Pt surface is proportional to the ratio $P_{O_2}/P_{C_2H_4}$.[58,131] Both of these observations had been interpreted quantitatively in terms of the kinetic scheme:

$$\frac{1}{2}\begin{Bmatrix} O_2(g) \xrightarrow{k_{ad}} O_2(a) \\ O_2(a) \rightarrow 2O(a) \end{Bmatrix} \qquad (71)$$

$$C_2H_4(g) + O(a) \xrightarrow{k} P \xrightarrow{\frac{5}{2}O_2} 2CO_2 + 2H_2O \qquad (72)$$

where $O_2(a)$ is a molecularly adsorbed precursor state with a negligible coverage,[131] and P is a reactive intermediate[131] which is rapidly oxidized by gaseous or adsorbed oxygen $O(a)$.

The rates of the steps given by Eqs. (71) and (72) can be written, respectively, as

$$r = k_{ad}P_{O_2}(1 - \theta_O) \tag{73}$$

$$r = kP_{C_2H_4}\theta_O \tag{74}$$

where θ_O is the coverage of chemisorbed oxygen. By equating these two expressions, one obtains both the experimental rate equation (70) and also, assuming a Langmuir isotherm for atomic oxygen chemisorption, the proportionality between a_O and $P_{O_2}/P_{C_2H_4}$.[58,131] The oxygen coverage θ_O can be computed as

$$\theta_O = 1/(1 + kP_{C_2H_4}/k_{ad}P_{O_2}) \tag{75}$$

On the fuel-lean side ($k_{ad}P_{O_2} \gg kP_{C_2H_4}$), the oxygen coverage is near unity, and, the step given by Eq. (72) is the rls. Thus, Eq. (70) reduces to

$$r_0 = kP_{C_2H_4} \tag{76}$$

On the fuel-rich side ($k_{ad}P_{O_2} \ll kP_{C_2H_4}$), θ_O is near zero, the oxygen adsorption step is the rls, and Eq. (70) reduces to

$$r_0 = k_{ad}P_{O_2} \tag{77}$$

It turns out[2,4] that varying V_{WR} and $e\Phi$ causes dramatic (up to 60-fold) increases in k but has practically no effect on k_{ad}. Thus, NEMCA is much more pronounced on the fuel-lean side, that is, when Eq. (76) is valid. This is shown in Fig. 27, which depicts the effect of the $P_{O_2}/P_{C_2H_4}$ ratio in the well-mixed reactor (CSTR) on the rate under open-circuit conditions and when V_{WR} is set at +1 V. There is a 60-fold increase in the rate for high $P_{O_2}/P_{C_2H_4}$ values.

It was found[2] that the kinetic constant k depends on $e\Phi$ according to

$$\ln(k/k_0) = \alpha e(\Phi - \Phi^*)/k_bT \tag{78}$$

and that the rate expression given by Eq. (70) remains valid under NEMCA conditions as well (Fig. 28), with k given by Eq. (78).

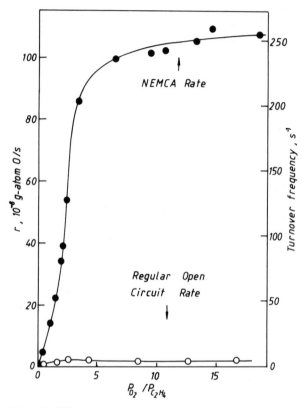

Figure 27. Effect of gaseous composition on the regular (open-circuit) catalytic rate of C_2H_4 oxidation on Pt/YSZ and on the NEMCA-induced catalytic rate on the same Pt catalyst film maintained at $V_{WR} = 1$ V. $T = 370°C$, $P_{C_2H_4} = 0.65 \times 10^{-2}$ bar. (Reprinted, with permission from Academic Press, from Ref. 2.)

Therefore, NEMCA does not induce a change in reaction mechanism, but only a pronounced increase in the Eley–Rideal rate constant k. For high $\Delta e\Phi$ values, when k has become sufficiently large, then NEMCA causes a change in the rls (Fig. 29); that is, NEMCA causes oxygen adsorption to become rate-limiting even under fuel-lean conditions. All this is described quantitatively by Eqs. (70) and (78).

The fact that the oxygen adsorption kinetic constant k_{ad} is insensitive to changes in V_{WR} and $e\Phi$ may provide an explanation for the oxygen

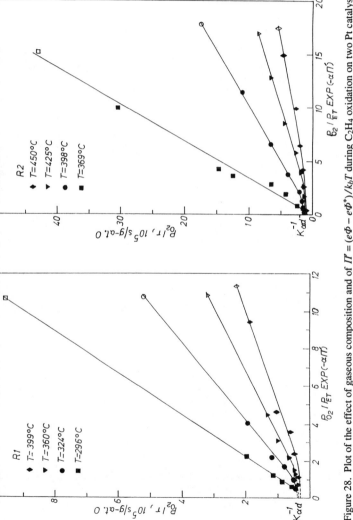

Figure 28. Plot of the effect of gaseous composition and of $\Pi = (e\Phi - e\Phi^*)/k_bT$ during C_2H_4 oxidation on two Pt catalyst films, labeled R1 and R2, showing that the rate expression given by Eq. (70) is valid both under open-circuit conditions (open symbols) and also under NEMCA conditions (filled symbols). (Reprinted, with permission from Academic Press, from Ref. 2.)

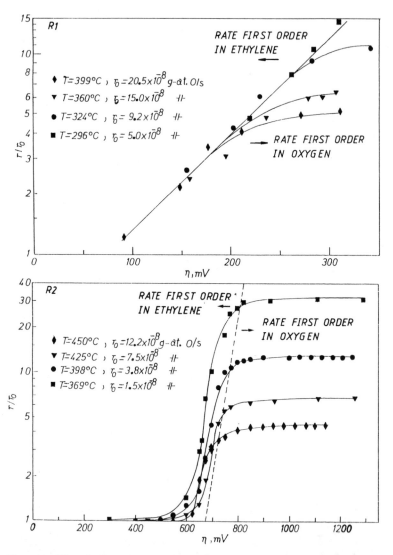

Figure 29. Effect of catalyst overpotential η on the rate and reaction order of C_2H_4 oxidation on two Pt catalyst films, labeled R1 and R2. For R_1, $P_{O_2} = 4.8 \times 10^{-2}$ bar and $P_{C_2H_4} = 0.4 \times 10^{-2}$ bar. For R_2, $P_{O_2} = 6.4 \times 10^{-2}$ bar and $P_{C_2H_4} = 0.4 \times 10^{-2}$ bar. (Reprinted, with permission from Academic Press, from Ref. 2.)

isotope exchange results of Sobyanin *et al.*[132] These authors reported that changing V_{WR} has no effect on the oxygen isotope exchange kinetics on Pt/YSZ at temperatures between 450 and 510°C and P_{O_2} up to 0.05 kPa.[132] On the basis of this, they questioned the variation in heats of adsorption with V_{WR} and $e\Phi$ and formulated a chain mechanism to explain NEMCA.[133] As shown in Section III.10 (Fig. 26), the binding strength of atomic oxygen on Pt/YSZ decreases by more than 0.6 eV with increasing V_{WR} and $e\Phi$. If, however, the rate-limiting step of the oxygen exchange process is oxygen adsorption rather than desorption, as suggested by the positive order dependence of the isotope exchange rate on P_{O_2},[132] then the insensitivity of the oxygen isotope exchange rate to changes in V_{WR} can be understood. The oxygen isotope exchange[132] is an interesting approach to studying NEMCA and is certainly worth investigating over a broader range of experimental conditions.

Thus, in order to rationalize the NEMCA behavior of the ethylene oxidation system, one needs only to concentrate on the kinetic constant k and on its dependence on $e\Phi$. As shown in Fig. 30, the exponential increase in k with $e\Phi$ is accompanied by a concomitant significant decrease in activation energy E and in the preexponential factor k^0, defined from

$$k = k^0 \exp(-E/k_b T) \tag{79}$$

As shown in Fig. 30,

$$E = E^\circ + \alpha_H \, \Delta e\Phi \tag{80}$$

and

$$k_b T \ln(k^0/k_o^0) = \alpha_S \, \Delta e\Phi \tag{81}$$

where α_S is a constant.[48] For the particular Pt catalyst film shown in the figure, $\alpha_H = -1$ and $\alpha_S = -0.4$. These values satisfy well the equality $\alpha = \alpha_S - \alpha_H$. Some significant variations in these values have been observed with other Pt films in the same study,[2,4] but with all films the equality $\alpha = \alpha_S - \alpha_H$ was satisfied within ±20%.

The linear decrease in activation energy with $e\Phi$ and, in fact, with a slope $\alpha_H = -1$ can be understood as follows. Since chemisorbed oxygen is an electron acceptor, increasing $e\Phi$ weakens the Pt= O chemisorptive bond, the cleavage of which is involved in the rls of the catalytic reaction (see Section III.10). Consequently, the activation energy decreases, in excellent agreement with the Polanyi principle,[127] and so does the preex-

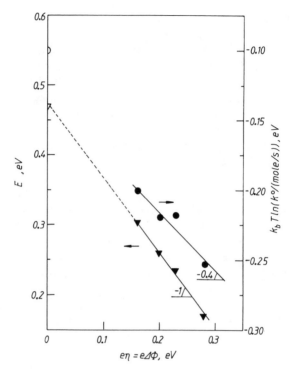

Figure 30. Effect of catalyst overpotential η and work function $e\Phi$ on the activation energy E and preexponential factor k^o of the kinetic constant k of C_2H_4 oxidation on Pt. T is the mean operating temperature. Conditions as in Fig. 21a. (Reprinted, with permission from Academic Press, from Ref. 2.)

ponential factor, because chemisorbed oxygen becomes more mobile and weakly bonded, and consequently its translational and vibrational entropy increases. This causes a pronounced decrease in the preexponential factor according to classical transition-state theory.[127]

In order to attempt a more quantitative description, one may start from the early theoretical considerations of Boudart,[134] who was first to tackle the problem of predicting the change in heats of adsorption with changing work function $e\Phi$. According to his early semiempirical electrostatic model, when the work function of a surface changes by $\Delta e\Phi$, then the heat of adsorption $-\Delta H_{ad}$ of covalently bonded adsorbed species should change by

$$\Delta(-\Delta H_{ad}) = -(n/2)\Delta(e\Phi) \tag{82}$$

where n is the number of valence electrons of the adatom taking part in the bonding. Since in the present case $n = 2$, Boudart's correlation reads

$$\Delta(-\Delta H_{ad}) = -\Delta(e\Phi) \tag{83}$$

Furthermore, to the extent that the nature of the activated complex of the catalytic step involving cleavage of the Pt=O bond (Eq. 72) does not change with varying $e\Phi$, one can use Polanyi's relationship,[127] i.e.,

$$\Delta E = \Delta(-\Delta H_{ad}) \tag{84}$$

which, in view of Eq. (83), yields

$$\Delta E = -\Delta(e\Phi) \tag{85}$$

which is in excellent agreement with experiment [Fig. 30 and Eq. (80)]. Boudart's early model has been criticized for its simplicity,[110] but it has been found on several occasions to provide a good fit to $-\Delta H_{ad}$ versus $e\Phi$ data. It does provide a quantitative fit to the data from the work discussed here[2] and from other NEMCA studies[4] where, as discussed below, linear variations in activation energy and heats of adsorption[129] with $e\Phi$ have been observed. It is worth noting that several workers have reported linear correlations between heats of adsorption of the same gas on different substrates as a function of substrate work function.[112,135]

Qualitatively, one can reach the same conclusions, but not a linear E versus $e\Phi$ relationship, via the semirigorous approach of Shustorovich,[136] who expresses the metal–adsorbate bond strength in terms of the differences $E_F - \varepsilon_A$ and $\varepsilon_A^* - E_F$, where E_F is the metal Fermi level (the variation in which is linear with $e\Phi$ in NEMCA experiments), ε_A is the energy of the highest occupied (σ or π) adsorbate orbital, and ε_A^* is the energy of the lowest unoccupied (σ^* or π^*) adsorbate orbital. Both ε_A and ε_A^* refer to the adsorbing molecule before adsorption and thus before the concomitant broadening of the bonding and antibonding energy levels. The recent rigorous *ab initio* calculations of Pacchioni *et al.*[137] for oxygen chemisorbed on Cu(100) and Pt(111) have shown similar linear trends.

In summary, the oxidation of C_2H_4 on Pt is one of the most thoroughly studied reactions from the point of view of NEMCA and, in view of its rather simple mechanistic scheme, one of the most thoroughly understood systems. The study of the NEMCA behavior of this system utilizing β''-Al_2O_3 as the ion donor leads to the same qualitative conclusions.[42]

(ii) Ethylene Oxidation on Rh

The oxidation of C_2H_4 to CO_2 on Rh has been investigated[48] at temperatures between 300 and 400°C. The reaction exhibits very strong electrophobic behavior, and the rate can be reversibly enhanced by up to 10,000% by supplying O^{2-} to the catalyst via positive potential application (up to 1.5 V). This ρ value (~100) is the highest obtained so far in NEMCA studies utilizing YSZ.

Figure 20 shows a typical galvanostatic transient. Positive current application ($I = 400\ \mu A$) causes an 88-fold increase in catalytic rate ($\rho = 88$). The rate increase is 770 times larger than the rate $I/2F$ of O^{2-} supply to the catalyst ($\Lambda = 770$). The NEMCA time constant τ is 40 s, in good qualitative agreement with the parameter $2FN/I = 18$ s.

Figure 31 shows the steady-state effect of P_{O_2} and imposed catalyst potential V_{WR} on the rate of C_2H_4 oxidation and compares the results with the open-circuit kinetics. The sharp rate decline for high P_{O_2} values is due to the formation of surface Rh oxide. Increasing V_{WR} causes a significant increase in the oxygen partial pressure at which oxide forms, $P_{O_2}^*$, and thus causes a dramatic increase in r for intermediate (1–2.5 kPa) P_{O_2} values.

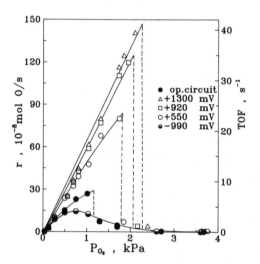

Figure 31. Effect of P_{O_2} and imposed catalyst potential V_{WR} on the rate of C_2H_4 oxidation on Rh/YSZ.[48] $T = 350°C$, $P_{C_2H_4} = 5$ kPa.

For low P_{O_2} values (reduced surface), the effect of V_{WR} is moderate, with ρ values up to 2. For high P_{O_2} values ($P_{O_2} > P_{O_2}^*$, oxidized surface), V_{WR} has practically no effect on the rate.

The effect of $P_{C_2H_4}$ and V_{WR} on the rate of C_2H_4 oxidation is shown in Fig. 32. Increasing V_{WR} causes a pronounced decrease in the ethylene partial pressure, $P_{C_2H_4}^*$, necessary to reduce the surface Rh oxide and thus a dramatic, up to 100-fold, increase in reaction rate for intermediate $P_{C_2H_4}$ values.

It is worth noting the change in the reaction order with respect to ethylene, from positive to negative, upon positive current application. This strongly suggests competitive adsorption of ethylene and the backspillover oxide ions $O^{\delta-}$.

Figures 33 and 34 show in detail the effect of V_{WR} and the corresponding rate, $I/2F$, of O^{2-} supply at a fixed gaseous composition. In the former case (Fig. 33), the gaseous composition has been chosen such that the surface is and remains reduced with positive and negative current application. The rate variation is moderate, as r/r_0 varies between 0.25 and 2.

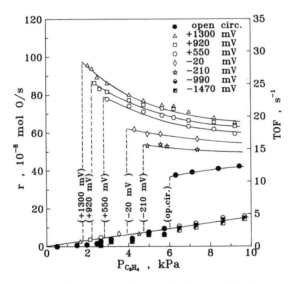

Figure 32. Effect of $P_{C_2H_4}$ and imposed catalyst potential V_{WR} on the rate of C_2H_4 oxidation on Rh/YSZ.[48] $T = 350°C$, $P_{O_2} = 1.3$ kPa.

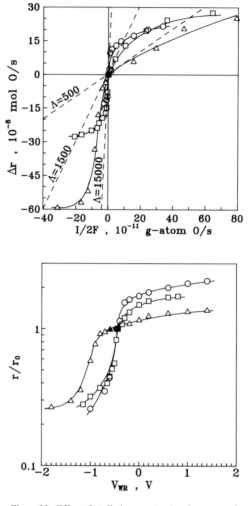

Figure 33. Effect of applied current (*top*) and corresponding catalyst potential V_{WR} (*bottom*) on the rate of C_2H_4
oxidation on a Rh surface which is reduced under open-
circuit conditions.[48] $P_{O_2} = 1.3$ kPa, $P_{C_2H_4} = 7.4$ kPa. O,
$T = 320°C$, $r_o = 1.74 \times 10^{-7}$ mol/s; □, $T = 350°Cr$, $r_o =
6.5 \times 10^{-7}$ mol/s; △, $T = 370°C$, $r_o = 8.4 \times 10^{-7}$ mol/s.
Filled symbols: open-circuit conditions.

The Λ values are up to 15,000. In the latter case (Fig. 34), the gas composition is such that under open-circuit conditions the surface is oxidized ($P_{O_2} > P_{O_2}^*$, $P_{C_2H_4} < P_{C_2H_4}^*$). At low temperatures ($T = 320°C$), the surface remains oxidized over the whole V_{WR} range, and the measured ρ and Λ values are moderate. At higher temperatures, positive currents and potentials lead to reduction of the surface Rh oxide, and the rate enhancement is very pronounced, with Λ values up to 50,000 and ρ values up to 100. The observed steady-state multiplicity with respect to V_{WR} (Fig. 34) is due to Rh oxide decomposition in conjunction with the galvanostatic operation. Potentiostatic operation leads to steady-state multiplicity with respect to the current. Figure 23a shows Arrhenius plots obtained at fixed V_{WR} values. Due to the reducing gas composition, all points in Fig. 23a correspond to a reduced surface. The open-circuit V_{WR}^0 is -500 to -600 mV. Increasing V_{WR} causes a dramatic decrease in activation energy from 0.92 eV (open-circuit conditions) to 0.3 eV for $V_{WR} = 0.8$ V with a concomitant pronounced decrease in the preexponential factor, r^o, defined from:

$$r = r^o \exp(-E/k_b T) \tag{86}$$

As a result of this, Fig. 23a presents a striking demonstration of the compensation effect with an isokinetic point at $T_\Theta = 372°C$. The compensation effect in heterogeneous catalysis has been the focal point of numerous studies and debates.[139,140] It is usually obtained from Arrhenius plots for several similar reactions on the same catalyst or for the same reaction on several similar catalysts. In the present case, it is obtained for one reaction and one catalyst by varying its potential or, equivalently, by varying the level of promoting species ($O^{\delta-}$) on its surface. Since the isokinetic point lies usually outside the temperature range of the kinetic investigation, several authors have even questioned the existence of a true compensation effect.[140] The results of Fig. 23 are, consequently, rather rare and show clearly that the compensation effect is a real one.

It is worth noting that below the isokinetic point ($T < T_\Theta$) the reaction exhibits electrophobic behavior, that is, $\partial r/\partial V_{WR} > 0$, while for $T > T_\Theta$ the reaction becomes electrophilic. At $T = T_\Theta$ the NEMCA effect disappears (see also the curve for $T = 370°C$ in Fig. 33).

As shown in Fig. 35, the decrease in activation energy E with eV_{WR} and $e\Phi$ is almost linear with a slope of -0.5. Also, the logarithm of the preexponential factor r^o decreases linearly with eV_{WR} (Fig. 22a), and upon

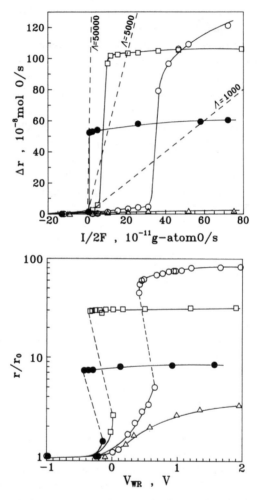

Figure 34. Effect of applied current (*top*) and corresponding catalyst potential V_{WR} (*bottom*) on the rate of C_2H_4
oxidation on a Rh surface which is oxidized under open-
circuit conditions.[48] $\triangle$, $P_{O_2} = 5$ kPa, $P_{C_2H_4} = 2.5$ kPa, $T =$
320°C, $r_0 = 0.87 \times 10^{-8}$ mol/s; ○, $P_{O_2} = 5$ kPa, $P_{C_2H_4} = 2.5$
kPa, $T = 350°C$, $r_0 = 1.8 \times 10^{-8}$ mol/s; □, $P_{O_2} = 5$ kPa,
$P_{C_2H_4} = 2.5$ kPa, $T = 370°C$, $r_0 = 3.67 \times 10^{-8}$ mol/s; ●,
$P_{O_2} = 1.2$ kPa, $P_{C_2H_4} = 3$ kPa, $T = 350°C$, $r_0 = 9.6 \times 10^{-8}$
mol/s.

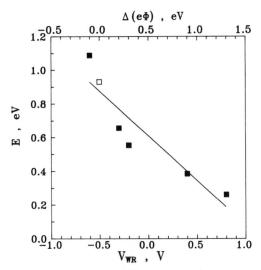

Figure 35. Effect of catalyst potential V_{WR} and corresponding work-function change $\Delta(e\Phi)$ on the activation energy of C_2H_4 oxidation on Rh.[48] $P_{O_2} = 1.3$ kPa, $P_{C_2H_4} = 7.4$ kPa.

plotting $k_b T_\Theta \ln(r^o/r_0^o)$, where r_0^o is the open-circuit preexponential factor, versus eV_{WR}, the slope is again -0.5. The difference between the two parallel lines equals the open-circuit activation energy E^o, as can be shown easily. As a result of Figs. 35 and 22a, the logarithm of r^o increases linearly with E with a slope of 18 eV^{-1} (Fig. 36).

The observed pronounced electrochemical promotion effect is due to the weakening of the Rh$=$O bond and the strengthening of the Rh$-C_2H_4$ bond with increasing V_{WR} and $e\Phi$. Increasing V_{WR} and $e\Phi$ destabilizes the Rh$=$O bond, owing to the increasing coverage of backspillover oxide ions, which exert repulsive through-the-metal and through-the-vacuum lateral interactions with chemisorbed oxygen.[137] Consequently, higher P_{O_2} values ($P_{O_2}^*$) are required to form surface Rh oxide (Figs. 31 and 32). Conversely, since increasing V_{WR} and coverage of $O^{\delta-}$ stabilizes the Rh$-C_2H_4$ bond via enhanced π-electron donation to the metal, it follows that smaller $P_{C_2H_4}$ values ($P_{C_2H_4}^*$) are required to reduce the surface Rh oxide, as experimentally observed.

These observations, which are in excellent agreement with the theory of electrochemical promotion regarding the effect of $e\Phi$ on the binding strength of electron acceptor (e.g., O) and electron donor (e.g., C_2H_4)

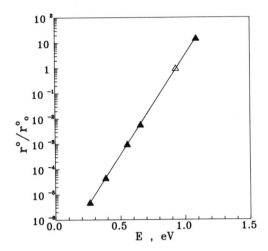

Figure 36. Dependence of the preexponential factor r^o on the activation energy E of C_2H_4 oxidation on Rh. Open symbol corresponds to open circuit.[48] $P_{O_2} = 1.3$ kPa, $P_{C_2H_4} = 7.4$ kPa.

adsorbates, are at a first glance counterintuitive since the surface Rh oxide is destabilized by supplying O^{2-} from the solid electrolyte *to* the catalyst. Alternatively, if one were to attempt to relate V_{WR} and oxygen activity, a_O, by using the Nernst equation for closed-circuit conditions, one would reach the conclusion that increasing a_O destabilizes the surface Rh oxide. This simply underlines, however, how erroneous the use of the Nernst equation can be outside its realm of applicability, that is, far from open-circuit conditions. It is worth emphasizing that ΔV_{WR} is an activation overpotential and not a concentration overpotential. In cases when it becomes the latter, then the Nernst equation can still be used.

(iii) C_2H_6 Oxidation on Pt

C_2H_6 oxidation on Pt has been investigated[49] in a "single-pellet" reactor at temperatures between 400 and 500°C. The reaction exhibits electrophobic behavior for positive currents ($V_{WR} > V^o_{WR}$) and electrophilic behavior for negative currents ($V_{WR} < V^o_{WR}$) (Figs. 37 and 38). In the former case, ρ is up to 20 and Λ up to 300. In the latter case, ρ is up to 7 and Λ up to −100. The open-circuit kinetic behavior indicates that the catalyst surface is predominantly covered with oxygen and that the cov-

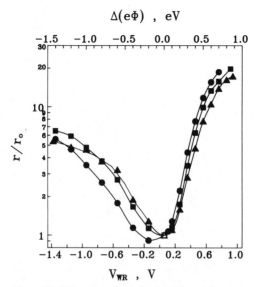

Figure 37. Effect of catalyst potential and work function on the rate of C_2H_6 oxidation on Pt.[49] P_{O_2} = 10.7 kPa, $P_{C_2H_6}$ = 1.65 kPa; ●, T = 500°C; ■, T = 460°C; ▲, T = 420°C.

erage of ethane is always low.[49] Consequently, the observed rate enhancement with $V_{WR} > V_{WR}^o$ can be attributed to enhanced ethane chemisorption and a weakening of the Pt=O bond. In order to explain the rate enhancement with negative currents ($V_{WR} < V_{WR}^o$), one must take into account that, since in this case no backspillover of ions can take place, the surface coverage of oxygen has to decrease so that Eq. (49) is obeyed, with $\Delta(e\Phi) = \Delta V_{WR}$. Consequently, the rate increase can be again attributed to the electrochemically induced decrease in oxygen coverage and enhancement in ethane chemisorption.

(iv) C_3H_6 Oxidation on Pt

C_3H_6 oxidation on Pt was studied[49] at temperatures between 250 and 400°C and was found to exhibit electrophilic behavior with negative currents, with Λ values of the order of –3000 and ρ values up to 6. The observed electrophilic behavior, under conditions where the rate is negative order in propene and positive order in oxygen, is due to the weakening

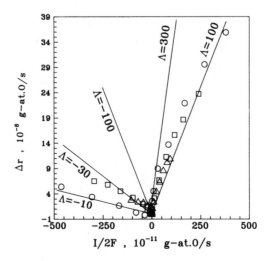

Figure 38. Effect of applied current on the increase in the rate of C_2H_6 oxidation.[49] Conditions as in Fig. 37. ○, $T = 500°C$; □, $T = 460°C$; △, $T = 420°C$.

of the chemisorptive bond of propene and concomitant enhanced chemisorption of oxygen with decreasing V_{WR} and $e\Phi$.

(v) CH₄ Oxidation on Pt

CH_4 oxidation on Pt is another reaction exhibiting electrophobic behavior for $V_{WR} > V_{WR}^o$ and electrophilic behavior for $V_{WR} < V_{WR}^o$. Figures 39a and 39b show typical galvanostatic transients for positive and negative currents.[4,44] The rate transients are normal first-order responses, but the V_{WR} transients are more complex. Figure 40 shows the steady-state effect of V_{WR} and $e\Phi$ on the rate of CH_4 oxidation for a low (1:1) CH_4-to-O_2 feed ratio ($P_{CH_4}^o = P_{O_2}^o = 2$ kPa). The rate increases exponentially with V_{WR} and $e\Phi$ both with increasing and decreasing $e\Phi$. The corresponding α values are 0.7 and –0.37.

As shown in Fig. 41, changing $e\Phi$ causes significant and linear variations in activation energy. The corresponding α_H values are –3.6 for $V_{WR} > V_{WR}^o$ and 1.06 for $V_{WR} < V_{WR}^o$. The former behavior is qualitatively similar to the case of C_2H_4 oxidation on Pt (Fig. 30) and thus can be explained by the same arguments. The latter is probably due to the fact

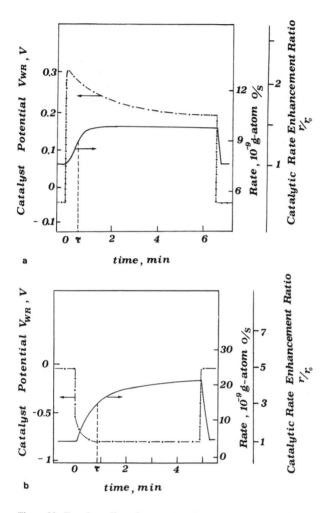

Figure 39. Transient effect of constant applied positive (a) and negative
(b) current on the rate of CH_4 oxidation to CO_2 on Pt. In (a) $T = 700°C$,
$I = 30$ mA, $P_{O_2} = P_{CH_4} = 0.02$ bar, in (b) $T = 700°C$, $I = -20$ mA,
$P_{O_2} = P_{CH_4} = 0.02$ bar. (Reprinted, with permission from Elsevier Science
Publishers, B.V., Amsterdam, from Ref. 4.)

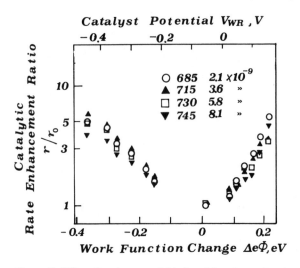

Figure 40. Effect of catalyst potential and work function on the rate
of CH_4 oxidation to CO_2 on Pt for a low (1:1) CH_4-to-O_2 feed ratio.
Maximum methane conversion is 4% $P_{CH_4} = P_{O_2} = 2$ kPa; T, °C; r_0,
g-atom O/s. (Reprinted, with permission from Elsevier Science Publishers, B.V., Amsterdam, from Ref. 4.)

that decreasing $e\Phi$ strengthens the Pt=O bond and thus enhances the
activation of the C–H bond in CH_4 to form CH_3· radicals, which can then
be rapidly oxidized in the gas phase.

The effect of V_{WR} and $e\Phi$ on catalytic rate is more pronounced when
high CH_4-to-O_2 ratios are used. Thus, for a 40:1 CH_4-to-O_2 feed ratio
($P_{CH_4} = 10$ kPa, $P_{O_2} = 0.25$ kPa, $T = 700$°C) a 70-fold increase in catalytic
rate is obtained for $e\Delta V_{WR} = \Delta e\Phi = 1$ eV (Fig. 42). This is due, again, to
the weakening of the Pt=O bond with increasing work function. Rate
enhancement factors Λ are low, typically less than five, due to the high
operating temperatures and concomitantly high I_0 values (Eq. 32).

In a recent study, Eng and Stoukides[85] also reported Λ values up to
five for this reaction and also detected the presence of trace C_2 hydrocarbons in the effluent stream. Since YSZ is known to promote catalytically
the oxidative coupling of CH_4,[141] the extent to which C_2 hydrocarbons can
be found in the products is dictated by the ratio of YSZ to Pt surfaces in
the reactor.

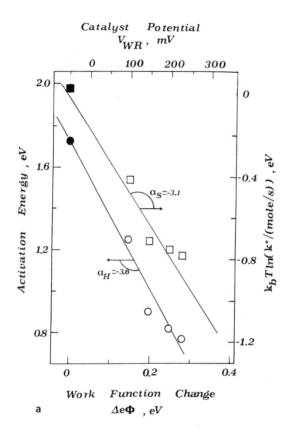

Figure 41. Effect of catalyst potential and work function on the activation energy E and preexponential factor of CH_4 oxidation to CO_2 on Pt for $V_{WR} > V_{WR}^o$ (a) and $V_{WR} < V_{WR}^o$ (b) $P_{O_2} = P_{CH_4} = 0.02$ bar. T is the mean operating temperature. (Reprinted, with permission from Elsevier Science Publishers, B.V., Amsterdam, from Ref. 4.)

(vi) CO Oxidation on Pt and Pd

The CO oxidation on Pt was the second reaction, after C_2H_4 oxidation on Ag, for which a nonfaradaic rate enhancement was observed.[14] Typical measured Λ values were of the order 10^2–10^3 while ρ was typically below

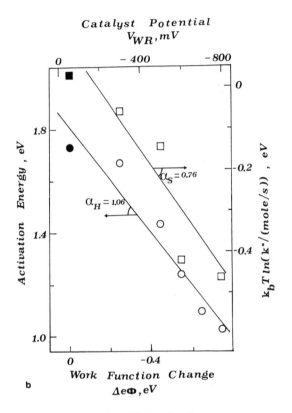

Figure 41. (continued)

five. In the earlier study of this system,[14] no storage oscilloscope was used in conjunction with the current interruption technique, and thus accurate determination of the ohmic-drop-free catalyst potential V_{WR} was not possible. Thus, results of this study were reported in terms of current rather than V_{WR}. Nevertheless, a more recent study involving accurate measurement of V_{WR} gives the same qualitative information.[143]

Figure 43 depicts some typical galvanostatic transients which show that, depending on the magnitude of the applied negative current (or overpotential), the reaction can exhibit either electrophobic behavior (for small negative currents) or electrophilic behavior for larger negative currents (in which case the magnitude of the rate enhancement depends

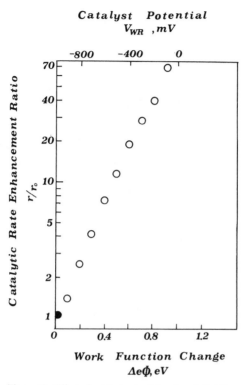

Figure 42. Effect of catalyst potential and work function on the rate of CH_4 oxidation to CO_2 on Pt for a high (40:1) CH_4-to-O_2 feed ratio. Maximum O_2 conversion is 62%. P_{CH_4} = 10 kPa, P_{O_2} = 0.25 kPa, T = 973 K. (Reprinted, with permission from Elsevier Science Publishers, B.V., Amsterdam, from Ref. 4.)

on previous catalyst history. Thus the curve labeled A in Fig. 43b was obtained when I = –600 μA was first applied to the cell. After several successive repetitions of negative current application, the rate increase stabilized to that labeled B in Fig. 43b.) The dual role of negative current is also shown by the steady-state r versus V_{WR} results presented in Fig. 44.

As shown in this figure and also in Fig. 45, increasing V_{WR} and $e\Phi$ above their open-circuit potential values leads to "volcano"-type behavior; that is, the rate goes through a maximum. Since CO chemisorption on

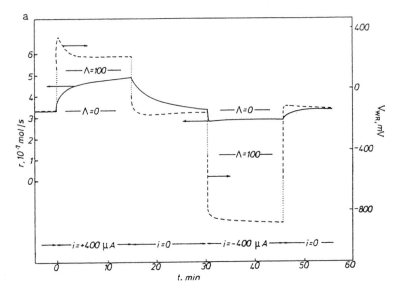

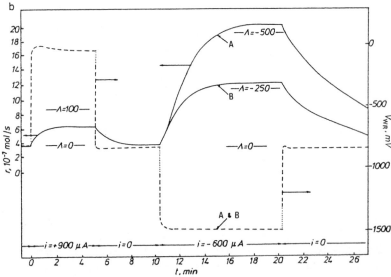

Figure 43. Transient effect of current on the rate of CO oxidation on Pt (—) and on catalyst potential (----). Inlet compositions and temperatures: (a) P_{CO} = 0.47 kPa, P_{O_2} = 10 kPa, T = 412°C; (b) P_{CO} = 2.9 kPa, P_{O_2} = 3.84 kPa, T = 555°C. (Reprinted, with permission from Academic Press, from Ref. 14.)

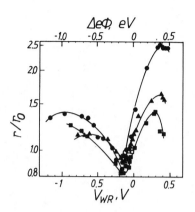

Figure 44. Steady-state effect of catalyst potential and work function on the rate of CO oxidation on Pt. Open symbols correspond to open-circuit conditions; ●, T = 485°C, r_o = 0.5 × 10⁻⁷ mol/s; ■, T = 505°C, r_o = 1.0 × 10⁻⁷ mol/s; ▲, T = 535°C, r_o = 1.5 × 10⁻⁷ mol/s. P_{CO} = 0.25 × 10⁻² bar, P_{O_2} = 11.3 × 10⁻² bar. (Reprinted, with permission from Trans Tech. Publications, from Ref. 143.)

Pt involves both donation and backdonation of electrons,[143] it is difficult to quantify the effect of changing $e\Phi$ on the strength of the CO chemisorption bond. Figure 46 shows that the location of the rate maximum is shifted to lower P_{CO} values with decreasing V_{WR}, which indicates a strengthening in the bond with decreasing $e\Phi$; that is, the electron acceptor character of CO dominates. Since the location of the rate maximum, which corresponds to roughly equal coverages of CO and oxygen, is only weakly affected with increasing $e\Phi$, this also indicates that both chemisorptive bonds are weakened. Consequently, one can interpret the observed "volcano" behavior in the usual way; that is, increasing $e\Phi$ decreases the binding strengths of chemisorbed reactants, thus enhancing their reaction rate, until the point is reached where the coverages of chemisorbed reactants become very low and the rate starts to decrease.

The observed rate increase with very negative V_{WR} values under reducing conditions has been attributed to a C=O disproportionation mechanism,[14] but this point requires further investigation utilizing $^{18}O_2$ in order to examine whether $C^{18}O_2$ is indeed formed.

As shown in Figs. 47–49, the rate and V_{WR} (or $e\Phi$) oscillations of CO oxidation can be started or stopped at will by imposition of appropriate currents.[14] Thus, in Fig. 47 the catalyst is initially at a stable steady state. Imposition of a negative current merely decreases the rate, but imposition of a positive current of 200 μA leads to an oscillatory state with a period of 80 s. The effect is completely reversible, and the catalyst returns to its initial steady state upon current interruption.

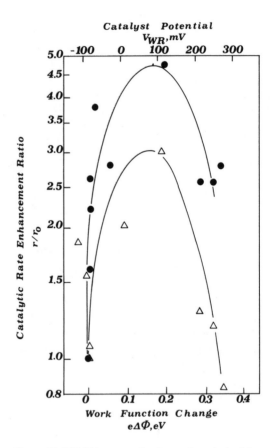

Figure 45. NEMCA-generated volcano plots obtained by increasing the catalyst work function above its open-circuit value during CO oxidation on Pt.[23] $P_{CO} = 0.2$ kPa, $P_{O_2} = 11$ kPa. ●, $T = 560°C$, $r_0 = 1.5 \times 10^{-9}$ g-atom O/s; △, $T = 538°C$, $r_0 = 0.9 \times 10^{-9}$ g-atom O/s. (Reprinted, with permission from Johnson Matthey Public Ltd., from Ref. 23.)

The opposite effect is depicted in Fig. 48, where the catalyst under open-circuit conditions exhibits stable limit cycle behavior with a period of 184 s. Imposition of a negative current of -400 μA leads to a steady state. Upon current interruption, the catalyst returns to its initial oscillatory state. Application of positive currents leads to higher frequency oscillatory states.

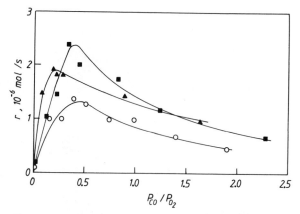

Figure 46. Effect of gaseous composition on the rate of CO oxidation on Pt under open-circuit conditions and for $V_{WR} = 0.5$ V and -0.5 V. $T = 535°C$; $\circ$, open circuit; $\blacksquare$, $V_{WR} = 500$ mV; $\blacktriangle$, $V_{WR} = -500$ mV. (Reprinted, with permission from Trans Tech Publications, from Ref. 143.)

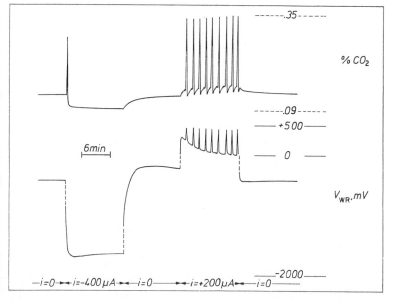

Figure 47. Induction of self-sustained rate and catalyst potential, or work-function, oscillations by NEMCA during CO oxidation on Pt. Inlet composition: $P_{CO} = 0.47$ kPa, $P_{O_2} = 16$ kPa; $T = 297°C$. (Reprinted, with permission from Academic Press, from Ref. 14.)

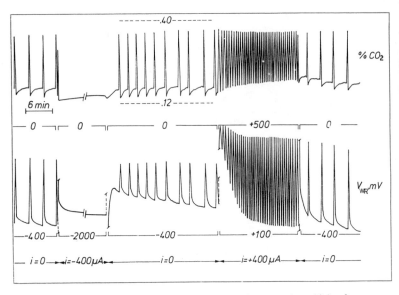

Figure 48. Transition from an oscillatory state to a steady state and to a higher frequency oscillatory state upon application of negative and positive current, respectively, during CO oxidation on Pt. Inlet composition: $P_{CO} = 0.47$ kPa, $P_{O_2} = 16$ kPa; $T = 332°C$. (Reprinted, with permission from Academic Press, from Ref. 14.)

A striking feature of the effect of current on the CO oxidation oscillations is shown in Fig. 49. It can be seen that the frequency of oscillations is a linear function of the applied current. This holds not only for intrinsically oscillatory states but also for those which do not exhibit oscillations under open-circuit conditions, such as the ones shown in Fig. 47. This behavior is consistent with earlier models developed to describe the oscillatory behavior of Pt-catalyzed oxidations under atmospheric pressure conditions which are due to surface PtO_2 formation,[58,131,145] as analyzed in detail elsewhere.[14]

The oxidation of CO on Pd is another reaction exhibiting NEMCA. In a preliminary study, values of the rate enhancement factor Λ of the order of 10^3 were measured at $T = 290°C$, $P_{CO} = 3 \times 10^{-2}$ kPa, and $P_{O_2} = 15$ kPa.[23] This reaction also is well known to exhibit oscillatory behavior[146] and deserves further examination.

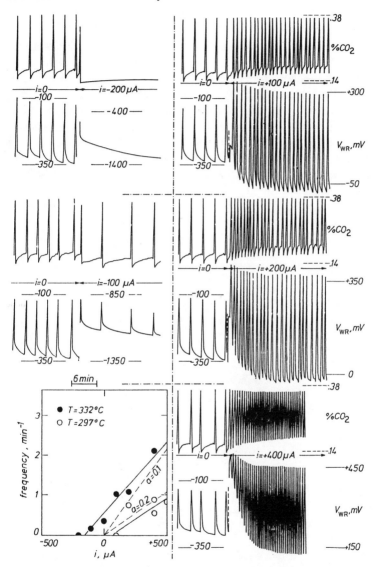

Figure 49. Effect of applied constant current on the frequency of the self-sustained rate and V_{WR} oscillations during CO oxidation on Pt. Conditions as in Fig. 48. Filled circles on the frequency versus current diagram are oscillatory states of this figure; open circles include states shown in Fig. 47. $P_{CO} = 0.47$ kPa, $P_{O_2} = 16$ kPa, $T = 297°C$. (Reprinted, with permission from Academic Press, from Ref. 14.)

(vii) CO Oxidation on Ag

Similarly to CO oxidation on Pt, this reaction on Ag exhibits electrophobic ($\Lambda > 0$) behavior for high V_{WR} and $e\Phi$ values and electrophilic ($\Lambda < 0$) behavior for low V_{WR} and $e\Phi$ values.[147] The rate dependence on V_{WR} and $e\Phi$ shown in Fig. 50 bears many similarities to that obtained with CO oxidation on Pt (Fig. 44) except that the "volcano" behavior at high $e\Phi$ is missing here, at least over the $e\Phi$ range investigated. Typically, Λ values are of the order of 20 in the electrophobic region and of the order of -800 in the electrophilic one.[147] Figures 51 and 52 show the rate dependence on reactant partial pressures under open-circuit and NEMCA conditions. Interestingly, the rate remains first-order in O_2 under all conditions, while the rate dependence on CO also retains its open-circuit qualitative features under NEMCA conditions.

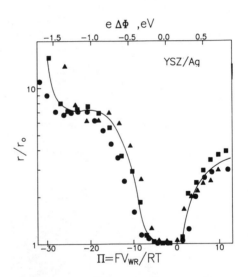

Figure 50.   Effect of catalyst potential and work function on the rate enhancement ratio during CO oxidation on Ag. $P_{O_2} = 3$ kPa, $P_{CO} = 5$ kPa, ●, $T = 363°C$, $r_0 = 2.7 \times 10^{-9}$ g-atom O/s; ▲, $T = 390°C$, $r_0 = 3.4 \times 10^{-9}$ g-atom O/s; ■, $T = 410°C$, $r_0 = 5.5 \times 10^{-9}$ g-atom O/s. (Reprinted, with permission from Trans Tech Publications, from Ref. 147.)

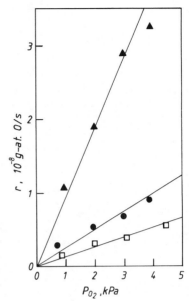

Figure 51. Effect of P_{O_2} on regular (open-circuit) and NEMCA-induced rate of CO oxidation on Ag. $T = 415°C$; $P_{CO} = 5$ kPa; $\square$, $I = 0$; $\bullet$, $V_{WR} = +475$ mV; $\blacktriangle$, $V_{WR} = -1300$ mV. (Reprinted, with permission from Trans Tech Publishers, from Ref. 147.)

(viii) C_2H_4 Oxidation on IrO_2

C_2H_4 oxidation on IrO_2 exhibits electrophobic behavior for positive currents and fuel-lean conditions.[21] Thus, at $T = 390°C$, $P_{C_2H_4} = 0.2$ kPa, and $P_{O_2} = 20$ kPa, $\rho \approx 6$ and $\Lambda = 100$. Figure 53 shows a typical galvanostatic transient. For negative currents, the reaction is weakly electrophilic.[21] This is the first demonstration of NEMCA utilizing a metal oxide catalyst, albeit IrO_2 is a metallic oxide, exhibiting a high density of states at the Fermi level.

2. Partial Oxidation Reactions

(i) C_2H_4 and C_3H_6 Epoxidation on Ag

The epoxidation of ethylene was the first reaction for which a non-faradaic rate enhancement was found.[13] As shown in Fig. 54, the observed increase in the rate of epoxidation, r_1, and oxidation to CO_2, r_2, were typically a factor of 300 higher than the rate $I/2F$ of supply of O^{2-} to the catalyst. A slight improvement in the selectivity to C_2H_4O (from 0.52 to 0.59) was observed with $I > 0$ under oxidizing conditions ($P_{C_2H_4} = 1.5$ kPa,

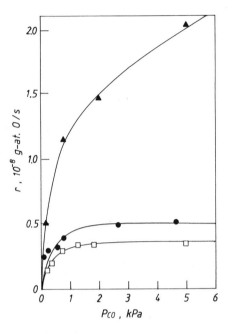

Figure 52. Effect of P_{CO} on regular (open-circuit) and NEMCA-induced rate of CO oxidation on Ag. $T = 415°C$; $P_{O_2} = 3$ kPa; $\square$, $T = 0$; $\bullet$, $V_{WR} = +475$ mV; $\blacktriangle$, $V_{WR} = -1300$ mV. (Reprinted with permission from Elsevier Science Publishers, B.V., Amsterdam, from Ref. 4.)

$P_{O_2} = 10$ kPa, $T = 400°C$), while negative currents caused a decrease in selectivity from 0.52 to 0.42 (Fig. 55). It was observed that the rate relaxation time constants during galvanostatic transients were of the order of $2FN/I$, strongly indicating that the change in catalytic properties was taking place over the entire catalyst surface. No reference electrode was used in earlier studies[13] so that η and V_{WR} could not be measured.

A qualitatively similar behavior was obtained during C_3H_6 epoxidation on Ag. Values of the enhancement factor Λ of the order of 150 were measured.[13] Both the rates of epoxidation and oxidation to CO_2 increased with $I > 0$ and decreased with $I < 0$. The intrinsic selectivity to propylene oxide was very low, typically 0.03, and could be increased only up to 0.04

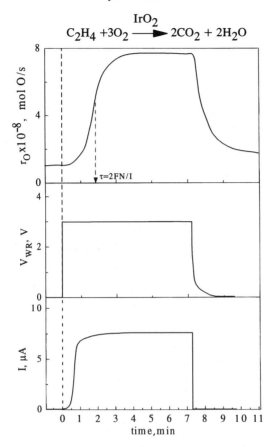

Figure 53. Potentiostatic and galvanostatic transient during C_2H_4 oxidation on $IrO_2/YSZ.^{21}$ $P_{C_2H_4} = 0.26$ kPa; $P_{O_2} = 20$ kPa; $T = 390°C$; $\Lambda \approx 100$.

by using positive currents. This was again an exploratory study, as no reference electrode was used, and thus η and V_{WR} could not be measured.

More recently, ethylene epoxidation on Ag has been investigated using a proper three-electrode system utilizing both doped ZrO_2^{43} and β''-Al_2O_3 solid electrolytes as the ion donor.[4] Very recently, these studies were extended[50,148] to examine the effect of higher operating pressures (5 bar) and of the addition of gas-phase $C_2H_4Cl_2$ "moderator," which is

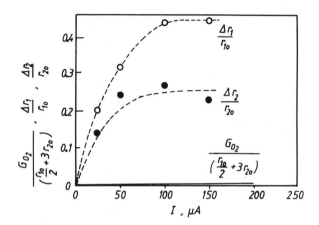

Figure 54. Steady-state effect of current on the increase in the rates of epoxidation (r_1) and deep oxidation to CO_2 (r_2) of C_2H_4 on Ag and comparison with the rate $G_{O_2} = I/4F$ of electrochemical oxygen supply.[13,28] $P_{C_2H_4} = 1.6$ kPa, $P_{O_2} = 10$ kPa; intrinsic ($I = 0$)) selectivity, 0.5. (Reprinted, with permission from Academic Press, from Ref. 28.)

known to promote the selectivity to ethylene oxide. In this section, we discuss the use of doped ZrO_2.

Figure 56 shows some typical galvanostatic transients, one for $I > 0$ and another for $I < 0$. In the former case (Fig. 56a), both rates of C_2H_4O and CO_2 formation increase, but both exhibit an initial "overshooting." The latter case (Fig. 56b) is more interesting. Initially, both rates decrease, but at steady state the rate of epoxidation has decreased while the rate of CO_2 formation has increased. Thus, epoxidation exhibits electrophobic behavior but oxidation to CO_2 exhibits electrophilic behavior.[43]

Figure 57 shows the effect of positive overpotential, that is, increasing work function, on the apparent activation energies E_i and preexponential factors k_i^0 of the epoxidation ($i = 1$) and deep oxidation ($i = 2$) reactions. After a slight initial increase, both E_1 and E_2 decrease substantially with increasing $e\Phi$ according to:

$$\Delta E_1 = -1.03 \, \Delta(e\Phi) \tag{87}$$

$$\Delta E_2 = -0.93 \, \Delta(e\Phi) \tag{88}$$

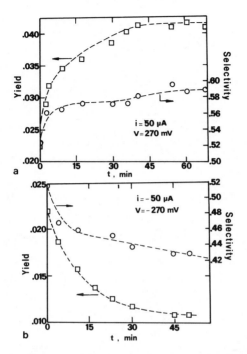

Figure 55. Transient effect of electrochemical O^{2-} pumping to (a) and from (b) an Ag catalyst film on selectivity to and yield of ethylene oxide.[13,28] Current applied at $t = 0$; $P_{C_2H_4} = 1.5$ kPa, $P_{O_2} = 10$ kPa; $T = 400°C$. (Reprinted, with permission from Academic Press, from Ref. 28.)

which is again strongly reminiscent of Boudart's correlation[134] and of the discussion for the case of C_2H_4 oxidation on Pt [Section IV.1(i)]. In fact, in view of Boudart's correlation, it would appear that the number of bonding electrons of the oxygen chemisorptive bond is $n = 2$, which corresponds to atomically chemisorbed oxygen, in agreement with the prevailing ideas about the C_2H_4 epoxidation system.[149,150]

Interestingly, the preexponential factors behave similarly to the activation energies, and $k_bT \ln k_1^0$ decreases linearly with increasing $e\Phi$[43]:

$$k_bT \Delta\ln k_1^0 = -0.93 \Delta e\Phi \qquad (89)$$

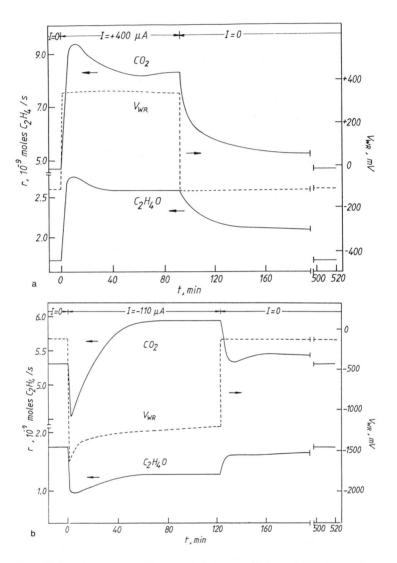

Figure 56. Transient response of the rates of ethylene epoxidation and oxidation to CO_2 on Ag upon O^{2-} supply to (a) and removal from (b) the catalyst. $P_{C_2H_4} = 2.2$ kPa, $P_{O_2} = 3.0$ kPa, $T = 435°C$. (Reprinted, with permission from Academic Press, from Ref. 43.)

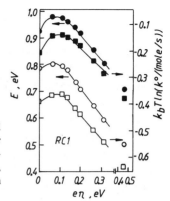

Figure 57. Effect of Ag/YSZ catalyst overpotential on the activation energy E and preexponential factor k^0 of ethylene epoxidation (open symbols) and oxidation to CO_2 (closed symbols). $P_{C_2H_4} = 2.48$ kPa, $P_{O_2} = 3.15$ kPa. (Reprinted, with permission from Academic Press, from Ref. 43.)

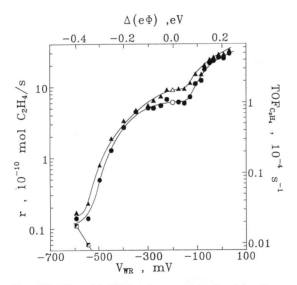

Figure 58. Effect of Ag/YSZ catalyst potential and work function on the rates of formation of ethylene oxide, acetaldehyde, and CO_2 at low $P_{O_2}/P_{C_2H_4}$ ratios. $T = 260°C$; $P = 500$ kPa; 3.5% O_2; 9.8% C_2H_4; ▲, C_2H_4O; ◨, CH_3CHO; ●, CO_2.[148]

$$k_b T \, \Delta \ln k_2^0 = -0.81 \, \Delta e \Phi$$

(90)

Similarly to the case of C_2H_4 oxidation on Pt [Section IV.1(i)], one can attribute this decrease to the increased entropy of chemisorbed atomic oxygen due to the weakening of the Ag=O bond with increasing work function $e\Phi$.

At lower temperatures (260°C), higher operating pressures (5 bar), and high C_2H_4-to-O_2 ratios (Fig. 58), ethylene oxide formation and CO_2 formation both exhibit electrophobic behavior over the entire V_{WR} range.[148] Both rates vary by a factor of 200 as V_{WR} is varied by 0.6 V (ρ varies between 3 and 0.015). The selectivity to ethylene oxide exhibits two local maxima. More interestingly, acetaldehyde appears as a new product. At higher oxygen pressures, acetaldehyde formation appears at higher potentials (Figs. 59 and 24), and the selectivity to acetaldehyde is up to 55% (Fig. 24).

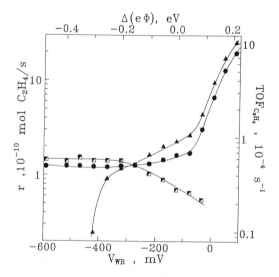

Figure 59. Effect of Ag/YSZ catalyst potential and work function on the rates of formation of ethylene oxide, acetaldehyde, and CO_2 at high $P_{O_2}/P_{C_2H_4}$ ratios. T = 270°C; P = 500 kPa; 8.5% O_2; 7.8% C_2H_4; ▲, C_2H_4O; ▨, CH_3CHO; ●, CO_2.[148]

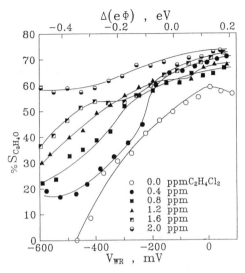

Figure 60. Effect of Ag/YSZ catalyst potential, work function, and feed partial pressure of dichloroethane on the selectivity to ethylene oxide; $T = 270°C$, $P = 500$ kPa, 8.5% O_2, 7.8% C_2H_4.[148]

Figures 60 and 61 show the effect of adding trace amounts of $C_2H_4Cl_2$ to the feed under NEMCA conditions. Dichloroethane suppresses the formation of acetaldehyde at negative potentials and leads, in conjunction with NEMCA, to ethylene oxide selectivity values of up to 75% for positive potentials (Fig. 62). As shown in the next section, even higher ethylene oxide selectivity values can be obtained using sodium, instead of $O^{\delta-}$, as the promoting ion.

(ii) Methanol Oxidation on Pt

Methanol oxidation on Pt was investigated at temperatures between 350° and 650°C, CH_3OH partial pressures, P_M, between 5×10^{-2} and 1 kPa, and oxygen partial pressures, P_{O_2}, between 1 and 20 kPa.[41] Formaldehyde and CO_2 were the only products detected in measurable concentrations. The open-circuit selectivity to H_2CO is of the order of 0.5 and is practically unaffected by gas residence time under the above conditions for methanol conversions below 30%. Consequently, the reactions of

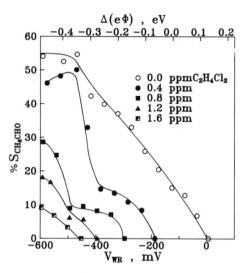

Figure 61. Effect of Ag/YSZ catalyst potential, work function, and feed partial pressure of dichloroethane on the selectivity to acetaldehyde.[148] Conditions as in Fig. 60.

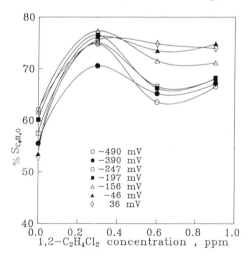

Figure 62. Effect of dichloroethane on the selectivity to ethylene oxide during ethylene oxidation on Ag/YSZ at various imposed catalyst potentials.[148] $V_{WR}^0 = -197$ mV, $T = 270°C$, $P = 500$ kPa, 3.5% O_2, 9.5% C_2H_4.

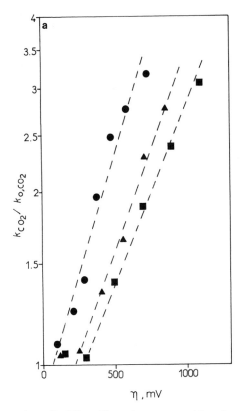

Figure 63. Effect of Pt catalyst overpotential on the kinetic constant of CH_3OH oxidation to CO_2 for positive (a) and negative (b) currents. $P_{CH_3OH} = 0.9$ kPa, $P_{O_2} = 19$ kPa; $T =$ ●, 698 K; ▲, 650 K; ■, 626 K. (Reprinted, with permission from Academic Press, from Ref. 41.)

H_2CO and CO_2 formation can be considered kinetically as two parallel reactions.

The effect of catalyst overpotential and potential on the rates of these two reactions is shown in Figs. 63 and 64. They both exhibit both electrophobic behavior for $V_{WR} > V_{WR}^o$ and electrophilic behavior for $V_{WR} < V_{WR}^o$.

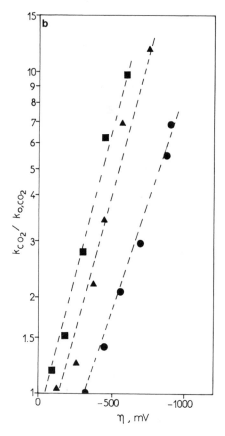

Figure 63. (continued)

Figure 65 shows the effect of the dimensionless potential $\Pi = FV_{WR}/RT$ on product selectivity, S, under constant feed conditions. The selectivity to H_2CO can be varied deliberately between 0.35 and 0.60 by varying the catalyst potential and goes through a maximum near the open-circuit potential value, V^o_{WR} . A similar study by Cavalca *et al.*[17] at lower temperatures and utilizing the single-pellet design gave qualitatively similar results. An important finding of that study was that NEMCA can also be induced by just short-circuiting the catalyst and counter electrodes and thus exploiting the open-circuit potential difference gener-

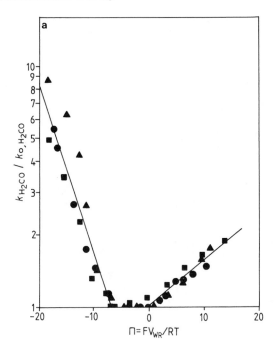

Figure 64. Effect of Pt catalyst dimensionless potential $\Pi = FV_{WR}/RT$ on the kinetic constants of formation of formaldehyde (a) and CO_2 (b). $T = \bullet$, 698 K; $\blacktriangle$, 650 K; $\blacksquare$, 626 K. (Reprinted, with permission from Academic Press, from Ref. 41.)

ated between the two electrodes due to the reaction-induced reduction in oxygen activity on the catalyst electrode.[17]

(iii) CH4 Oxidative Coupling on Ag

The oxidative coupling of CH_4 (OCM) in solid oxide fuel cells has attracted considerable attention in recent years because of the enormous interest in the production of C_2 hydrocarbons from natural gas. Work in this area utilizing solid electrolytes prior to 1990 has been reviewed.[84]

It was recently reported that the solid electrolyte itself, that is, Y_2O_3-doped ZrO_2, is a reasonably selective catalyst for CH_4 conversion to C_2 hydrocarbons, that is, ethane and ethylene,[141] and this should be taken into account in studies employing stabilized ZrO_2 cells. At the same

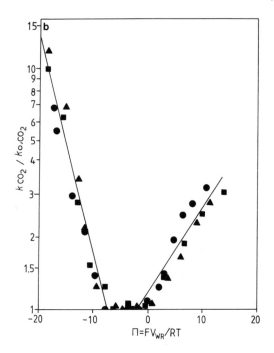

Figure 64. (continued)

time, it was found[4,23,45] that the use of Ag catalyst films leads to C_2 selectivities above 0.6 for low methane conversions.

As shown in Fig. 66, one can influence dramatically both the total CH_4 conversion as well as product selectivity by varying the Ag catalyst potential. Thus, under open-circuit conditions ($V_{WR} = V^o_{WR}$), the CH_4 conversion is near 0.04 with a C_2 selectivity (methane molecules reacting to form C_2H_4 and C_2H_6 per total number of reacting CH_4 molecules) near 0.6. Increasing V_{WR} increases the methane conversion to 0.3 and decreases the selectivity to 0.23, while decreasing V_{WR} decreases the conversion to 0.02 and increases the selectivity to 0.75. The reaction exhibits electrophobic NEMCA behavior both for $I > 0$ and for $I < 0$, with Λ values up to 5. The low Λ values are due to the, unavoidably, high I_0 values (Eq. 53) because of the high operating temperature. Rate enhancement ratios up to 30, 7, and 3.5 were observed for the three main products, CO_2, C_2H_4, and C_2H_6, respectively.[45]

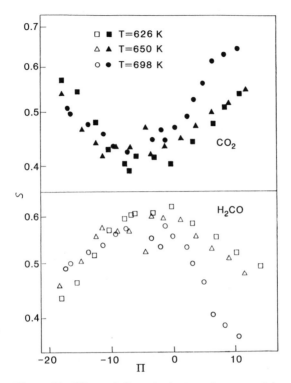

Figure 65. Effect of dimensionless catalyst potential $\Pi = FV_{WR}/RT$ on the selectivity to CO_2 and formaldehyde during CH_3OH oxidation on Pt. Conditions as in Fig. 63. (Reprinted, with permission from Academic Press, from Ref. 41.)

The rates of C_2H_4, C_2H_6, and CO_2 formation depend exponentially on V_{WR} and $e\Phi$ according to Eq. (66) with α values of 1.0, 0.65, and 1.2, respectively, for $I > 0$ and 0.15, 0.08, and 0.3, respectively, for $I < 0$. Linear decreases in activation energy with increasing $e\Phi$ have been found for all three reactions.[45] It should be emphasized, however, that, due to the high operating temperatures, Λ is near unity, and electrocatalysis, rather than NEMCA, plays the dominant role.

Nevertheless, this system is quite interesting from a technological point of view. As recently shown,[60] Ag and Ag–Sm_2O_3 anodes operating with pure CH_4 (no oxygen addition, chemical cogeneration mode) lead to

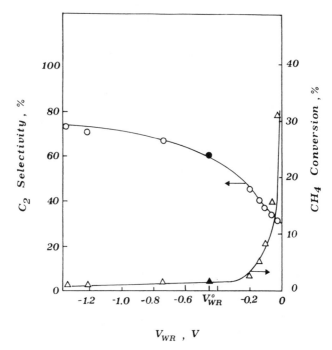

Figure 66. Effect of Ag/YSZ catalyst potential on CH_4 conversion and on selectivity to C_2 hydrocarbons. $T = 800°C$, $P_{O_2} = 0.25$ kPa, $P_{CH_4} = 10.13$ kPa, $V_{WR}^o = -0.45$ V. (Reprinted, with permission from Elsevier Science Publ., B.V. Amsterdam from Ref. 4.)

C_2H_4 selectivities up to 100% and total C_2H_4 yields up to 85%, when the zirconia reactor is used in a gas-recycle loop containing a molecular sieve adsorbent (molecular sieve 5A) for C_2 hydrocarbon trapping.[60] This is by far the highest ethylene or C_2 hydrocarbon yield obtained so far with the OCM reaction.

Similar studies utilizing Au electrodes on YSZ showed again that the selectivity and yield of C_2 hydrocarbons can be significantly affected by applying currents or potentials to the cell.[16,54,84] The behavior with Au appears to be qualitatively similar to that obtained with Ag electrodes although electrophilic behavior has also been reported.[16,54] The latter study[84] also reports on the catalytic properties of the YSZ electrolyte, which, as previously mentioned, is also active, to some extent, for the oxidative coupling reaction.[141]

3. Dehydrogenation and Hydrogenation Reactions

(i) Methanol Dehydrogenation on Ag and Pt

The dehydrogenation and decomposition of CH_3OH on Ag was one of the first catalytic systems for which NEMCA was studied in detail.[40] This investigation was carried out at temperatures between 600 and 680°C and CH_3OH partial pressures up to 7 kPa. Total CH_3OH conversion was kept below 20% to avoid consecutive reactions. Under open-circuit conditions, the main product is H_2CO (typical selectivity S_{H_2CO} is 0.85–0.90) with lesser amounts of CO ($S_{CO} \approx 0.07$–0.11) and CH_4 ($S_{CH_4} \approx 0.03$–0.05) Residence time in the CSTR has practically no effect on selectivity, so that the three reactions

$$CH_3OH \rightarrow H_2CO + H_2 \tag{91a}$$

$$CH_3OH \rightarrow CO + 2H_2 \tag{91b}$$

$$CH_3OH + H_2 \rightarrow CH_4 + H_2O \tag{91c}$$

can be viewed macroscopically and kinetically as parallel reactions obeying simple Langmuir-type rate expressions.[40]

Figure 67 shows a typical galvanostatic transient. At the start of the experiment the circuit is open, and the rates of H_2CO, CO, and CH_4 production are $r_{0,H_2CO} = 4.9 \times 10^{-8}$ mol/s, $r_{0,CO} = 7.6 \times 10^{-9}$ mol/s, and $r_{0,CH_4} = 1.5 \times 10^{-9}$ mol/s. At $t = 0$ a galvanostat is used to apply a current $I = -2mA$, with a corresponding rate of O^{2-} removal from the tpb of $-I/2F = 1.04 \times 10^{-8}$ g-atom/s. This causes a 380% increase in r_{H_2CO} and a 413% increase in r_{CO}. The corresponding enhancement factors are $\Lambda_{H_2CO} = -17.5$ and $\Lambda_{CO} = -3$. There is also a 190% increase in r_{CH_4} with an enhancement factor $\Lambda_{CH_4} = -0.3$, but this rate increase has been shown[40] to be faradaic and due to the electrocatalytic reaction

$$CH_3OH + 2e^- \rightarrow CH_4 + O^{2-} \tag{92}$$

taking place at the tpb.

Figure 68 shows the effect of the dimensionless catalyst potential Π on the rates of formation of H_2CO, CO, and CH_4. The α values are −0.14, −0.30, and −0.65, respectively. However, as shown previously,[40] the α_{CO} value is the cathodic transfer coefficient for the reaction in Eq. (92) and

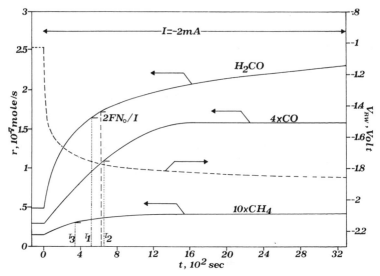

Figure 67. Rate and catalyst potential response to a step change in applied current during CH₃OH dehydrogenation and decomposition on Ag. The experimental time constants τ are compared with $2FN/I$. $T = 660°C$, $P_{CH_3OH} = 5.2$ kPa. (Reprinted, with permission from Academic Press, from Ref. 40.)

is not a true NEMCA coefficient. The corresponding effect of Π on product selectivity is shown in Fig. 69.

The effect of overpotential η on the apparent activation energies E of this reaction system is shown in Fig. 70. The apparent activation energies are extracted from $\ln r$ versus T^{-1} plots and cannot be attributed specifically to any single reaction step. The observed decrease in E_{CH_4} with increasing $|\eta|$ with a slope near -1 is in good qualitative agreement with the classical theory of electrocatalytic reactions.[100]

The observed electrophilic behavior of the formation of H₂CO and CO has been interpreted by taking into account the strengthening of the chemisorptive bond of methoxy intermediates[40] with decreasing catalyst work function. This strengthening in the chemisorptive bond causes, from classical bond energy/bond order conservation considerations, a weakening in the intra-adsorbate C-H bonds and thus facilitates dehydrogenation. More negative potentials, that is, lower work-function values, promote complete dehydrogenation, and thus the selectivity to CO increases (Fig. 69).

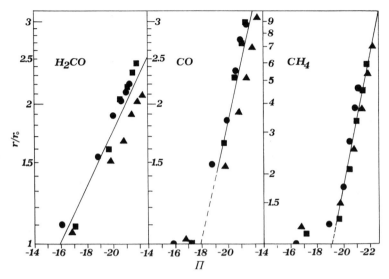

Figure 68. Effect of dimensionless catalyst potential $\Pi = FV_{WR}/RT$ on the rates of formation of H_2CO, CO, and CH_4 during CH_3OH dehydrogenation and decomposition on Ag/YSZ. $P_{CH_3OH} = 5$ kPa. ▲, $T = 620°C$; ■, $T = 643°C$; ●, $T = 663°C$. (Reprinted, with permission from Academic Press, from Ref. 40.)

A qualitatively similar behavior has been observed in a preliminary study of CH_3OH dehydrogenation on Pt at temperatures between 400 and 500°C, where enhancement factors Λ of the order of −10 were measured.[4]

(ii) CO_2 Hydrogenation on Rh

The hydrogenation of CO and CO_2 on transition-metal surfaces is a promising area for using NEMCA to affect rates and selectivities. In a recent study of CO_2 hydrogenation on Rh,[51,151] where the products were CH_4 and CO, under atmospheric pressure and at temperatures between 300 and 500°C it was found that CH_4 formation is electrophobic (Fig. 71a) while CO formation is electrophilic (Fig. 71b). Values of the enhancement factor Λ on the order of 100 up to 220 were measured (Fig. 72), in good agreement with the values calculated from Eq. (32).

The observed increase in CH_4 formation with increasing catalyst potential and work function can be attributed to the preferential formation on the Rh surface of electron donor hydrogenated carbonylic species

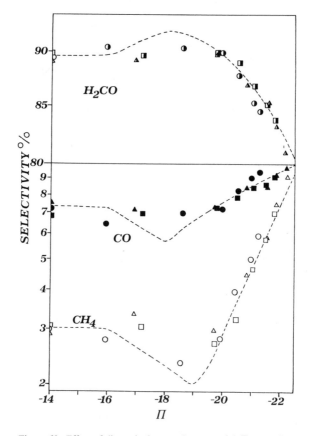

Figure 69. Effect of dimensionless catalyst potential Π on product selectivity to H_2CO, CO, and CH_4 during CH_3OH dehydrogenation and decomposition on Ag. Conditions as in Fig. 68. Dashed lines from Eq. (24) in Ref. 40. (Reprinted, with permission from Academic Press, from Ref. 40.)

leading to formation of CH_4 and to the decreasing coverage of electron acceptor carbonylic species resulting in CO formation.[151]

(iii) CO_2 Hydrogenation on Pd

CO_2 hydrogenation on Pd was investigated under atmospheric pressure and at temperatures between 500 and 600°C, where the only product

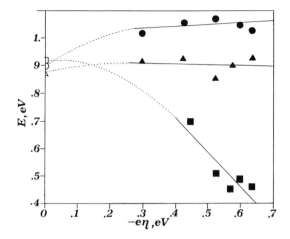

Figure 70. Effect of catalyst overpotential on the apparent activation energies of formation of H_2CO (●), CO (▲), and CH_4 (■) during CH_3OH dehydrogenation and decomposition on Ag. (Reprinted, with permission from Academic Press, from Ref. 40.)

was CO, because Pd, contrary to Rh, does not adsorb CO_2 dissociatively.[151] This difference in reaction pathway is also reflected in the NEMCA behavior of the system, since in the present case CO formation is not only enhanced (by up to 600%) with decreasing catalyst potential and work function, but also enhanced, although to a minor extent, via catalyst potential increase (Fig. 73). Absolute values of the enhancement factor Λ up to 150 were measured.

(iv) CO Hydrogenation on Pd

CO hydrogenation on Pd is of considerable technological interest as Pd catalysts have been investigated thoroughly in recent years for the production of alcohols and other oxygenated products from synthesis gas.[151]

The hydrogenation of CO on Pd was studied under NEMCA conditions in a single-pellet YSZ cell reactor at a total pressure of 12.5 bar and at temperatures between 330 and 370°C. Under these conditions, a variety of products are obtained, including hydrocarbons (CH_4, C_2H_4, C_2H_6), alcohols (CH_3OH), and aldehydes (HCHO, CH_3CHO). The distribution

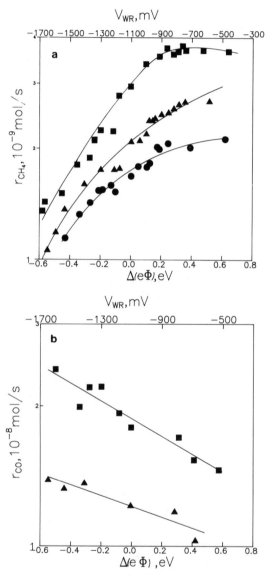

Figure 71. Effect of catalyst potential and work function on the rate of CH$_4$ (a) and CO formation (b) during CO$_2$ hydrogenation on Rh/YSZ.[151] P_{CO_2} = 1 kPa; P_{H_2} = 1.5 kPa; ●, T = 400°C; ▲, T = 451°C; ■, T = 468°C.

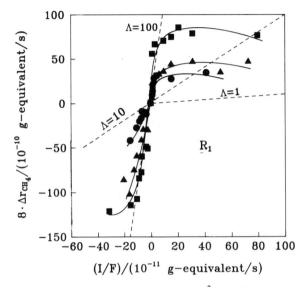

Figure 72. Effect of current and concomitant O^{2-} supply/removal on the change in the rate of CH_4 production during CO_2 hydrogenation on Rh/YSZ.[151] R_1; $P_{CO_2} = 1$ kPa; $P_{H_2} = 1.5$ kPa; •, $T = 400°C$; ▲, $T = 451°C$; ■, $T = 468°C$.

of the reaction products depends on temperature and reactor space time.[51,151]

It was found that both the catalytic rates and the selectivity to the various products can be altered significantly (rate changes up to 250% were observed) and reversibly under NEMCA conditions. Depending on the product, electrophobic or electrophilic behavior is observed, as shown in Fig. 74. In addition to the selectivity modification due to the different NEMCA effect for the rate of formation of each product, acetaldehyde, which is not produced under open-circuit conditions, is formed at negative overpotentials (Fig. 75). Values of the enhancement factor Λ up to 10 were observed in this complex system.

(v) Methane Reforming on Ni

Methane reforming on Ni is of great technological interest in the area of solid oxide fuel cells (SOFCs) because methane reforming is catalyzed

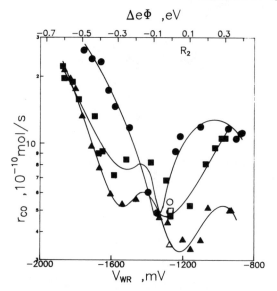

Figure 73. Effect of catalyst potential and work function on the rate of CO_2 hydrogenation on Pd/YSZ (reverse water–gas shift reaction).[151] $P_{H_2} = 73$ kPa; $P_{CO_2} = 22.5$ kPa; $\blacktriangle$, $T = 546°C$; $\blacksquare$, $T = 559°C$; $\bullet$, $T = 575°C$.

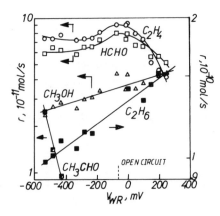

Figure 74. Effect of catalyst potential on the rates of formation of C_2H_6, C_2H_4, H_2CO, CH_3OH, and CH_3CHO during CO hydrogenation on Pd/YSZ. The rate of CH_4 formation is of the order of 10^{-9} mol/s and is only weakly affected by V_{WR}. Single-pellet design; $P = 12.5$ bar, $T = 350°C$; $P_{H_2}/P_{CO} = 1.8$; flow rate, 85 cm^3 STP/min.[151]

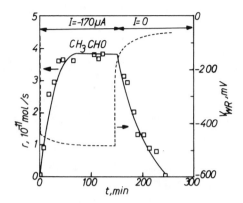

Figure 75. Transient effect of applied negative current on the rate of formation of CH_3CHO. Acetaldehyde does not form under open-circuit conditions; thus, ρ is nominally infinite. $P = 12.5$ bar, $T = 350°C$; $P_{H_2}/P_{CO} = 1.8$; flow rate, 85 cm³ STP/min.[151]

by the Ni surface of the Ni-stabilized ZrO_2 cermet used as the anode material in power-producing SOFC units.[38,63] The ability of SOFC units to reform methane "internally," that is, in the anode compartment, permits the direct use of methane or natural gas as the fuel, without a separate external reformer, and thus constitutes a significant advantage of SOFCs in relation to low-temperature fuel cells.

The extent to which anode polarization affects the catalytic properties of the Ni surface for the methane–steam reforming reaction via NEMCA is of considerable practical interest. In a very recent investigation,[52] a 70 wt % Ni–YSZ cermet was used at temperatures between 800 and 900°C with low steam-to-methane ratios (0.2–0.35). At 900°C, the anode characteristics were $i_0 = 0.2$ mA/cm², $\alpha_a = 2$, and $\alpha_c = 1.5$. Under these conditions, spontaneously generated currents were of the order of 60 mA/cm², and catalyst overpotentials were as high as 250 mV. It was found that the rate of CH_4 consumption due to the reforming reaction increases with increasing catalyst potential; that is, the reaction exhibits overall electrophobic NEMCA behavior with $\alpha \sim 0.13$. Measured Λ and ρ values were of the order of 12 and 2, respectively.[52] These results show that NEMCA can play an important role in anode performance even when the anode/solid electrolyte interface is nonpolarizable (high I_0 values) as is the case in fuel cell applications.

(vi) H₂S Dehydrogenation on Pt

H_2S dehydrogenation on Pt was investigated by Stoukides and co-workers[152] at temperatures between 600 and 750°C and was found to be electrophobic, with ρ values up to 11. The counter and reference electrodes were placed either in a separate compartment or in the reactor (single-pellet design[4,47]). The rate of H_2S decomposition was found to increase exponentially with V_{WR} and to increase slowly during galvanostatic transients ($\tau \approx 10$ min) as in typical NEMCA experiments. However, no Λ values were reported,[152] and it is likely that, owing to the high operating temperature, electrocatalysis, in addition to NEMCA, plays an important role.

V. ELECTROCHEMICAL PROMOTION WITH Na⁺-CONDUCTING SOLID ELECTROLYTES

It has been shown[3,7,42] that β''-Al_2O_3, a Na^+ conductor,[34–36] can also induce NEMCA on metal surfaces. Here, the dominant electrocatalytic reaction is

$$Na^+(\beta''\text{-}Al_2O_3) + e^- \leftrightharpoons Na(M) \qquad (93)$$

where Na(M) stands for Na adsorbed on the metal catalyst film surface. An attractive feature of using β''-Al_2O_3 or other cation donors is that one can compute coulometrically, via Faraday's law, the amount and coverage of alkali dopant introduced onto the catalyst surface. Also, as described in Section III, one can then compare computed dipole moments with literature values, in view of the very rich literature which exists for the interaction of alkali dopants with transition-metal surfaces.[117–121]

1. Ethylene Oxidation on Pt/β″-Al₂O₃

Ethylene oxidation on Pt/β''-Al_2O_3 was studied[42] at temperatures between 150 and 300°C, but most of the NEMCA experiments were carried out at 290°C. Experimental details about the cell can be found in Ref. 42. The open-circuit kinetic behavior was found to be similar to the case of C_2H_4 oxidation on Pt/doped ZrO_2 (Section IV.1); that is, the rate expression

$$r_0 = kk_{ad}P_{C_2H_4}P_{O_2}/(kP_{C_2H_4} + k_{ad}P_{O_2}) \qquad (70)$$

was found to provide a quantitative fit to the data at temperatures above 250°C, that is, when the coverage of C_2H_4 becomes negligible.[42] However, and despite the similarity in turnover frequency values measured under similar T and gas composition conditions, it was found that the open-circuit k values are significantly higher, by a factor of 20, and the k_{ad} values are significantly lower, by a factor of 100, than in the case[2,4] of Pt/doped ZrO_2. Experimentally, this makes it much more difficult to work on the fuel-lean side ($kP_{C_2H_4} \ll k_{ad}P_{O_2}$), where the NEMCA effect is very pronounced, with Pt/β''-Al_2O_3 than with Pt/$ZrO_2(Y_2O_3)$. The origin of this difference in the kinetic constant values is not obvious but may be related to the large systematic difference (~0.5 eV) observed[4,42] between the work functions of Pt/β''-Al_2O_3 and Pt/$ZrO_2(Y_2O_3)$. It was also noticed that, during catalyst film preparation, the Pt surface gets contaminated by Na, presumably due to Na diffusion during calcination, but that this Na contamination can then be removed electrochemically.[42]

Figure 76 shows a typical galvanostatic and potentiostatic transient. At the start of the experiment, the circuit is open and the steady-state (regular) catalytic rate r_0 is 5.7×10^{-7} g-atom O/s with a corresponding V^0_{WR} of -430 mV. At $t = 0$ a galvanostat is used to apply a constant current

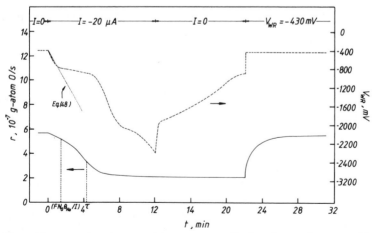

Figure 76. Rate and catalyst potential response to application of a negative current (Na supply to the catalyst) during C_2H_4 oxidation on Pt/β''-Al_2O_3, followed by potentiostatic restoration of the initial state; $T = 291$°C, $P_{O_2} = 5.0$ kPa, $P_{C_2H_4} = 2.1 \times 10^{-2}$ kPa. (Reprinted, with permission from Academic Press, from Ref. 42.)

$I = -20 \mu$A between the catalyst and the counter electrode. Sodium cations are pumped onto the catalyst surface at a rate I/F of 2.1×10^{-10} g-atom/s. This causes a 66% decrease in the catalytic rate, which drops to 2.1×10^{-7} g-atom O/s. The steady-state change in catalytic rate, $\Delta r = -3.6 \times 10^{-7}$ g-atom O/s, is 1720 times larger than the rate of supply of Na^+. The corresponding enhancement factor Λ is 3440; that is, the reaction exhibits again electrophobic behavior as in the case of $Pt/ZrO_2(Y_2O_3)$. At the same time, the catalyst potential and work function decrease in a complex manner. The catalytic rate transient is complete before the appearance of the second break in the V_{WR} transient at -900 mV, in agreement with the steady-state behavior described below. The rate relaxation time constant τ is 250 s, in reasonable agreement with

$$\tau = FN\theta_{Na}/I \tag{94}$$

where N is the catalyst surface area in moles of Pt, and θ_{Na} can be computed either from Faraday's law or from the induced $e\Phi$ change, using literature values for the initial dipole moment P_0 of Na on Pt.[42] The latter is possible in view of the excellent agreement in the P_0 values computed from the initial slopes of $e\Phi$ versus I galvanostatic transient plots, such as Figs. 16 and 76, with literature P_0 values, as described in detail in Section III. Thus, the dashed line in Fig. 76 results from Eq. (48) (Section III) with $A_C = 4.25 \times 10^{-2}$ m^2,[42] $I = -20 \mu$A, and the literature value of $P_0 = 1.75 \times 10^{-29}$ C·m for the initial dipole moment of Na/Pt(111).[111]

As shown in Fig. 76, when the circuit is opened ($I = 0$), the catalyst potential starts increasing but the reaction rate stays constant. This is different from the behavior observed with O^{2-}-conducting solid electrolytes and is due to the fact that the spillover oxygen anions can react with the fuel (e.g., C_2H_4, CO), albeit at a slow rate,[4] whereas Na(Pt) can be scavenged from the surface only by electrochemical means. Thus, as shown in Fig. 76, when the potentiostat is used to impose the initial catalyst potential, $V_{WR}^0 = -430$ mV, then the catalytic rate is restored within 100–150 s to its initial value, since Na(Pt) is now pumped electrochemically as Na^+ back into the β''-Al_2O_3 lattice.

The steady-state effect of work function on catalytic rate is shown in Fig. 77. As in the case of using doped ZrO_2 as the solid electrolyte,[2,4] there is a V_{WR} and corresponding $e\Phi$ range in which the rate increases exponentially with $e\Phi$. At higher $e\Phi$ values, the rate levels off because in this region oxygen chemisorption becomes rate-limiting and the rate constant

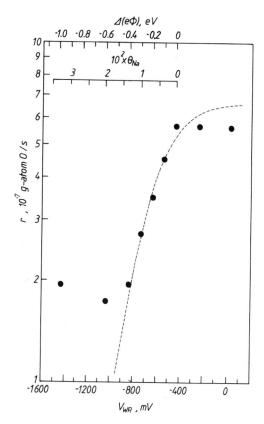

Figure 77. Effect of catalyst potential V_{WR}, work function $e\Phi$, and corresponding Na coverage on the rate of C_2H_4 oxidation on Pt/β''-Al$_2$O$_3$. The dashed line is from kinetic model discussed in Ref. 42. (Reprinted, with permission from Academic Press, from Ref. 42.) P_{O_2} = 5.0 kPa, $P_{C_2H_4} = 2.1 \times 10^{-2}$ kPa, T = 291°C, k_{ad} = 12.5 s^{-1}.

k_{ad} is, similarly to the case of ZrO$_2$(Y$_2$O$_3$) solid electrolyte ion donor, rather insensitive to changing $e\Phi$.

Also, the rate plateau at low $e\Phi$ values is strongly reminiscent of the observed behavior with Pt/ZrO$_2$(Y$_2$O$_3$). Thus, the behavior is qualitatively very similar. The inserted θ_{Na} abscissa in Fig. 77 is constructed on the basis of the following form of the Helmholtz equation:

$$\Delta\Phi = -(N_{Pt}P_0/\varepsilon_0)\theta_{Na} \tag{95}$$

and is also in very good agreement with θ_{Na} computed coulometrically via Faraday's law.[42] As shown in Fig. 77, a θ_{Na} value of the order of 0.01 suffices to decrease the catalytic rate by a factor of 3.

Figure 78 shows the dependence of the Eley–Rideal kinetic constant k on V_{WR} and $e\Phi$. For -830 mV $< V_{WR} < -430$ mV, k is exponentially dependent on $e\Phi$ according to

$$\ln(k/k_0) = \alpha e(\Phi - \Phi^*)/k_b T \tag{96}$$

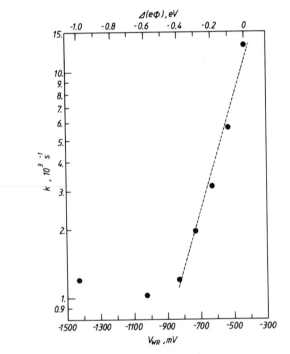

Figure 78. Effect of catalyst potential and work function on the kinetic constant k of C_2H_4 oxidation on Pt/β''-Al_2O_3. The dashed line is from the kinetic model of Ref. 42. (Reprinted, with permission from Academic Press, from Ref. 42.) $P_{O_2} = 5.0$ kPa, $P_{C_2H_4} = 2.1 \times 10^{-2}$ kPa, $T = 291°C$, $k_{ad} = 12.5$ s^{-1}.

with $\alpha = 0.28$. Thus, despite the qualitative similarities in the NEMCA behavior of C_2H_4 oxidation using $Pt/ZrO_2(Y_2O_3)$ and Pt/β''-Al_2O_3, there do exist quantitative differences, as α is between 0.5 and 1.0 for $Pt/ZrO_2(Y_2O_3)$ (Ref. 4).

It is worth emphasizing, however, that in both cases C_2H_4 oxidation exhibits electrophobic behavior, that the relaxation time constants τ can be estimated from similar formulas [Eqs. (51) and (94)], and that the enhancement factors Λ can again be estimated from the same formula (Eq. 53).

The fact that very small Na coverages ($\theta_{Na} = 0.015$) suffice to induce pronounced (up to 70%) decreases in catalytic rate (i.e, the Na "toxic-

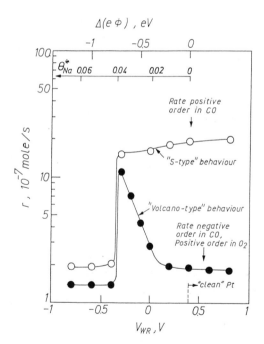

Figure 79. Effect of catalyst potential V_{WR}, corresponding work-function change $\Delta(e\Phi)$, and approximate linearized Na coverage θ_{Na}^* on the rate of CO oxidation on Pt/β''-Al_2O_3. Conditions: $T = 350°C$, $P_{O_2} = 6$ kPa; ●, P_{CO} = 5.3 kPa; ○, $P_{CO} = 2.8$ kPa. (Reprinted, with permission from Academic Press, from Ref. 7.)

ity"[153,7] is 0.70/0.015 = 47) rules out the possibility of any "geometric" interpretation of NEMCA. It also provides strong evidence for some "long-range" electronic interactions. For if the effect of each Na atom were localized to those Pt atoms immediately adjacent to it, then for $\theta_{Na} = 0.015$ one would expect at most a 10% rate decrease, a factor of 7 smaller than the observed one. Although surface heterogeneity could play a role, the observed behavior[42] is strongly reminiscent of the infrared spectroscopic work of Yates and co-workers,[121] who studied the CO + K/Ni(111) system and found evidence that a single K atom can influence as many as 27 coadsorbed CO molecules.

One can explain the observed exponential decrease in the kinetic constant k with decreasing $e\Phi$ [Fig. 78 and Eq. (96)] by the same physical reasoning used to explain the k dependence on $e\Phi$ with ZrO_2 solid electrolytes: Spillover $Na^{\delta+}$-compensating charge dipoles cause a more or less uniform decrease in $e\Phi$ and a concomitant increase in the binding strength of the Pt=O bond, cleavage of which is rate-limiting.[4,42] To the extent that the heat of adsorption of oxygen increases linearly with decreasing $e\Phi$ (Section III.10), one can then directly explain the observed exponential dependence of k on $e\Phi$.

It is also worth emphasizing the excellent agreement between the initial dipole moment P_0 values for Na/Pt computed in Ref. 42 and literature values for Na/Pt(111).[116] This agreement shows that Na introduced on the catalyst surface via β''-Al_2O_3 to induce NEMCA is in the same binding state as Na introduced from the gas phase.[117–121] The advantage of β''-Al_2O_3 is, as previously discussed, the possibility of *in situ* monitoring and control of the dopant coverage.

2. CO Oxidation on Pt/β''-Al_2O_3

CO oxidation on Pt/β''-Al_2O_3 was investigated[7] at temperatures in the range 300–430°C. Figures 79–81 show the effect of catalyst potential and corresponding Na coverage θ_{Na}, gaseous composition, and temperature on the reaction rate. Figure 82 shows the corresponding effect on activation energy. For fuel-rich conditions, the reaction is electrophilic, with ρ values up to 8. This is due to enhanced oxygen chemisorption with increasing θ_{Na}. For fuel-lean conditions, the reaction is weakly electrophobic. Then, at $V_{WR} \approx -0.3$, which corresponds to $\theta_{Na} \approx 0.05$, the rate decreases abruptly due to formation of a surface CO–Na–Pt complex that poisons the rate.[7]

Figure 83 shows typical galvanostatic transients. Computed dipole moments of Na/Pt are again in reasonable agreement with the literature

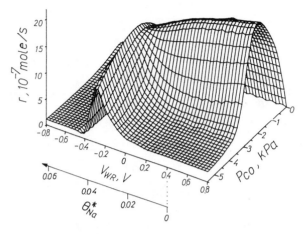

Figure 80. Effect of catalyst potential and corresponding linearized Na coverage θ_{Na}^{*} and P_{CO} on the rate of CO oxidation on Pt/β''-Al$_2$O$_3$. $T = 350°C$, $P_{O_2} = 6$ kPa. (Reprinted, with permission from Academic Press, from Ref. 7.)

value of 1.75×10^{-29} C·m (5.3 D).[116] The promotion index P_{Na} is up to 250 under fuel-rich conditions, which is the highest P_i value reported so far in electrochemical promotion studies (Table 3). In the region of CO–Na–Pt complex formation, P_{Na} is as low as –30 (Fig. 83b). This complex system provides an excellent example of the importance of promotional effects in catalysis.

3. Ethylene Epoxidation on Ag/β''-Al$_2$O$_3$

Earlier exploratory studies of ethylene epoxidation on Ag/β''-Al$_2$O$_3$ at atmospheric pressure and $T = 410°C$ had shown that both the activity and selectivity of Ag can be markedly affected via NEMCA.[4]

In a very recent study[148] at temperatures between 250 and 300°C and higher pressures (5 bar), it was found that technologically important ethylene oxide selectivity values can be obtained in the presence of C$_2$H$_4$Cl$_2$ moderators. Figure 84 shows a typical galvanostatic experiment. Negative currents, that is, Na supply to the catalyst, enhances the rate of epoxidation without affecting the rate of CO$_2$ formation. Consequently, the selectivity to ethylene oxide increases substantially (Fig. 85).

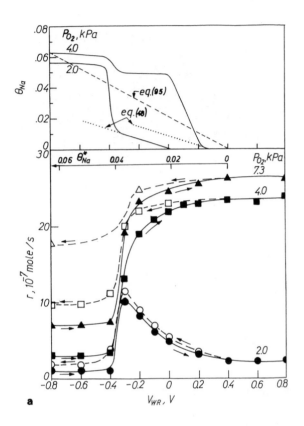

Figure 81. (a) Effect of V_{WR} and linearized (Ref. 7) Na coverage θ^*_{Na} on the rate of CO oxidation on Pt/β''-Al$_2$O$_3$ at varying P_{O_2}. Other conditions: $P_{CO} = 2$ kPa, $T = 350°C$. The top part of the figure shows the corresponding variation of θ_{Na} with V_{WR}. (b) Effect of Na coverage θ_{Na} on the rate of CO oxidation. Conditions as in (a). (Reprinted, with permission from Academic Press, from Ref. 7.)

Figure 86 shows the effect of catalyst potential V_{WR} corresponding linearized[148] sodium coverage θ^*_{Na}, and C$_2$H$_4$Cl$_2$ partial pressure on the selectivity to ethylene oxide. For $V_{WR} = -0.25$ V and $P_{C_2H_4Cl_2} = 1.0$ ppm, the selectivity to ethylene oxide is 88%, which is one of the highest values reported for this important reaction.

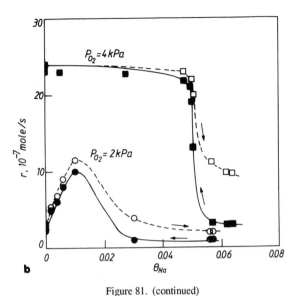

b

Figure 81. (continued)

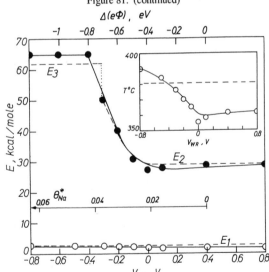

Figure 82. Effect of catalyst potential V_{WR} on the apparent activation energy and on the temperature (inset) at which the transition occurs from a high (●) to a low (○) E value. The dashed lines and predicted asymptotic E_1, E_2, E_3 activation energy values are from the kinetic model discussed in Ref. 7. Conditions: P_{O_2} = 5.8 kPa, P_{CO} = 3.5 kPa. (Reprinted, with permission from Academic Press, from Ref. 7.)

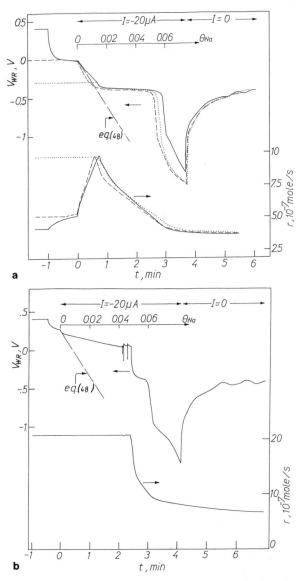

Figure 83. Rate and catalyst potential response to application of negative currents (a,b), for the case of "volcano-type" behavior (a) and "S-type" behavior (b) of the reaction rate, and to application of positive currents (c). See text for discussion. Conditions: (a) $P_{CO} = 2$ kPa, $P_{O_2} = 2$ kPa, $T = 350°C$, catalyst C1;

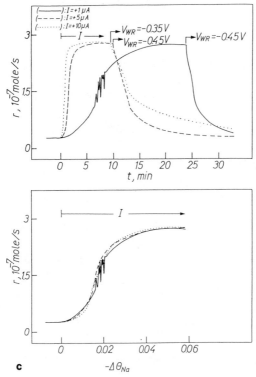

Figure 83. (continued) (b) P_{CO} = 2 kPa, P_{O_2} = 4 kPa, T = 350°C, catalyst C1; (c) P_{CO} = 0.73 kPa, P_{O_2} = 0.86 kPa, T = 402°C, catalyst C2. (Reprinted, with permission from Academic Press, from Ref. 7.)

4. NO Reduction by C_2H_4 on Pt/β''-Al$_2$O$_3$

NO reduction by C_2H_4 on Pt/β''-Al$_2$O$_3$ was investigated at temperatures between 250 and 400°C by Harkness and Lambert[19] both under atmospheric pressure and under high-vacuum conditions. The reaction exhibits pronounced electrophilic behavior (Fig. 87). The authors concluded that this is due to enhanced oxygen chemisorption rather than enhanced NO dissociation. The results are very spectacular in that a formally infinite ρ value was obtained[19]; that is, r_o was immeasurably low on the unpromoted surface (θ_{Na} = 0, Fig. 87). At higher Na coverages (0.03 to 0.3), the differential promotion index p_i is up to 500.

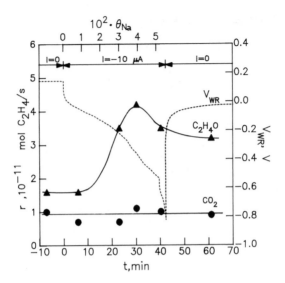

Figure 84. Ethylene epoxidation on Ag/β''-Al$_2$O$_3$: Transient effect of a negative applied current (Na supply to the catalyst) on the rates of ethylene oxide and CO$_2$ formation and on catalyst potential, and Na coverage.[148] $T = 260°C$, $P = 5$ atm, $P_{O_2} = 17.5$ kPa, $P_{C_2H_4} = 49$ kPa.

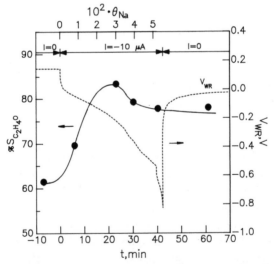

Figure 85. Ethylene epoxidation on Ag/β''-Al$_2$O$_3$: Transient effect of a negative applied current (Na supply to the catalyst) on catalyst potential, Na coverage, and selectivity to ethylene oxide.[148] Conditions as in Fig. 84.

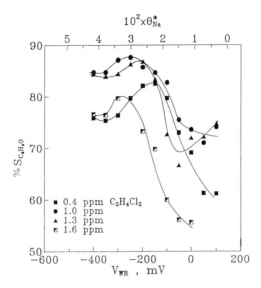

Figure 86. Effect of catalyst potential and of gas-phase 1,2-$C_2H_4Cl_2$ partial pressure on the selectivity of ethylene epoxidation on Ag/β''-Al$_2$O$_3$.[50] $T = 260°C$; $P = 500$ kPa, 4% O_2, 13% C_2H_4.

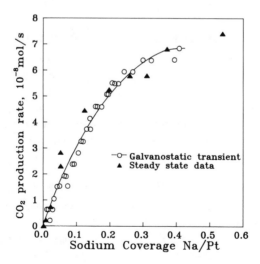

Figure 87. Effect of Na coverage on the rate of NO reduction by C_2H_4 on Pt/β''-Al$_2$O$_3$. The rate enhancement ratio ρ is nominally infinite; 23 Torr C_2H_4, 5.0 Torr NO, 753 K. (Reprinted, with permission from Academic Press, from Ref. 19.)

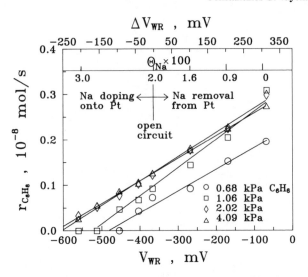

Figure 88. Effect of catalyst potential, Na coverage, and benzene partial pressure on the rate of benzene hydrogenation on Pt/β''-Al$_2$O$_3$.[18] T = 130°C, P_{H_2} = 33.35 kPa, $P_{C_6H_6}$ = 2.02 kPa, P = 1 atm, flow rate = 81 cm^3(STP)/min.

5. Benzene Hydrogenation on Pt/β''-Al$_2$O$_3$

Benzene hydrogenation on Pt/β'''-Al$_2$O$_3$ was investigated at temperatures between 100 and 150°C by Cavalca and Haller.[18] The reaction is electrophobic; that is, the rate decreases dramatically with increasing Na coverage (Fig. 88). The toxicity index $-P_i$ is up to 50. The effect is due to decreased benzene chemisorption with decreasing V_{WR} and $e\Phi$ owing to reduced donation of π-electrons to the metal. These results are also quite spectacular, since ρ values approaching zero were obtained.[18]

6. CO$_2$ Hydrogenation on Pd

CO$_2$ hydrogenation on Pd was investigated[151] under atmospheric pressure and at temperatures between 540 and 605°C. The CO formation rate (reverse water–gas shift reaction) exhibits electrophilic behavior over the entire potential range examined; that is, the rate increases by up to 600% with increasing sodium coverage (Fig. 89). This behavior can be explained by the enhancement of CO$_2$ adsorption on the Pd surface with increasing sodium coverage.

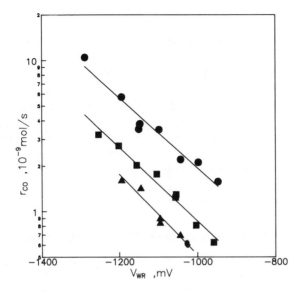

Figure 89. Effect of catalyst potential on the rate of CO_2 hydrogenation on Pd/β''-Al_2O_3.[151] $P_{H_2} = 67.4$ kPa; $P_{CO_2} = 19.3$ kPa; ▲, $T = 545°C$; ■, $T = 568°C$; ●, $T = 605°C$.

VI. ELECTROCHEMICAL PROMOTION WITH H^+-, F^--, AND OTHER ION-CONDUCTING SOLID ELECTROLYTES

The very wide applicability of electrochemical promotion in heterogeneous catalysis has been demonstrated recently by using a variety of solid electrolytes as the promoting ion donor. The group of Sobyanin pioneered the use of $CsHSO_4$ as a proton conductor to induce NEMCA during C_2H_4 hydrogenation on Pt.[15] This was followed by the elegant work of Stoukides and co-workers, who used $SrCe_{0.95}Yb_{0.05}O_3$ as a proton conductor at 750°C to enhance the nonoxidative dimerization of CH_4.[20] Very recently, Nafion was used as a proton conductor to obtain a spectacular nonfaradaic rate enhancement of H_2 oxidation on Pt at room temperature.[55] Parallel work in Patras has shown that F^--conducting solid electrolytes can induce NEMCA on Pt at elevated temperatures,[56] while a striking demonstration of NEMCA was recently obtained with C_2H_4 oxidation on Pt supported on TiO_2.[154] This is very likely related to the well-known effect of strong metal–support interactions (SMSI).[155,156]

1. H_2 Oxidation on Pt/Nafion

H_2 oxidation was studied[55] at 25°C using a Nafion membrane on which a porous Pt electrode had been deposited. A mixture of H_2 (1–2 kPa) and O_2 (1–5 kPa) was bubbled on the Pt surface, and current densities of 10–20 mA/cm^2 ($\Delta V_{WR} \approx 0.5$–1V) were applied. Values of the rate enhancement ratio ρ up to 6 and of the faradaic efficiency Λ up to 20 were obtained. The H_2 oxidation was found to be electrophobic, and the observed shifts in rate maxima with respect to P_{H_2} and P_{O_2} as well as the rate enhancement were attributed to the weakening of the Pt=O and strengthening of the Pt–H bond with increasing catalyst potential V_{WR} and work function $e\Phi$. The measured Λ values were in good agreement with $2Fr_0/I_0$ as in NEMCA studies at higher temperatures.

2. Methane Dimerization Using Proton Conductors

The reaction of nonoxidative CH_4 dimerization to ethane and ethylene was investigated by Stoukides and co-workers[20] at 750°C on Ag electrodes in a single-pellet NEMCA reactor arrangement.[4,47] $SrCe_{0.95}Yb_{0.05}O_3$ was used as the solid electrolyte. This material is known to exhibit both protonic (H^+) and oxide ion (O^{2-}) conductivity, the former dominating at temperatures below 750°C.[20] The reaction was found to be electrophobic, with ρ values up to 8. The total selectivity to C_2H_4 and C_2H_6 was near 100%. Thermodynamics places very stringent limits on the maximum equilibrium conversion of this reaction, provided $\Lambda > 1$. When $\Lambda < 1$, however, these limitations vanish, as the process is similar to an electrolytic one. No Λ values were reported, unfortunately, in this interesting study,[20] which showed that r increases exponentially with V_{WR}

3. Ethylene Hydrogenation on Ni/CsHSO$_4$

Ethylene hydrogenation was investigated[15] at 150–170°C using Ni as the catalyst supported on CsHSO$_4$, a protonic conductor.[15] The reaction was found to be electrophobic; that is, proton supply to the Ni catalyst was found to decrease the rate by a factor of 6 ($\rho = 0.16$), and proton removal was found to increase the rate by a factor of 2 ($\rho = 2$). The corresponding Λ values are 6 for hydrogen removal and 300 for hydrogen supply. These results are intriguing in that proton supply and removal have the opposite effects from what would be anticipated from mass action kinetic considerations. It is very likely that increasing V_{WR} and $e\Phi$, which corresponds to proton removal, enhances the binding of ethylene and hydrogen on the

Ni surface, since both are electron donors, and thus enhances the rate of hydrogenation.

4. CO Oxidation on Pt/CaF$_2$

CO oxidation on Pt/CaF$_2$ reaction was investigated[56] at 500–700°C and was found to be electrophobic, with ρ values up to 2.5 and Λ values up to 200. Increasing V_{WR} was found to increase the activation energy and preexponential factor, leading to the appearance of the compensation effect, with an isokinetic point at 650°C. The promoting role of F$^-$ was found to be qualitatively similar to that of O^{2-}, although the ρ and P_i values are significantly smaller than in the case of O^{2-}.

5. C$_2$H$_4$ Oxidation on Pt/TiO$_2$

C$_2$H$_4$ oxidation on Pt/TiO$_2$ was studied[154] at temperatures between 450 and 600°C and, as in the case of O^{2-} and Na$^+$-conducting support,[2,4,42] was found to be electrophobic. Rate enhancement ρ values were up to 20, with Λ values up to 5000. Observing NEMCA with a TiO$_2$ support is surprising, since TiO$_2$ (rutile) is an n-type semiconductor and its ionic (O^{2-}) conductivity is quite low, so at best it can be considered a mixed conductor. In the catalytic literature, however, TiO$_2$ is a very well-known support, owing to the effect of strong metal–support interactions (SMSI),[155,156] discovered by Tauster *et al.* in the seventies[156] and subsequently shown to be due to decoration[155] of the surface of the well-dispersed metal (e.g., Pt) surface with TiO$_2$ moieties. At this point, it is not clear whether the observed pronounced NEMCA behavior is due to backspillover of O^{2-} or of negatively charged TiO$_x$ moieties, or of both. *In situ* or *ex situ* XPS could easily clarify this point. In any case, however, it is very likely that the observed NEMCA behavior is an electrochemically induced and controlled SMSI. This could be very valuable for the systematic and controlled *in situ* study of the SMSI effect.

VII. ELECTROCHEMICAL PROMOTION USING AQUEOUS ELECTROLYTES

The recent discovery of NEMCA in aqueous systems[8] is of considerable theoretical and practical importance. In this case, no ion migration (backspillover) is necessary to account for the observed behavior, which appears again to be due to the effect of changing potential and work function on the binding strength of adsorbates.[8] Here, there appears to be only one double layer of interest, that at the electrode/electrolyte interface, although, since gases are produced and/or consumed, the electrode/gas interface may also have a role.

1. H_2 Evolution and Aldehyde Oxidation at Ib Metals in Alkaline Solutions

Heitbaum, Anastasijevic, Baltruschat, and co-workers were the first to report a nonfaradaic enhancement in the rate of H_2 evolution on Cu[157] and Ag[57] electrodes during formaldehyde oxidation. They used differential electrochemical mass spectroscopy[157] to measure the rate of H_2 evolution during formaldehyde oxidation on Cu and Ag in weakly alkaline solutions and found values of the faradaic efficiency Λ up to 2; Λ was found to increase with decreasing catalyst potential. They attributed their interesting findings to the heterogeneously catalyzed reaction

$$H_2CO + OH^- \rightarrow HCOO^- + H_2 \qquad (97)$$

the rate of which was proposed to be potential-dependent (electrophilic behavior). The authors concluded that the observed phenomenon is due to the interaction of the electric field of the double layer with the adsorbed H_2CO molecule (water dipole orientation, change in the surface concentration of OH^-) and that the effect is similar to the NEMCA effect of solid-state electrochemistry.

2. Hydrogen Oxidation on Pt in Aqueous Solutions

The first clear demonstration of NEMCA in aqueous electrochemistry was reported recently[8] for the oxidation of H_2 on Pt electrodes in $0.1M$ KOH solutions at temperatures between 25 and 50°C. The original work was carried out using finely dispersed Pt supported on graphite,[8] but identical results were later obtained using Pt black supported on a Teflon frit.[158] In both cases, the H_2–O_2 mixture (P_{O_2} and P_{H_2} were between 0.1 and 2 kPa) was bubbled through a Teflon frit, and the rates r_{H_2} (mol H_2/s) and r_O ($\frac{1}{2}$ mol O_2/s) of H_2 and O_2 consumption were measured via on-line gas chromatography and mass spectrometry. The gas flow rates were chosen such that the conversion of H_2 and O_2 was maintained under all conditions below 40%. The absence of diffusional limitations was verified by varying the total gas flow rate between 100 and 600 cm^3 STP/min and observing no significant change in the rates of H_2 and O_2 consumption. No problems with deterioration of catalyst or support were encountered over prolonged periods (~200 h) of operation at potentials below 1.5 V with respect to a reference H_2 electrode (r.h.e.) immersed in the same solution. Substitution of the $0.1M$ KOH solution with a $0.1M$ LiOH solution led to the same results. The Pt counter electrode was situated in a separate compartment, electrolytically connected via a Flemion membrane, so that the rates of H_2 and O_2 consumption on the working catalyst electrode could be accurately measured without any interference from the gases produced or consumed at the counter electrode.

Hydrogen and oxygen are consumed on the Pt surface at a rate r_c by the catalytic reaction

$$H_2 + \frac{1}{2} O_2 \xrightarrow{r_c} H_2O \tag{98}$$

Under open-circuit conditions, the catalyst potential $V_{WR} \equiv E$ (r.h.e.) takes values of the order 0.4–0.85 V, that is, –0.35 to +0.1 V on the normal hydrogen electrode scale (n.h.e.), depending on the hydrogen-to-oxygen ratio.

When a positive current I is applied between the catalyst electrode and the Pt counter electrode, then the catalyst potential E changes to more positive values (Fig. 90), and the following electrochemical (net charge-transfer) reactions take place at the surface of the Pt catalyst electrode:

$$H_2 + 2OH^- \xrightarrow{r_{e,1}} 2H_2O + 2e^- \tag{99}$$

$$2OH^- \xrightarrow{r_{e,2}} \frac{1}{2} O_2 + H_2O + 2e^- \tag{100}$$

where the forward reaction given by Eq. (100), that is, $r_{e,2} > 0$, takes place when E (r.h.e.) is above the oxygen reduction potential (1.23 V), and the reverse reaction, that is, $r_{e,2} < 0$, occurs otherwise. It follows from simple mass-balance considerations that, in general,

$$r_{H_2} = r_c + r_{e,1} \tag{101}$$

$$r_O = r_c - r_{e,2} \tag{102}$$

$$I/2F = r_{e,1} + r_{e,2} \tag{103}$$

and, thus,

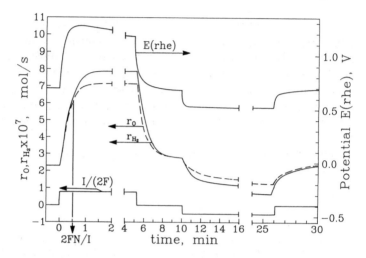

Figure 90. NEMCA in aqueous media (0.1M KOH): Transient effect of applied positive and negative currents (I = 15 and −10 mA) on the rates of consumption of hydrogen (r_{H_2}) and oxygen (r_O). P_{H_2} = 0.75 kPa, P_{O_2} = 1.06 kPa; gas flow rate Q = 280 cm^3/min at STP. (Reprinted, with permission from *Nature*, Macmillan Magazines Ltd., from Ref. 8.)

$$r_{H_2} - r_O = I/2F \qquad (104)$$

Consequently, if r_c were to remain constant, application of a positive current would increase r_{H_2} by less than $I/2F$ ($\Delta r_{H_2} \leq I/2F$) and would decrease or increase r_O again by less than $I/2F(-I/2F \leq \Delta r_O \leq I/2F)$.

Surprisingly, as shown in the galvanostatic transient of Fig. 90, Δr_{H_2} is 720% higher than $I/2F$, and Δr_O is 620% higher than $I/2F$. The increase, Δr_{H_2}, in r_{H_2} is 344% relative to the open-circuit value $r_{H_2}^o = r_O^o = r_c^o$. The nonfaradaic behavior [$\Lambda_{H_2} = \Delta r_{H_2}/(I/2F) = 7.2$] is due to the electrochemical activation of the catalytic reaction given by Eq. (98), the rate (r_c) of which increases by between 310 and 344%. Figure 90 also shows that:

(a) The rate relaxation time constant τ upon constant current application is again ~$2FN/I$, as in solid-state electrochemical promotion studies.

(b) The effect is reversible; that is, r_{H_2}, r_O, and catalyst potential E all return to their open-circuit values upon current interruption.

(c) Negative currents also cause a nonfaradaic decrease in r_{H_2} and r_O.

(d) The reaction exhibits electrophobic behavior; that is, $\Lambda > 1$ and $\partial r/\partial E > 0$.

(e) At steady state, the difference $r_{H_2} - r_O$ always equals $I/2F$, in accordance with Eq. (104).

The steady-state effect of positive current on r_{H_2} and r_O is shown in Fig. 91. The faradaic efficiency Λ exceeds 20 (2000%) for low currents. Figure 92 shows the corresponding effect of catalyst potential $V_{WR} \equiv E$ on r_{H_2} and r_O, together with the dependence of I on E. The break in the plot of $\log I$ versus E coincides with the observed inflection in r_{H_2} and r_O and corresponds to the onset of Pt oxide formation.[8,100,101,159,160] As shown in Fig. 92, the (predominantly catalytic) rates r_{H_2} and r_O depend exponentially on catalyst potential E, as in studies with solid electrolytes with slopes comparable to the Tafel slopes seen here. This explains why the

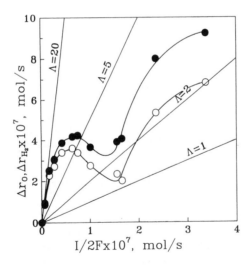

Figure 91. NEMCA in H_2 oxidation on Pt in 0.1M KOH: Steady-state effect of applied positive (anodic) current (I) on the increase in the rates of hydrogen (●) and oxygen (○) consumption. $P_{H_2} = 0.8$ kPa, $P_{O_2} = 1.25$ kPa; $r_{H_2}(= r_O^0 = r_c^0) = 2.38 \times 10^{-7}$ mol/s is the open-circuit catalytic rate; $Q = 540$ cm³/min at STP. (Reprinted, with permission from *Nature*, Macmillan Magazines Ltd., from Ref. 8.)

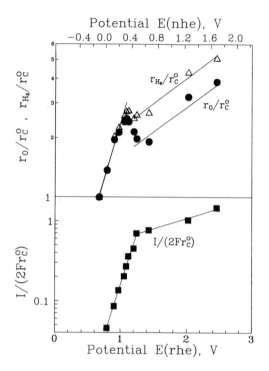

Figure 92. Steady-state effect of ohmic-drop-free catalyst potential on current (*bottom*) and on the rates of
hydrogen (△) and oxygen (●) consumption (*top*). Condi-
tions as in Fig. 91: $r_{H_2}(= r_O^o = r_c^o) = 2.38 \times 10^{-7}$ mol/s is
the open-circuit catalytic rate. (Reprinted, with permission
from *Nature*, Macmillan Magazines Ltd., from Ref. 8.)

observed magnitude of the faradaic efficiency Λ (~2–20) is in good
agreement with $2Fr_c^o/I_0$ (r_c^o is the open-circuit catalytic rate, and I_0 is the
exchange current), which is known to predict the expected magnitude of
$|\Lambda|$ in solid-electrolyte studies.[4]

Figure 93 shows the effect of P_{O_2}, P_{H_2}, and catalyst potential on the
rate of H_2 oxidation. The reaction mechanism is of the Langmuir–
Hinshelwood type, as also manifested by the observed rate maxima
upon varying P_{O_2} and P_{H_2}. Increasing catalyst potential causes both a
pronounced increase in r_{H_2} and a shift of the rate maximum to higher

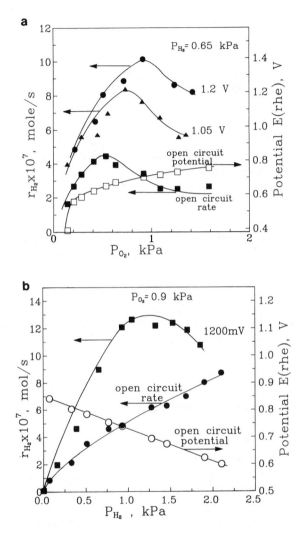

Figure 93. Effect of electrode-catalyst potential and oxygen (a) and hydrogen (b) partial pressure on the rate of hydrogen oxidation on Pt in 0.1M KOH (a) and 0.1M LiOH (b). $Q = 500$ cm^3 STP/min. (Reprinted, with permission from *Nature*, Macmillan Magazines Ltd., from Ref. 8.)

P_{O_2} and lower P_{H_2} values. This shift indicates a weakening of the Pt=O chemisorptive bond and a strengthening of the Pt–H bond, both consistent with the anticipated effect of increased potential and work function[4] on the binding strength of electron-acceptor (oxygen) and electron-donor (hydrogen) adsorbates. These considerations can also account for the appearance of the local rate maximum with respect to potential on the left of the break in the log *I–E* curve (Fig. 92).

Thus, similarly to the case of solid-state electrochemistry,[4] the observed nonfaradaic electrochemical modification of catalytic activity (NEMCA) appears to be due to the effect of changing potential and work function on the binding strength of the adsorbates. Changing catalyst potential affects the electric field in the metal–solution double layer with a concomitant change in the surface concentrations of OH$^-$ and K$^+$ and in the orientation of H_2O dipoles. These changes will then affect the strength of the Pt=O and Pt–H covalent bonds, both of which have a certain ionic character, via direct electrostatic or through-the-metal[137] interactions.

The NEMCA effect in aqueous electrochemistry may be of considerable technological value, for example, in the electrochemical treatment of toxic organics[161] or the production of useful industrial chemicals.[66]

It must be emphasized, however, that since the faradaic efficiency Λ is on the order of $2Fr_0/I_0$, one anticipates to observe NEMCA behavior only for those systems in which there is a measurable open-circuit catalytic activity r_0. Consequently, the low operating temperatures of aqueous electrochemistry may severely limit the number of reactions where nonfaradaic Λ values can be obtained.

It is also worth noting that the recent findings about the one-to-one correspondence between change in (ohmic drop-free) catalyst potential and work function in solid-state electrochemistry[3,4,59] are also applicable to the work function of liquid-free, gas-exposed electrode surfaces in aqueous electrochemistry. Such surfaces, created when gases are consumed or produced on an electrode surface, may also play a role in the observed NEMCA behavior. The one-to-one correspondence between $e\Delta V_{WR}$ and $\Delta(e\Phi)$ is strongly reminiscent of the similar one-to-one relationship established with emersed electrodes previously polarized in aqueous solutions.[162,163]

VIII. SUMMARY OF THE THEORY OF NEMCA

Throughout this chapter, the reader has encountered several aspects of NEMCA, or electrochemical promotion in catalysis. The phenomena have been rationalized on the basis of some simple electrochemical and catalytic rules, all confirmed by independent experimental studies. Here we summarize these rules, which, we believe, can explain, at least semiquantitatively, and without any exception, all the reported aspects of electrochemical promotion.

(1) The electrochemical promotion effect, or nonfaradaic electrochemical modification of catalytic activity (NEMCA), is due to the controlled migration (backspillover) of ions from the solid electrolyte to the gas-exposed catalyst-electrode surface under the influence of the applied current or potential.[4,125]

(2) The backspillover ions ($O^{\delta-}$, $Na^{\delta+}$, etc.) are each accompanied by their compensating (screening) charge in the metal, thus forming surface dipoles.[4] Consequently, these surface dipoles form an *"effective electrochemical double layer"* on the gas-exposed, that is, catalytically active, catalyst surface, in addition to the classical double layer that exists at the metal/solid electrolyte interface (Fig. 1).

(3) Due to the strong lateral repulsive interactions between the parallel-oriented surface dipoles, the migration of these dipoles on the catalyst surface is fast. The rate of migration is not limited by surface diffusion but rather by the rate, I/nF, of creation of the surface dipoles at the metal–solid electrolyte–gas three-phase boundaries (tpb). Consequently, the time, τ, required to form the "effective electrochemical double layer" during galvanostatic transients is of the order of

$$\tau \approx nFN\theta_i/I \tag{105}$$

where n is the absolute value of the ion charge, N is the number of metal moles of the gas-exposed catalyst surface, θ_i is the established steady-state surface coverage of the migrating ions (or dipoles), and I is the applied current.

For example, in the case of metals deposited on YSZ, where *in situ* XPS has shown that $\theta_{O^{\delta-}}$ can be of the order of unity,[125] τ is on the order of $2FN/I$. In the case of metals deposited on β''-Al_2O_3, τ is on the order of $FN\theta_{Na}/I$, where θ_{Na} is the coulometrically measured Na coverage.[42]

(4) The electrochemically (potentiostatically or galvanostatically) induced migration of backspillover dipoles on the catalyst surface is

accompanied by a concomitant change $\Delta(e\Phi)$ in the work function, $e\Phi$, of the gas-exposed electrode surface. It is[3,4]

$$\Delta(e\Phi) = e\Delta V_{WR} \tag{106}$$

where ΔV_{WR} is the change in the (ohmic-drop-free) catalyst-electrode potential. Equation (106) is valid even when no migration (backspillover) of ions can take place (e.g., negative current application to metal/YSZ interfaces, which also frequently causes NEMCA). In this case, the coverages and dipole moments (therefore also chemisorptive binding strengths) of covalently bonded adsorbates change to satisfy Eq. (106).

(5) The effective electrochemical double layer established on the gas-exposed catalyst-electrode surface affects the binding strength of covalently bonded adsorbates, that is, chemisorbed reactants and reaction intermediates. These may or may not[48,125] occupy the same type of surface sites as the backspillover ions.

The binding strength of adsorbates is affected due to direct electrostatic interactions in the effective double layer (usually termed through-the-vacuum interactions by the catalytic community) and also due to through-the-metal interactions.[137]

(6) Increasing V_{WR} and $e\Phi$ via positive current application, that is, increasing the coverage of negatively charged backspillover ions on the gas-exposed catalyst surface, weakens the chemisorptive bond of electron-acceptor adsorbates (e.g., normally chemisorbed atomic oxygen) and strengthens the chemisorptive bond of electron-donor adsorbates (e.g., olefins or dissociatively chemisorbed hydrogen).

(7) Decreasing V_{WR} and $e\Phi$ (negative current application), that is, increasing the coverage of positively charged backspillover ions on the gas-exposed catalyst surface, strengthens the chemisorptive bond of electron-acceptor adsorbates and weakens the chemisorptive bond of electron-donor adsorbates.

(8) The variation in chemisorptive bond strengths, or heats of adsorption, with $\Delta(e\Phi)$ discussed in (6) and (7) is often linear,[129] with slopes of the order -1 to $+1$. This fits nicely, via the Polanyi principle,[127] with the frequently observed linear variation in the activation energies E of the *catalytic* reactions with changing $e\Phi$ observed in NEMCA studies.

(9) The change in chemisorptive bond strengths with changing catalyst potential and work function, described in (6), (7), and (8), is the cause of NEMCA, or electrochemical promotion in catalysis, and leads to the

observed dramatic nonfaradaic variations in catalytic rates with $\Delta(e\Phi)$. Owing to the linear variation of heats of adsorption, activation energies, and logarithms of preexponential factors with $\Delta(e\Phi)$, catalytic rates, r, are often found to depend exponentially on $e\Phi$ over wide (0.3–1 eV) $e\Phi$ ranges:

$$\ln(r/r_0) = \alpha\Delta(e\Phi)/k_b T \qquad (107)$$

$$\ln(r/r_0) = \alpha F\, \Delta V_{WR}/RT \qquad (108)$$

(10) The NEMCA coefficient α is positive for electrophobic reactions and negative for electrophilic ones. The sign and magnitude of α (typically -1 to $+1$) depends primarily on the polarity of the (partially ionic) chemisorptive bonds broken and formed in the rate-limiting step of the catalytic reaction.

(11) The enhancement factor or faradaic efficiency, Λ, defined by

$$\Lambda \equiv \Delta r_{\text{catalytic}} / (I/2F) \qquad (109)$$

is positive for electrophobic reactions and negative for electrophilic ones. The order of magnitude of the absolute value, $|\Lambda|$, of the faradaic efficiency Λ can be estimated for any reaction from

$$|\Lambda| \approx 2Fr_0/I_0 \qquad (110)$$

where r_0 is the open-circuit catalytic rate (expressed in moles of O), and I_0 is the exchange current of the metal/solid electrolyte interface during the catalytic reaction, extracted from the usual Tafel plots.[4,100] The derivation of Eq. (110), which is in good agreement with measured Λ values of some 30 catalytic reactions (ranging from 1 to 10^5), is simple and has been presented elsewhere.[4] It stems from the exponential dependence of the rates of catalytic, r, and electrocatalytic, $I/2F$, reactions on catalyst-electrode potential. The predictions of Eq. (110) are almost quantitative when the absolute value, $|\alpha|$, of the NEMCA coefficient, α, is close to the values of the anodic ($I > 0$) or cathodic ($I < 0$) transfer coefficients α_a and α_c.

The parameter I_0 is usually[4,46] proportional to the tpb length and can be, largely, controlled by appropriate choice of the sintering temperature during catalyst film preparation.[4]

Equation (110) is quite important, as it defines the limits of applicability of NEMCA ($|\Lambda| > 1$). In order to observe a nonfaradaic rate enhancement (NEMCA), the open-circuit rate, r_o, of the catalytic reaction must be larger than the electrocatalytic exchange rate $I_0/2F$. In simple terms, the catalytic reaction must be *faster* than the electrocatalytic one.

It is also worth underlining that $|\Lambda|$ has an additional important physical meaning provided it is sufficiently larger than one. It is the ratio of the lifetimes (on the catalyst surface) of the promoting ion and of the reactants involved in the catalytic reaction. Equivalently, it is the ratio of the NEMCA-promoted catalytic rate to the rate of consumption of the promoting ion on the catalyst surface (due to desorption or side reaction with one of the reactants). This latter rate, which at steady state equals $I/2F$, can also be conveniently extracted from log rate versus time curves upon current interruption.[56]

(12) The NEMCA effect does not appear to be limited to any specific type of catalytic reaction, metal catalyst, or electrolyte, particularly in view of the recent demonstration of NEMCA using aqueous electrolytes.[8] The catalyst, however, must be electronically conductive, and the only report of NEMCA on an oxide catalyst is for the case of IrO_2, which is a metallic oxide. It remains to be seen if NEMCA can be induced on semiconductor catalysts.

IX. CONCLUDING REMARKS

The nonfaradaic activation of heterogeneous catalytic reactions via the NEMCA effect is a novel and promising application of electrochemistry.[6] There is a lot of new surface chemistry to be explored[5] with several promising technological possibilities, primarily in product selectivity modification. At the very least, the new phenomenon or electrochemical promotion will allow for a systematic study and enhanced understanding of the role of promoters in heterogeneous catalysis.

ACKNOWLEDGMENTS

We thank the EEC JOULE and STRIDE-HELLAS programs for financial support during the last years. The first author is also grateful to the École Polytechnique Fédérale de Lausanne (EPFL), Lausanne, for an invited Professorship which enabled him to concentrate on the writing of this chapter.

REFERENCES

[1]C. G. Vayenas, S. Bebelis, and S. Neophytides, *J. Phys. Chem.* **92** (1988) 5083.

[2]S. Bebelis and C. G. Vayenas, *J. Catal.* **118** (1989) 125.

[3]C. G. Vayenas, S. Bebelis, and S. Ladas, *Nature (London)* **343** (1990) 625.

[4]C. G. Vayenas, S. Bebelis, I. V. Yentekakis, and H.-G. Lintz, *Catal. Today* **11** (1992) 303.

[5]J. Pritchard, *Nature (London)* **343** (1990) 592.

[6]J. O'M. Bockris and Z. S. Minevski, *Elektrochim. Acta* **39** (1994) 1471.

[7]I. V. Yentekakis, G. Moggridge, C. G. Vayenas, and R. M. Lambert, *J. Catal.* **146** (1994) 293.

[8]S. Neophytides, D. Tsiplakides, P. Stonehart, M. M. Jaksic, and C. G. Vayenas, *Nature (London)* **370** (1994) 45.

[9]C. Wagner, *Adv. Catal.* **21** (1970) 323.

[10]C. G. Vayenas and H. M. Saltsburg, *J. Catal.* **57** (1979) 296.

[11]S. Pancharatnam, R. A. Huggins, and D. M. Mason, *J. Electrochem. Soc.* **122** (1975) 869.

[12]C. G. Vayenas and R. D. Farr, *Science* **208** (1980) 593.

[13]M. Stoukides and C. G. Vayenas, *J. Catal.* **70** (1981) 137; *J. Electrochem. Soc.* **131** (1984) 839.

[14]I. V. Yentekakis and C. G. Vayenas, *J. Catal.* **111** (1988) 170.

[15]T. I. Politova, V. A. Sobyanin, and V. D. Belyaev, *React. Kinet. Catal. Lett.* **41** (1990) 321.

[16]O. A. Marina and V. A. Sobyanin, *Catal. Lett.* **13** (1992) 61.

[17]C. A. Cavalca, G. Larsen, C. G. Vayenas, and G. L. Haller, *J. Phys. Chem.* **97** (1993) 6115.

[18]L. Basini, C. A. Cavalca and G. L. Haller, *J. Phys. Chem.* **98** (1994) 10853; C. A. Cavalca, Ph.D. Thesis, Yale Univ. (1995).

[19]I. Harkness and R. M. Lambert, *J. Catal.* **152** (1995) 211.

[20]P. C. Chiang, D. Eng, and M. Stoukides, *J. Catal.* **139** (1993) 683.

[21]E. Varkaraki, J. Nicole, E. Plattner, Ch. Comninellis, and C. G. Vayenas, *J. of Applied Electrochem.* **25** (1995) 978.

[22]C. G. Vayenas, S. Bebelis, S. Neophytides, and I. V. Yentekakis, *Appl. Phys. (A)* **49** (1989) 95.

[23]C. G. Vayenas, S. Bebelis, I. V. Yentekakis, P. Tsiakaras, and H. Karasali, *Platinum Met. Rev.* **34**(3) (1990) 122.

[24]C. G. Vayenas, S. Bebelis, and C. Kyriazis, *Chemtech* **21** (1991) 500.

[25]C. G. Vayenas, in *Elementary Reaction Steps in Heterogeneous Catalysis,* Ed. by R. W. Joyner and R. A. van Santen, Kluwer Academic Publishers, Dordrecht, The Netherlands, 1993, pp. 73–92.

[26]C. G. Vayenas, S. Bebelis, I. V. Yentekakis, Ch. Karavasilis, and Y. Jiang, *Solid State Ionics* **72** (1994) 321.

[27]C. G. Vayenas, S. Ladas, S. Bebelis, I. V. Yentekakis, S. Neophytides, Y. Jiang, Ch. Karavasilis, and C. Pliangos, *Electrochim. Acta* **39** (1994) 1849.

[28]C. G. Vayenas, *Solid State Ionics* **28–30** (1988) 1521.

[29]M. Stoukides, *Ind. Eng. Chem. Res.* **27** (1988) 1745.

[30]P. J. Gellings, H. S. A. Koopmans, and A. J. Burgraaf, *Appl. Catal.* **39** (1988) 1.

[31]H.-G. Lintz and C. G. Vayenas, *Angew. Chem.* **101** (1989) 725; *Angew. Chem., Int. Ed. Engl.* **28** (1989) 708.

[32]C. Wagner and W. Schottky, *Z. Phys. Chem.* **B11** (1930) 163.

[33]H. Rickert, *Electrochemistry of Solids*, Springer-Verlag, Berlin, 1982.

[34]G. C. Farrington, B. Dunn, and J. C. Thomas, in *High Conductivity Solid Ionic Conductors*, Ed. by T. Takahashi, World Scientific, Singapore, 1989.

[35]E. C. Subbarao and H. S. Maiti, *Solid State Ionics* **11** (1984) 317.

[36]D. F. Schriver and G. C. Farrington, *Chem. Eng. News* **1985** (May 20).

[37]W. Göpel, *Sensors and Actuators B* **18–19** (1994) 1.

[38] F. Grosz, in *Proceedings of the 2nd International Symposium on Solid Oxide Fuel Cells*, Athens, Greece, CEC Publ., Luxembourg, 1991, pp. 7–24.

[39] H. Iwahara, T. Esaka, H. Uchida, and N. Maeda, *Solid State Ionics* **3–4** (1981) 359.

[40] S. Neophytides and C. G. Vayenas, *J. Catal.* **118** (1989) 147.

[41] C. G. Vayenas and S. Neophytides, *J. Catal.* **127** (1991) 645.

[42] C. G. Vayenas, S. Bebelis, and M. Despotopoulou, *J. Catal.* **128** (1991) 415.

[43] S. Bebelis and C. G. Vayenas, *J. Catal.* **138** (1992) 588; **138** (1992) 570.

[44] P. Tsiakaras and C. G. Vayenas, *J. Catal.* **140** (1993) 53.

[45] P. Tsiakaras and C. G. Vayenas, *J. Catal.* **144** (1993) 333.

[46] C. G. Vayenas, A. Ioannides, and S. Bebelis, *J. Catal.* **129** (1991) 67.

[47] I. V. Yentekakis and S. Bebelis, *J. Catal.* **137** (1992) 278.

[48] C. Pliangos, I. V. Yentekakis, X. E. Verykios, and C. G. Vayenas, *J. Catal.* **154** (1995) 124.

[49] A. Kaloyannis and C. G. Vayenas, *J. Catal.,* submitted.

[50] Ch. Karavassilis, S. Bebelis, and C. G. Vayenas, *J. Catal.,* submitted.

[51] H. Karasali, S. Bebelis, and C. G. Vayenas, *J. Catal.* manuscript in preparation.

[52] S. Neophytides, S. Bebelis, and C. G. Vayenas, *Proceedings of the 1st European Solid Oxide Fuel Cell Forum*, Ed. U. Bossel, J. Kinzel Publ., Lucerne, Switzerland (1994), Vol 1, pp. 197–206.

[53] O. A. Marina, V. A. Sobyanin, and V. D. Belyaev, *Materials Sci. and Engineering* **B13** (1992) 153.

[54] O. A. Mar'ina, V. A. Sobyanin, V. D. Belyaev, and V. N. Parmon, *Catal. Today* **13** (1992) 567.

[55] S. Neophytides, D. Tsiplakides, O. Enea, M. M. Jaksic, and C. G. Vayenas, *Electrochim. Acta*, manuscript in preparation.

[56] I. V. Yentekakis and C. G.Vayenas, *J. Catal.* **149** (1994) 238.

[57] N. A. Anastasijevic, H. Baltruschat, and J. Heitbaum, *Electrochim. Acta* **38** (1993) 1067.

[58] C. G. Vayenas, C. Georgakis, J. N. Michaels, and J. Tormo, *J. Catal.* **67** (1981) 348.

[59] S. Ladas, S. Bebelis, and C. G. Vayenas, *Surf. Sci.* **251/252** (1991) 1062–1069.

[60] Y. Jiang, I. V. Yentekakis, and C. G. Vayenas, *J. Catal.* **148** (1994) 240.

[61] B. C. H. Steele, in *Electrode Processes in Solid State Ionics*, Ed. by M. Kleitz and J. Dupuy, Reidel, Dordrecht, The Netherlands, 1976.

[62] J. N. Michaels, C. G. Vayenas, and L. L. Hegedus, *J. Electrochem. Soc.* **133** (1985) 552.

[63] S. C. Singhal and H. Iwahara, eds., *Proceedings of the 3rd International Symposium on Solid Oxide Fuel Cells*, Vol. 93–94, The Electrochemical Society, Pennington, New Jersey, 1993.

[64] R. D. Farr and C. G. Vayenas, *J. Electrochem. Soc.* **127**, (1980) 1478.

[65] C. Sigal and C. G. Vayenas, *Solid States Ionics* **5** (1981) 567.

[66] C. G. Vayenas, S. Bebelis, and C. C. Kyriazis, *Chemtech* **21** (1991) 422; **21** (1991) 500.

[67] I. V. Yentekakis and C. G. Vayenas, *J. Electrochem. Soc.* **136** (1989) 996.

[68] S. Neophytides and C. G. Vayenas, *J. Electrochem. Soc.* **137** (1990) 834.

[69] Y. Jiang, I. V. Yentekakis, and C. G. Vayenas, *Science* **264** (1994) 1563.

[70] N. Kiratzis and M. Stoukides, *J. Electrochem. Soc.* **134** (1987) 1925.

[71] J. N. Michaels and C. G. Vayenas, *J. Catal.* **85** (1987) 477.

[72] J. N. Michaels and C. G. Vayenas, *J. Electrochem. Soc.* **131** (1984) 2544.

[73] R. DiCosimo, J. D. Burrington, and R. K. Grasselli, *J. Catal.* **102** (1986) 234.

[74] E. J. L. Schouler, M. Kleitz, E. Forest, E. Fernandez, and P. Fabry, *Solid State Ionics* **3–4** (1981) 431.

[75] T. M. Gür and R. A. Huggins, *J. Electrochem. Soc.* **126** (1979) 1067.

[76] T. M. Gür and R. A. Huggins, *Science* **219** (1983) 967.

[77] T. M. Gür and R. A. Huggins, *J. Catal.* **102** (1986) 443.

[78] T. Hayakawa, T. Tsunoda, H. Orita, T. Kameyama, H. Takahashi, K. Takehira, and K. Fukuda, *J. Chem. Soc. Jpn. Chem. Commun.* **1986**, 961.

[79]K. Otsuka, S. Yokoyama, and A. Morikawa, *Chem. Lett. Chem. Soc. Jpn.* **1985**, 319.

[80]S. Seimanides and M. Stoukides, *J. Electrochem. Soc.* **133** (1986) 1535.

[81]K. Otsuka, K. Suga, and I. Yamanaka, *Catal. Lett.* **1** (1988) 423.

[82]K. Otsuka, K. Suga, and I. Yamanaka, *Chem. Lett. Jpn.* **1988**, 317.

[83]V. D. Belyaev, O. V. Bazhan, V. A. Sobyanin, and V. N. Parmon, in *New Developments in Selective Oxidation*, Ed. by G. Centi and F. Trifiro, Elsevier, Amsterdam, 1990, p. 469.

[84]D. Eng and M. Stoukides, *Catal.-Rev.-Sci.-Eng.* **33** (1991) 375.

[85]D. Eng and M. Stoukides, *J. Catal.* **30** (1991) 306.

[86]P. H. Chiang, D. Eng, and M. Stoukides, *J. Electrochem. Soc.* **138** (1991) L11.

[87]W. Dönitz and E. Erdle, *Int. J. Hydrogen Energy* **10** (1985) 291.

[88]D. Eng and M. Stoukides, *Catal. Lett.* **9** (1991) 47.

[89]P. Tsiakaras, Ph.D. Thesis, University of Patras, 1993.

[90]D. Y. Wang and A. S. Nowick, *J. Electrochem. Soc.* **126** (1979) 1155.

[91]D. Y. Wang and A. S. Nowick, *J. Electrochem. Soc.* **126** (1979) 1166.

[92]D. Y. Wang and A. S. Nowick, *J. Electrochem. Soc.* **128** (1981) 55.

[93]E. J. L. Schouler and M. Kleitz, *J. Electrochem. Soc.* **134** (1987) 1045.

[94]H. H. Hildenbrand and H.-G. Lintz, *Catal. Today* **9** (1991) 153.

[95]H. Okamoto, G. Kawamura, and T. Kudo, *J. Catal.* **82** (1983) 322.

[96]C. G. Vayenas, *J. Catal.* **90** (1984) 371.

[97]J. Tafel, *Z. Phys. Chem. (Leipzig)* **50** (1905) 641.

[98]J. A. V. Butler, *Trans. Faraday Soc.* **19** (1924) 729.

[99]T. Erdey-Gruz and M. Volmer, *Z. Phys. Chem. (Leipzig)* **150** (1930) 203.

[100]J. O'M. Bockris and A. K. N. Reddy, *Modern Electrochemistry*, Vol. 2, Plenum Press, New York, 1970.

[101]J. O'M. Bockris and S. U. M. Khan, *Surface Electrochemistry: A Molecular Level Approach*, Plenum Press, New York, 1993.

[102]J. S. Newman, Electrochemical Systems, Prentice-Hall, Englewood Cliffs, New Jersey, 1973.

[103]M. Manton, Ph.D. Thesis, MIT, 1986.

[104]C. G. Vayenas and J. N. Michaels, *Surf. Sci.* **120** (1982) L405.

[105]M. Peukert and H. P. Bonzel, *Surf. Sci.* **145** (1984) 239.

[106]M. Peukert and H. Ibach, *Surf. Sci.* **136** (1983) 319.

[107]Y. Jiang, A. Kaloyannis, and C. G. Vayenas, *Electrochim. Acta* **38** (1993) 2533.

[108]H. J. Reiss, *J. Phys. Chem.* **89** (1985) 3783.

[109]H. J. Reiss, *J. Electrochem. Soc.* **135** (1988) 2476.

[110]P. M. Gundry and F. C. Tompkins, in *Experimental Methods in Catalyst Research*, Ed. by R. B. Anderson, Academic Press, New York, 1968, pp. 100–168.

[111]J. Hölzl and F. K. Schulte, in *Solid Surface Physics*, Ed. by G. Höhler, and E. Niekisch, Springer-Verlag, Berlin, 1979, pp. 1–150.

[112]S. Trasatti, in *Advances in Electrochemistry and Electrochemical Engineering*, Vol. 10, Ed. by H. Gerischer and C. W. Tobias, John Wiley & Sons, New York, 1977.

[113]H. Amariglio, *J. Chim. Phys.* **64** (1967) 1391.

[114]N. W. Ashcroft and N. D. Mermin, in *Solid State Physics*, Hott-Saunders Intl. Eds., Philadelphia, USA, 1976, pp. 354–371.

[115]C. Lamy, in *Propriétés Électriques des Interfaces Chargées*, Ed. by D. Schuhmann, Masson, Paris, 1978, pp. 210–241.

[116]W. Schröder and J. Hölzl, *Solid State Commun.* **24** (1977) 777.

[117]C. A. Papageorgopoulos and J. M. Chen, *Surf. Sci.* **52** (1975) 40.

[118]H. P. Bonzel, *Surf. Sci. Rep.* **8** (1987) 43.

[119]D. Heskett, *Surf. Sci.* **199** (1988) 67.

[120]T. Aruga and Y. Murata, *Prog. Surf. Sci.* **31** (1989) 61.

[121]K. J. Uram, L. Ng, and J. R. Yates, Jr., *Surf. Sci.* **177** (1986) 253.

[122]T. Arakawa, A. Saito, and J. Shiokawa, *Appl. Surf. Sci.* **16** (1983) 365.

[123]T. Arakawa, A. Saito, and J. Shiokawa, *Chem. Phys. Lett.* **94** (1983) 250.

[124]U. Vöhrer, Ph.D. Thesis, University of Tübingen, 1992.

[125]S. Ladas, S. Kennou, S. Bebelis, and C. G. Vayenas, *J. Phys. Chem.* **97** (1993) 8845.

[126]I. V. Yentekakis and C. G. Vayenas, manuscript in preparation.

[127]M. Boudart and G. Djéga-Mariadassou, *Kinetics of Heterogeneous Catalytic Reactions*, Princeton University Press, Princeton, New Jersey, 1984.

[128]Z. Xu, J. T. Yates, Jr., L. C. Wang, and H. J. Kreuzer, *J. Chem. Phys.* **96** (1991) 1628.

[129]S. Neophytides and C. G. Vayenas, *Ionics* **1** (1995) 80; *J. Phys. Chem.* 1995, in press.

[130]J. L. Falconer and R. J. Madix, *Surf. Sci.* **48** (1975) 393.

[131]C. G. Vayenas, B. Lee, and J. N. Michaels, *J. Catal.* **66** (1980) 36.

[132]V. A. Sobyanin, V. I. Sobolev, V. D. Belyaev, O. A. Mar'ina, A. K. Demin, and A. S. Lipilin, *Catal. Lett.* **18** (1993) 153.

[133]V. A. Sobyanin and V. D. Belyaev, *React. Kinet. Catal. Lett.* **51**(2) (1993) 373.

[134]M. Boudart, *J. Am. Chem. Soc.* **74** (1952) 3556.

[135]S. Cerny and V. Ponec, *Catal. Rev.* **2** (1968) 249.

[136]E. Shustorovich, *J. Mol. Catal.* **54** (1989) 307.

[137]G. Pacchioni, F. Illas, S. Neophytides, and C.G. Vayenas, manuscript in preparation.

[138]S. Neophytides, D. Tsiplakides, P. Stonehart, M. M. Jaksic, and C. G. Vayenas, *Electrochim. Acta*, submitted, 1995.

[139]E. Cremer, *Adv. Catal.* **7** (1955) 75.

[140]G.-M. Schwab, *J. Catal.* **84** (1983) 1.

[141]S. Seimanides, P. Tsiakaras, X. E. Verykios, and C. G. Vayenas, *Appl. Catal.* **68** (1991) 41.

[142]V. D. Belyaev, V. A. Sobyanin, A. K. Demin, A. S. Lipilin, and V. A. Zapesotski, *Mendeleev Commun.* **1991**, 53.

[143]H. Karasali and C. G. Vayenas, *Mater. Sci. Forum* **76** (1991) 171.

[144]D. Lackey and P. A. King, *J. Chem. Soc., Faraday Trans. 1* **1987**, 83.

[145]I. V. Yentekakis, S. Neophytides, and C. G. Vayenas, *J. Catal.* **67** (1981) 348.

[146]S. Ladas, R. Imbihl, and G. Ertl, *Surf. Sci.* **219** (1989) 88.

[147]Ch. Karavasilis, S. Bebelis, and C. G. Vayenas, *Mater. Sci. Forum* **76** (1991) 175.

[148]Ch. Karavasilis, S. Bebelis, and C. G. Vayenas, *J. Catal.*, submitted, 1995, in press.

[149]R. A. Van Santen and H. P. C. E. Kuipers, *Adv. Catal.* **35** (1987) 265.

[150]R. B. Grant and R. M. Lambert, *J. Catal.* **92** (1985) 364.

[151]H. Karasali, Ph.D. Thesis, University of Patras, 1994.

[152]H. Alqahtany, P.-H. Chiang, D. Eng, M. Stoukides, and A. R. Robbat, *Catal. Lett.* **13** (1992) 289.

[153]E. Lamy-Pitara, L. Bencharif, and J. Barbier, *Appl. Catal.* **18** (1985) 117.

[154]C. Pliangos, I. V. Yentekakis, X. E. Verykios, and C. G. Vayenas, *J. Catal.* **154** (1995) 124.

[155]G. L. Haller and D. E. Resasco, *Adv. Catal.* **36** (1989) 173.

[156]S. J. Tauster, S. C. Fung, and R. L. Garten, *J. Am. Chem. Soc.* **100** (1978) 170.

[157]H. Baltruschat, N. A. Anastasijevic, M. Beltowska-Brzezinska, G. Hambitzer, and J. Heitbaum, *Ber. Bunsenges. Phys. Chem.* **94** (1990) 996.

[158]S. Neophytides, D. Tsiplakides, P. Stonehart, M. Jaksic, and C. G. Vayenas, *Electrochim. Acta*, submitted, 1995.

[159]B. E. Conway, in *Electrodes of Conductive Metallic Oxides*, Ed. by S. Trassatti, Elsevier, Amsterdam, 1981, Chapter 9.

[160]B. E. Conway and B. K. Tilak, *Adv. Catal.* **38** (1992) 1.

[161]E. Plattner and Ch. Comninellis, in *Process Technologies for Water Treatment*, Ed. by S. Stucki, Plenum Press, New York, 1988, pp. 205–217.

[162]D. L. Rath and D. M. Kolb, *Surface Science* **109** (1981) 641.

[163]E. R. Kötz, H. Neff, and K. Müller, *J. Electroanal. Chem.* **215** (1986) 33.

3

Effect of Surface Structure and Adsorption Phenomena on the Active Dissolution of Iron in Acid Media

Ileana-Hania Plonski

Institute of Atomic Physics, IFTM, RO-76900 Bucharest, Romania

I. INTRODUCTION

An understanding at the molecular level of the processes leading to iron corrosion and of the complex factors influencing it have long been considered a *sine qua non* condition for progress in the field of corrosion research. During its long history,[1-240] this research has provided the scientific background for attempts to master, to prevent, or at least to minimize all kinds of corrosion phenomena as well as to elaborate new highly corrosion-resistant materials. Uniform corrosion in aqueous acid electrolytes is one of the most important corrosion phenomena in practice, being encountered in many branches of the chemical and related industries, during the cleaning of heat-transfer surfaces in traditional power plants, and in the reprocessing of nuclear fuel and decontamination of heavy water reactors. Insight into the processes leading to iron corrosion has mainly come from the study of electrode kinetics using electrochemical methods in conjunction with surface-morphological and crystallographic investigations and chemical analysis of the electrolyte.

1. Corrosion in the Active State

Electrochemical iron dissolution is a process in which metal atoms are transferred as ions from the lattice surface sites into the interstices of the

Modern Aspects of Electrochemistry, Number 29, edited by John O'M. Bockris *et al.* Plenum Press, New York, 1996.

short-range-ordered structure of water molecules and solute ions in the reaction layer, forming dissolved hydrated or complexed compounds. Except for the adsorbed intermediate species, the dissolved ions leave the reaction layer by transport phenomena.

It is a well-known fact that iron dissolution occurs in four different states, namely, the active, passive, transpassive, and brightening states, determined by the nature and kinetics of the reactions involved, which depend in turn on the potential and electrolyte composition. A schematic polarization curve for anodic dissolution of iron in acid solutions is given in Fig. 1. The shape of the curve depends on the nature of the electrolyte, the polarization program, and hydrodynamics. The Fe/Fe^{2+}, H^+/H_2, and Fe/Fe_3O_4 equilibrium potentials are calculated for $a_{Fe^{2+}} = 1M$, $p_{H_2} = 1$ atm, and pH = 2 in the bulk of the solution. The potentials delimiting the four different states and the nature of the reactions in each range of potential are still a matter of debate. It can also be seen that the total current–anodic potential curve displays a nonmonotonous but continuous shape representing the gentle interference of various processes, each having its own particular kinetics.

The overall active range of iron dissolution was subdivided by Lorenz et al.[16,24–29] into the following ranges:

(i) The *active range* (I_1 reaction according to Bech-Nielsen[176,235]), extending between the Fe/Fe^{2+} equilibrium potential and the onset of the first hump (see Fig. 1). However, this last boundary is ambiguous because some authors have placed it at the potential corresponding to the first current maximum, and others at the equilibrium potential of the Fe/Fe_3O_4 electrode.[37] Nevertheless, it should be noted that in the region of high current densities, due to the expulsion of H^+ by the high Fe^{2+} flux and the concentration or possible crystallization of ferrous salts, the pH near the electrode surface may be increased so that the E^{eq}_{Fe/Fe_3O_4} theoretically calculated for the pH in the bulk of the solution is in fact shifted toward more negative values.

(ii) The *transition range* between the onset of the first hump and the onset of the second one, a region which, according to Fig. 1, is delimited by the E^{eq}_{Fe/Fe_3O_4} theoretically calculated for the pH in the bulk of the solution.

Irrespective of the placement of E^{eq}_{Fe/Fe_3O_4} and its neighbor E^{eq}_{Fe/Fe_2O_3} in our overview schemes, by shifting the potential toward positive values, parallel reactions leading to magnetite, γ-ferric oxyhydroxide, hematite, and their intermediary products become thermodynamically possible.

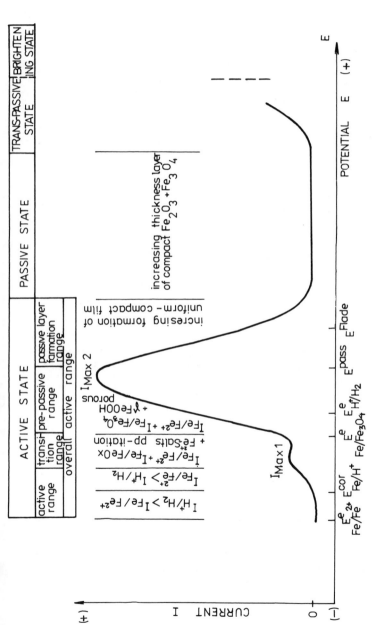

Figure 1. Schematic anodic polarization curve of iron in acid solution between the active and the transpassive states of the metal.

(iii) The *prepassive range* (I_2 reaction according to Bech-Niel-sen[176,235]), extending from E^{eq}_{Fe/Fe_3O_4} and delimited by or extending slightly past the passivation potential.

(iv) The *passive layer formation range*, between the passivation potential and the Flade potential, where two-dimensional and three-dimensional compact films consisting of iron hydroxides and oxides are progressively formed.[*]

Thus, iron is considered to be in an active state between the equilibrium potential of the Fe = Fe^{2+} + $2e^-$ reaction and the onset of passivation (i.e., the Flade potential), a region termed "the overall active range," in which oxides or hydroxides may be present locally or as porous layers allowing direct contact between the metal and the electrolyte solution.

The polarization curve in the active state displays at least two current maxima, I_{Max1} and I_{Max2}, in the range $1 < pH < 6$.[176,235] Because the current density at the minimum in between the two humps increases with the increase of the rotation speed of the electrode, and with the decrease of the pH and of the potential sweep rate, at pH > 3 the first maximum degenerates into a plateau and at pH < 2 becomes an inflection point. Practically, I_{Max1} is not detectable at pH < 1. The active corrosion of iron proceeds usually by several series of parallel elementary processes, the main reactions being the anodic oxidation of the metal and reduction of an oxidant (hydronium or dissolved oxygen). These processes, ferrous ion formation and hydrogen evolution or oxygen reduction, described by the stoichiometric reactions

$$Fe = Fe^{2+} + 2e^- \qquad (1)$$

$$2H^+ (aq) + 2e^- = H_2 \qquad (2)$$

$$O_2 (aq) + 4H^+ + 4e^- = 2H_2O \qquad (3)$$

occur randomly at the metal surface and also at specific sites with different probabilities which depend on the surface heterogeneity from chemical, structural, morphological, and energetic standpoints.

Unfortunately, the overvoltage region in which these reactions can be studied is very narrow; as will be seen below, not too far from the corrosion potential in both directions, additional processes interfere. Figures 2 and

---

[*]The passivation potential and the Flade potential are defined in slightly different ways by various authors.

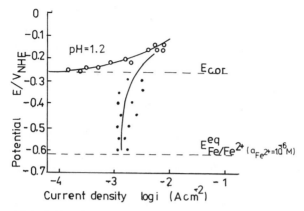

Figure 2. Experimental steady-state polarization curve, ($\circ$) and photocolorimetrically determined rate of Armco iron dissolution ($\bullet$) in 0.05M H_2SO_4 in the region of potential $E_{Fe/Fe^{2+}}^{eq} < E < E_{Fe/H^+}^{cor}$. (Experimental points taken from Ref. 1.)

3 show typical anodic polarization curves illustrating the two extreme situations of active iron dissolution. In Fig. 2, the polarization curve and the potential dependence of the rate of iron dissolution, determined photocolorimetrically, taken from Ref. 1, are depicted. The first extreme situation pertains to iron dissolution in very acidic solutions, especially after cathodical prepolarization.

Thus, in 0.05M H_2SO_4, pH 1.6, an anomalous dissolution rate of iron was found experimentally, independent of the diffusion of hydrogen from inside the metal, of the cathodic potential, and of hydrodynamics. The dissolution rate slightly increased with hydronium concentration with a reaction order close to 0.3 and with iron impurity content. In an attempt to explain this phenomenon, Kolotyrkin and Florianovich[2] advanced the hypothesis of a chemical dissolution via the intermediate formation of hydroxo ferrous ions:

$$Fe + H_3O^+ \rightarrow FeOH^+ + H_2 \qquad (4)$$

An alternative explanation was proposed by Vorkapic and Drazic.[12] As was observed a long time ago, during the cathodic polarization of iron, hydrogen evolution is accompanied by hydrogen penetration into the metal lattice and its accumulation in a *ca.* 0.1-mm surface layer. As a

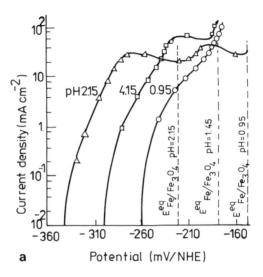

Figure 3. Experimental steady-state polarization curves, showing the active and the transition ranges of the "overall active dissolution range," as a function of pH. Equilibrium potential lines for the iron/magnetite reaction are superimposed. (a) Iron in solutions of $0.5M$ FeClO$_4$ + $0.5M$ BaClO$_4$; pH values of the solutions are labeled on the curves (from Ref. 15). (b) Iron (rotating disk electrode) in $1M$ ClO$_4^-$ solution, pH 2.93 (from Ref. 177). (c) Iron (rotating disk electrode) in Na$_2$SO$_4$ and H$_2$SO$_4$ solution, pH 3.45 (from Ref. 22).

consequence, there is a change in the lattice parameters resulting in fragility of the metal, which favors breaking off of small pieces of metal, a phenomenon called the "chunk effect."[3-11] Vorkapic and Drazic[12] performed a decisive experiment in which the anomalous dissolution rate was found to decrease by nearly one order of magnitude when iron was introduced into a magnetic field. As a result, they attributed the increase in the concentration of Fe^{2+} ions in the solution to the mechanical disintegration of the surface layer under the influence of the cathodically evolving hydrogen; as the hydrogen penetrates into the lattice, it breaks off chunks of particles.

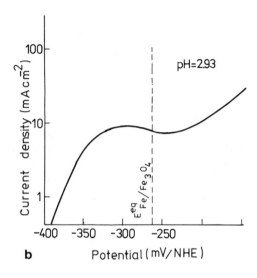

b

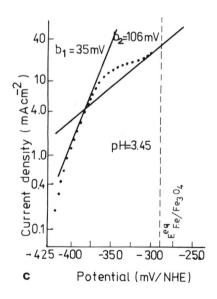

c

Figure 3. (Continued)

In fact, the two explanations are complementary because, in order not to be influenced by the electrode potential, the electron-exchange reaction in Eq. (4) must occur in the solution outside the double layer. Moreover, an additional effect consisting in surface screening by hydrogen bubbles collecting at certain sites on the surface is superimposed[13] (see also Ref. 6 in Ref. 13). Consequently, the data obtained from experiments performed in very acidic solutions, after a long waiting time under open-circuit conditions, and especially with cathodically prepolarized iron, can be affected by absorbed hydrogen and its consequences in relation to the change of surface structure and anomalous dissolution.

The other extreme situation refers to the transition range of anodic active iron dissolution[14] and is illustrated in Fig. 3, in which the experimental steady-state polarization curves of iron in acid solutions of different pH are reproduced from the literature[15,22,177]; the equilibrium potential lines for the iron/magnetite reaction are superimposed. The transition range of active iron dissolution begins with an increase in the steady-state Tafel slope and a decrease in the reaction order with respect to hydroxyl ion; thus, for the upper part of the polarization curve, usually encompassing the equilibrium potential of the iron/magnetite reaction, in moderately acidic solutions, higher anodic Tafel slopes of 100–120 mV dec^{-1} and reaction orders in pH close to zero were reported by Bech-Nielsen and co-workers[19–23] and by Lorenz and co-workers.[16,24–28] Further on, a first current maximum, I_{max1}, and a subsequent minimum in the anodic polarization curve appear. Under certain conditions, the current maximum 1 degenerates to a flat plateau or to an inflection point, as can be seen for pH 0.95 in Fig. 3A. Both I_{max1} and its corresponding potential, E_{imax1}, decrease as the pH increases (with a reaction order dI_{max1}/dpH = –0.5) but are independent of anion activity in the electrolyte.[16,17] According to Bech-Nielsen,[18] in this region of potential the anodic dissolution of iron takes place via two distinct parallel electron source reactions, leading to two currents I_1 and I_2. Different reaction schemes based on a formally divalent iron oxide or hydroxide species adsorbed at the electrode surface were assumed, and the corresponding kinetic equations were established.[16,19,24,28–31] Support of this idea was provided by Drazic and Hao's[32] calculations that hydroxo compounds of iron (in their case ferrous hydroxide) in the adsorbed state are thermodynamically stable even in acidic solutions because their chemical potential is changed by adsorption. In addition to this, as a rule E_{imax1} is situated at potentials that are only about 50 mV more negative than the equilibrium potential of the Fe/Fe$_3$O$_4$

system, suggesting that the current density does not grow further due to the adsorbed intermediates involved in the parallel magnetite formation, which partially block the dissolution surface. In supporting this idea, Morel[33] observed that at high current densities, the anodic dissolution of iron in acidic sulfate solutions was diffusion-limited. Thus, the author studied the influence of the rotation speed of a polycrystalline Armco iron electrode in $0.5M$ H_2SO_4, pH 0.5, on the anodic polarization curves and on the dissolution products crystallized on the metal surface at the end of the experiment. At potentials between $E_{Fe/Fe_3O_4}^{eq} + 50$ mV and $E_{Fe/Fe_3O_4}^{eq} + 100$ mV, under static conditions, a crystalline layer containing salts such as $FeSO_4 \cdot 7H_2O$ was noticed as a result of supersaturation of the solution near the electrode. This layer completely disappeared when a rotating disk electrode was used, proving that it resulted from a slow transport phenomenon. Concomitantly, a crystalline layer of Fe_3O_4 and γ-FeOOH was observed by electron diffraction and microscopy. In contrast with iron sulfate, magnetite and γ-ferric oxyhydroxide persisted locally even with a disk electrode rotating at a high speed, suggesting their formation, in the adsorbed state, directly from the metal. It was also noted by Allgaier and Heusler[14] that the spacing of kinks at monatomic steps approaches its minimum value, comparable to interatomic spacing, just where the transition range begins, confirming a change in the mechanism of the iron dissolution reaction at more positive potentials.[34] Irrespective of the adopted mechanism, one must keep in mind that starting with an anodic overvoltage greater than 100 mV (with respect to the corrosion potential), the polarization curves are influenced by the accumulation of ferrous salts and by the formation of some solid ferrous–ferric oxides and hydroxides in the adsorbed state which account for the increase in the steady-state Tafel slope.

Therefore, the active region of iron dissolution in acid media can be subdivided in turn into three parts, depending on the pH, potential, and hydrodynamics.

(i) In very acidic solutions near the corrosion potential, supplementary effects of hydrogen evolution interfere, such as lattice surface damage, anomalous chemical dissolution, surface screening by hydrogen bubbles; catalytic species resulting from sulfate ion reduction were also observed by Bala.[35]

(ii) In moderately acidic solutions ($0 <$ pH < 4), in a rather narrow range of potential, $E^{cor} < E < E^{cor} + 80$ mV (which can be widened by working under strong hydrodynamic conditions), the anodic ferrous ion formation and hydrogen evolution prevail and control the electrochemical kinetics. It is supposed that in this region, when a rational polarization program is used, parallel processes such as the above-mentioned hydrogen effects, ferrous salt accumulation, and/or oxide and hydroxide formation do not interfere. This part of the anodic curve displays linear regions characterized by low steady-state Tafel slopes, <60 mV dec^{-1} (frequently 30–40 mV dec^{-1}), and a rather positive reaction order with respect to OH$^-$ ions (between 1 and 2).

(iii) At $E > E^{cor} + (80$ to $120)$ mV (depending on pH), an increase in Tafel slope and a decrease in the reaction order with respect to OH$^-$ have been observed and mark the start of processes specific to the transition range of the overall active range of iron dissolution; among them, the formation of crystallized ferrous and ferric solid species including anions and their blocking effect on the metal dissolution superimpose and change the mechanism and the kinetics.

The insight that has been gained to date into metal dissolution and corrosion has mainly come from the study of electrode kinetics in the second region. The aim of this chapter is to give an up-to-date account of advances in knowledge about dissolution and corrosion in this second region of active anodic polarization, that is, for $0 < \eta^{an} < 100$ mV versus E^{cor}. The extreme regions will not be encompassed in the following treatment, but it must be kept in mind that before the additional processes specific to the extreme regions prevail, they interfere with the main corrosion reactions, changing the kinetic data to a small extent but sufficiently to become a source of misinterpretations.

Because comprehensive surveys and extensive review articles related to iron/acid interface electrochemistry and including new experimental results and critical interpretations of the proposed mechanisms have been published recently,[34,36–40] a lot of the information included in these papers will not be repeated here; the reader is advised to consult this literature.

2. Applying Statistics to Corrosion Data

The main insight into the kinetics and mechanism of iron dissolution and corrosion has come from electrochemical studies using complex E–t or I–t perturbation programs. Some of the conclusions that have emerged from this large body of work are summarized here.

(i) Iron dissolution in aqueous acidic solutions free of oxygen, under hydrodynamic conditions, is under the main control of a charge-transfer step, a conclusion drawn from the Tafel-like linear current–potential dependence.

(ii) The cathodic reduction of hydronium ions on iron is controlled to a large extent by the atomic hydrogen formation step followed by recombination, as indicated by the Tafel slope close to -120 mV dec^{-1}, a reaction order in H$^+$ ion equal to 1, and a low degree of coverage by adsorbed hydrogen near the corrosion potential, averaged over the entire surface area, $\Theta_H^{corr} \ll 1$.

(iii) A negative reaction order in hydronium ion between -1 and -2 of anodic polarization was generally observed. It was concluded that either the OH$^-$ ion participates directly in some stages leading to intermediate formation of iron hydroxo compounds or H atom exerts an indirect influence by blocking the active sites for metal dissolution, or both.

The electrochemical quantities relevant for the kinetics and mechanism, obtained from experimental measurements, are the interdependence between potential, current, and species in the bulk of the solution (especially hydronium ions and anions) and time. Theoretically, one needs all the characteristic slopes $\delta E/\delta \log i_i$, $\delta \eta/\delta \log i_i$, $\delta E/\delta \log i_s$, $\delta \eta/\delta \log i_s$, $\delta \log i_i/\delta \log a_j$, $\delta \log i_s/\delta \log a_j$, $\delta a_j/\delta t$, $\delta \log i_i/\delta$ pH, $\delta \log i_s/\delta$ pH, and $\delta E/\delta t$, as well as the characteristic times, that is, the waiting, peak, and polarization times τ_w, τ_{peak}, and τ_p, respectively. Practically, one cannot maintain all the conditions under control, but some of them are important, and others irrelevant.

In spite of numerous investigations, a high variability in the data characterizes the electrochemical measurements, leading to the conclusion that important variables have been totally ignored or believed to be constant. Only a few examples are given below.

Lorenz and co-workers[44–47] reported for Fe in HCl + NaCl solutions a steady-state Tafel slope, b_s^a, close to 60 mV dec^{-1} for pH 0–1.5 and 30 mV dec^{-1} for pH 1.5–3, whereas Oftedal,[48] apparently working under the same conditions, obtained $b_s^a = 30$ mV dec^{-1} in solutions of pH < 1.5 and 60–70 mV dec^{-1} for pH > 1.5. Atkinson and Marshall[49] found for Fe in $1M$ HCl a steady-state Tafel slope close to 44 mV dec^{-1} at lower current density and 85 mV dec^{-1} at higher current density, at variance with Drazic and Drazic,[50] who reported 120 mV dec^{-1} for chlorine ion concentrations in the range between 10^{-1} and $2M$ at low current densities. Moreover, if the steady-state Tafel slope values for iron in sulfate acidic solutions are plotted as a function of the number of authors reporting a certain value, a distribution with two wide maxima is obtained, as seen in the histogram given in Fig. 4B, suggesting both poor reproducibility *and* at least one uncontrolled variable.

During the past 10 years, there has been much effort to find out the causes of the irreproducibility of the data. The possible explanations are as follows.

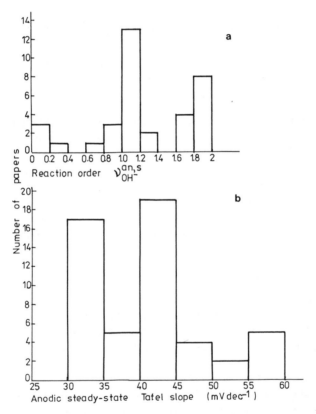

Figure 4.   Histograms showing the distribution of values reported in the literature for the steady-state Tafel slope and the reaction order in OH⁻ for anodic dissolution of iron in acid media. The literature values are taken from Refs. 1, 19, 33, 35, 37, 38, 43, 46, 50–52, 79, 80, 109, 123, 125, 136, 137, 155, 158–162, 173, 174, 176, 177, 184, 185, 200, 205, 212, and 214.

(i) The Time Effect

The time effect is the influence exerted by the length of time elapsed under certain experimental conditions prior to a change in these conditions and was first described in various ways a long time ago. Thus, one can distinguish the influence of the following characteristic times:

- The *waiting time*, τ_w, termed also exposure, or contact or immersion time, is the time during which the electrode is in contact with the solution under open-circuit conditions from the moment of immersion up to the measurement or moment of setting on the polarization.
- The *polarization time*, τ_p, is the time elapsed from the moment of setting on the polarization up to the measurement.
- The *interruption time*, τ_{int}, is the period of time under open-circuit conditions between two measurements under polarization.

As will be seen below, these times influence the chemistry of both the solution and the metal sides of the reaction layer.

(ii) Effect of Surface Structure and Morphology

Surface structure and morphology refers to the nature and number of specific sites at the metal surface, such as kink and step atoms, atoms in edges, dislocations, grain boundaries, corners or etch pyramids, and grooves, from which the dissolution can start with different probabilities and also according to different mechanisms. The surface structure and morphology is highly dependent on the metal history prior to immersion and can be partially controlled by crystallographic, metallurgical, chemical, or mechanical treatments. As will be seen below, these two effects influence each other, the mechanism and kinetics of the corrosion process, as well as the degree of coverage with adsorbed intermediates and species from the solution.

(iii) Mishandling of the Experimental Data

Three kinds of mistakes can result from mishandling of the data. The first arises when the standard deviation of the slopes is estimated from the formulas derived for data obtained directly from measurements instead of from specific relationships established for slopes. As a result of this type of error, apparently different reproducible data may, in reality, be randomly distributed and perhaps even overlapped. The second mistake arises from the extrapolation of observations made in the extreme region of polarization, for instance, at high current densities, to the corrosion potential, where the phenomena are not the same. The third mistake relates especially to data obtained for the reduction of ferrous ions in the cathodic region of polarization. It is a general consensus of the desire to obtain a unified mechanism for the iron reaction in both the anodic and the cathodic direction. Unfortunately, a lot of difficulties are encountered in attempts to do so:

(i) In acidic solutions, iron dissolution and deposition cannot be accurately studied near the equilibrium potential, due to the simultaneous evolution of hydrogen at a high rate. Thus, it is necessary either to subtract the hydrogen current from the total current or to work in a potential range far away from the equilibrium potential, where other processes interfere and the mechanism can change.

(ii) Near the equilibrium potential of the iron electrode, hydrogen codeposition gives rise to additional effects, namely, penetration of hydrogen into the metal, destroying its structure and morphology and causing the chunk effect, and significant changes in pH near the electrode surface.

(iii) Another difficulty emerges from a certain liberty in handling the data. For example, in Ref. 51, after subtraction of the hydrogen current from the experimental cathodic polarization versus log i curve, a steady-state cathodic Tafel slope for iron close to -60 mV dec^{-1} was obtained. If one accepts *a priori* that ferrous ion reduction proceeds with the assistance of hydroxyl ions through an intermediate hydroxo species, as is the case in Ref. 51, the iron Tafel-like line must be corrected for the pH change near the surface, yielding $b_{Fe}^{c} = -120$ mV dec^{-1}, consistent with stepwise electron transfer. On the other hand, if one presumes *a priori* that OH$^-$ ions do not take part directly in the reaction of Fe^{2+} reduction, no correction is necessary, and the experimental slope of -60 mV dec^{-1} attests a two-electron-transfer reaction in one step. Consequently, an iron cathodic Tafel slope derived after a set of operations, based on hypothetical assertions, leads to ambiguous conclusions concerning the mechanism.

As a result of mishandling of the data as well as of the time and surface effects discussed above, a discouraging amount of variability in the data is seen in Fig. 5, in which histograms have been plotted to show the distributions of values for the cathodic steady-state Tafel slope and reaction order in OH$^-$ in the published data.

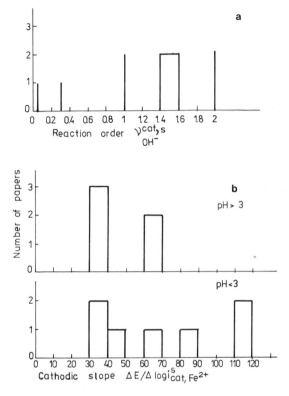

Figure 5. Histograms showing the distribution of values reported
in the literature for the steady-state Tafel slope and the reaction
order in OH^- for cathodic deposition of iron in acid media. The
literature values are taken from Refs. 37, 38, 47, 51, 70, 200, 214,
and 220–224.

Fundamental studies on the kinetics of iron dissolution have been
carried out in oxygen-free acid aqueous solutions containing sulfate,
perchlorate, and halogenide ions under static or hydrodynamic conditions.
Iron single-crystal faces and polycrystalline iron of different purities,
previously exposed to heat treatments and chemical or electrochemical
attack of the metal surface, as well as cathodically deposited iron have
been used. Within the framework of this chapter, a summary of typical
new experimental results will be presented, along with recently proposed
mechanisms and interpretations.

II. THE OPEN CIRCUIT

There are three reasons for the importance of the investigation of the iron/acid interface under open-circuit conditions. First, it is of practical significance because the main corrosion processes encountered in real life involve free corrosion. Second, it supplies diagnostic criteria for elucidating the reaction mechanism such as the steady-state corrosion potential and corrosion current as a function of time, pH, anions, hydrodynamics, and metal surface characteristics. In this section, special attention will be given to the kinetic quantities $\Delta \log i_s^{cor}/\Delta$ pH and $\Delta E_s^{cor}/\Delta$ pH, where the superscript "cor" and the subscript "s" refer to steady-state open-circuit conditions. Finally, a number of experiments, especially recent ones, carried out under well-controlled conditions have revealed the influence of the waiting time on both open-circuit and polarization kinetic quantities.

1. The Transitory Regime of the Open Circuit

Despite the fact that as early as 1970 Akiyama et al.[52] observed that the corrosion potential, corrosion current, and differential capacity are time-dependent, there are not many papers reporting on the transitory regime of the open circuit, and only one[53] in which the potential was recorded as a function of time from the moment of immersion. Figure 6 represents such a typical potential transient for iron in $H_2SO_4 + K_2SO_4$ solutions, under nitrogen bubbling and a flow rate of 1 dm^3 min^{-1}. Equilibrium potentials of the thermodynamically possible reactions for $a_{Fe^{2+}} = 10^{-6}M$ and $p_{H_2} = 1$ atm are indicated on the right-hand side of the plot. However, because $p_{H_2} < 1$ atm during the first part of the transient, the equilibrium potential of the hydrogen electrode is temporarily situated at more positive values.

The following remarks may be made regarding the potential–time trend:

(i) The trace always starts with a jump in the potential region in which both iron dissolution and oxide formation are thermodynamically possible and then sharply drops. The observed negative slope dE^{cor}/dt corroborates the predominance of electron source reactions in this region, the most probable being $Fe \rightarrow Fe^{2+} + 2e^-$.

(ii) The slope remains steep until the potential reaches a value close to the equilibrium potential of the hydrogen electrode, at which point there

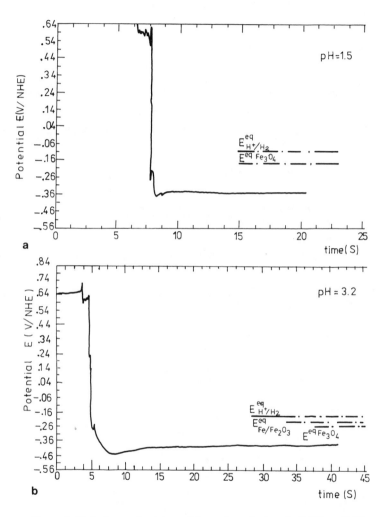

Figure 6. Typical open-circuit potential transients for iron in $H_2SO_4 + K_2SO_4$ solutions under hydrodynamic conditions (flow rate 1 $dm^3 min^{-1}$) at the moment of immersion. The equilibrium potential lines of the thermodynamically possible reactions are superimposed. (Taken from Ref. 53.)

is a bend in the trace, evidently due to the contribution of the $H^+ + e^- \rightarrow$ $H \rightarrow \frac{1}{2}H_2$ process.

(iii) The shift of the potential in the negative direction continues up to a value which lies in the region of the equilibrium potential of the iron/iron oxide systems. A small wave and/or a short arrest seems to mark the reductive dissolution of oxides on the metal surface.

(iv) As one moves further along the curve, the potential shifts more slowly in the negative direction, passes through a minimum, and then changes direction, increasing monotonously toward the quasi-steady state of active iron corrosion. The decrease in the absolute value of the slope dE^{cor}/dt can be explained by the decrease of the anodic current associated with the process $Fe \rightarrow Fe^{2+} + 2e^-$ owing to the shift of the electrode potential to more negative values and to the enhanced contribution of the cathodic current produced by the reaction $H^+ + e^- \rightarrow H \rightarrow \frac{1}{2}H_2$.

(v) At the minimum, we have $dE^{cor}/dt = O$; that is, the electron source and sink intensities balance each other. Beyond the minimum, the slope dE^{cor}/dt is positive suggesting the predominance of the cathodic hydrogen current, probably because of a decrease in the number of active sites for metal dissolution via the blocking effect of adsorbed hydrogen intermediates or of other inhibitory species.

(vi) At longer times, the pH near the electrode surface slowly increases (owing to H^+ depletion and Fe^{2+} accumulation[54]), so that the electrode potential moves again in the negative direction. When the circulation was halted after 30 min, a rapid shift of about -50 mV occurred, demonstrating an influence of transport phenomena.

The following relationships were found for the Fe/SO_4^{2-} system[53]:

$$\hat{E}_{min}^{cor} = -318 - 43(\pm 12)pH \ (mV/NHE)$$

$$t_{min} = -9.7 + 4(\pm 1)pH \ (s) \quad (valid \ for \ pH \ 2.3-3.8)$$

$$\hat{E}_{30}^{cor} = -267 - 40(\pm 15)pH \ (mV/NHE)$$

Experiments performed with complexing buffer solutions containing $0.1M$ HCl + $0.1M$ disodium citrate (Na_2HCit) in the pH range 2.2–4.6 under the same conditions led to

$$\hat{E}_{min}^{cor} = -271 - 41(\pm 10)pH \ (mV/NHE)$$

$$t_{min} = 0.43 + 0.17(\pm 14)pH \ (s)$$

$$\hat{E}_{30}^{cor} = -216 - 41(\pm 2)\text{pH (mV/NHE)}$$

Here, the subscripts "min" and "30" denote the value at the minimum and after 30 min, respectively, the symbol "^" means the least-squares line, and the value in parentheses is the estimated standard deviation of the slope calculated according to ASTM G16/74. A comparison of the two systems revealed that the E^{cor}–t transients lie in a region of potentials about 50–60 mV more negative and the time necessary to reach the minimum is longer and much more influenced by pH in sulfate than in chloride–citrate solutions. In our opinion, this behavior is explained by the buffering capacity of the $HCl–Na_2HCit$ solution and the high adsorbability of the Cl^- and Cit^{3-} anions, which have the ability to inhibit the iron dissolution and/or the oxide formation.

Figure 7 shows the time evolution of the corrosion potential of pure iron (99.9%) in concentrated sulfuric acid (pH 0.0–0.7), under static conditions, reported by Mansfeld[55] and by Drazic and Zecevic.[56] In both cases, experimental points for the first two minutes after immersion are lacking; thereafter, the open-circuit potential shifts in the positive direction, reaches a maximum value, and then decreases monotonously toward a value 15–20 mV more negative. Concomitantly (see Fig. 8), the rate of iron corrosion increases during the first hours and, after an oscillation, decreases, reaching a stationary value.[57] An increase in the corrosion rate

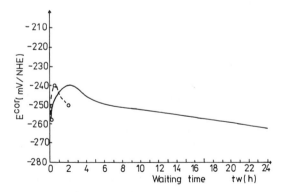

Figure 7. The time evolution of the open-circuit potential of pure iron (99.9%) in sulfuric acid solution, pH 0.03–0.7, under static conditions. [Solid (pH 0.4–0.7) and dashed (pH 0.03) lines taken from Refs. 55 and Ref. 56, respectively.]

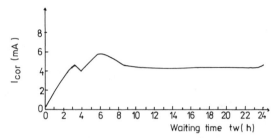

Figure 8. The time dependence of the corrosion current for iron (99.9%) in $0.5M$ H_2SO_4 solution, under static conditions. (Taken from Ref. 55.)

of iron in sulfuric acid, pH 0.4–0.7, with time, over the course of an hour, was also reported by von Frenzel[58] (see Fig. 9); like Akiyama et al.,[52] von Frenzel found a simultaneous increase in the double-layer capacity. It was suggested that this behavior may be explained by an increase in surface roughness.

According to Drazic,[38] waiting at open circuit produces a surface layer of hydroxo particles including SO_4^{2-} whose structure changes with time and which increasingly inhibits the main dissolution reaction. In fact, a certain adsorption of sulfate ions on an iron surface from dilute deaerated aqueous Na_2SO_4 solutions at open circuit had been previously observed by Hackerman and Stephens.[59]

Some time later, Drazic et al.[60,61] found that the corrosion potential of Armco iron in $0.5M$ H_2SO_4 in the presence of $0.1M$ NaCl became more positive with time over a period of during 24 hours. At the same time, the corrosion rate (determined from linear current–potential cathodic plots after a 5-min anodic pulse of 100 mA cm^{-2}) decreased. For this system, the authors concluded that sulfate and halide anions adsorbed at a slow rate and inhibited the metal dissolution.

In order to derive the kinetic equations describing the open-circuit potential transients plotted in Fig. 6, Plonski[62,63] proposed an equivalent electric circuit, depicted in Fig. 10. Here, G_A is an anodic current generator characterized by R_A, the equivalent internal resistance, and I_A, the sum of all faradaic currents generated by the electron source reactions, and G_C is a cathodic current generator characterized by R_C, its equivalent internal resistance, and I_C, the overall cathodic current generated by faradaic electron sink reactions. I_A and I_C are functions of time, potential, and the

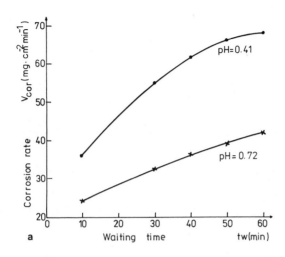

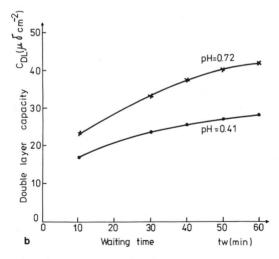

Figure 9. The time dependence of the corrosion rate (a) and of the double-layer capacity (b) of iron in sulfuric acid solutions. (Taken from Ref. 58.)

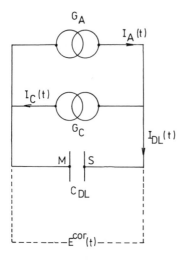

Figure 10. Electric circuit equivalent to the metal/solution (M/S) interface under open-circuit conditions. G_A, Anodic current generator (electron source); I_A, sum of anodic currents generated by the local electron source reactions; G_C, cathodic current generator (electron sink); I_C, sum of cathodic currents generated by the local electron sink reactions; $I_{DL} = I_C - I_A$ = the local current charging the equivalent double-layer capacitor, C_{DL}. (From Refs. 62 and 63.)

concentrations of reactants and products and of the free active surface for the respective reaction, obeying the laws of electrochemical kinetics. C_{DL} is the double-layer capacitor formed by the metal plate, M, and the equivalent solution plate, S. According to this equivalent electric circuit,

$$I_{DL} = -I_F = -I_A + I_C \tag{5}$$

where the charging current intensity, I_{DL}, is supplied by the net current resulting from the local faradaic reactions. (It is known that the circuit is closed through a current generator only via its internal resistance.) Therefore, the time dependence of the open-circuit potential, $E^{cor}(t)$, measured against a reference electrode, E_{ref}, is given by

$$dE^{cor}/dt = d(E_{Fe} - E_{ref})/dt = -I_F/C_{DL} = -(I_A - I_C)/C_{DL} \tag{6}$$

where E_{Fe} is the potential drop across the metal–solution interface. The minus sign in front of I_F in Eq. (5) means that the metal plate accumulates negative charges when, in open circuit, local electron source reactions prevail. At the steady-state corrosion potential, E_s^{cor}, the two local current generators balance each other.

On the basis of the equivalent circuit adopted, the slope dE^{cor}/dt, controlled by the iron dissolution, $I_{Fe/Fe^{2+}}$, oxide formation, $I_{Fe/FeOx}$, and hydrogen evolution, I_{H^+/H_2}, is given by the expression

$$-dE^{cor}/dt = (i_{Fe} \cdot \Theta_{Fe} + i_{FeOx} \cdot \Theta_{Ox} - i_H \cdot \Theta_H)A_{Fe}/C_{DL} \qquad (7)$$

where A_{Fe} is the metal surface area, and Θ_{Fe}, Θ_{Ox}, and Θ_H represent the free actual surface fraction on which the above reactions take place.

The initial slope, $(dE^{cor}/dt)_{t\to 0}$ (experimentally distorted by the non-instantaneous contact between all parts of the metal surface and the solution), is mainly determined by the ferrous ion formation, $Fe \to Fe^{2+} + 2e^-$, occurring at maximum bare surface area for this reaction. As a result, the potential shifts rapidly in the negative direction concomitantly with a decrease in i_{Fe}, and therefore in the absolute value of dE^{cor}/dt, as can be seen from the decay of the E^{cor}–t transient depicted in Fig. 6.

After a small wave, denoted "wv," presumed to be caused by the formation and dissolution of a superficial iron oxide, the hydrogen evolution reaction interferes, entailing a large decrease in the absolute value of dE^{cor}/dt according to

$$-C_{DL} \, dE^{cor}/dt =$$

$$FA_{Fe}[2k_{Fe} \cdot \Theta_{Fe}^{wv} \cdot \exp(2\alpha \, E_{wv}) - k_H \cdot \Theta_H^{wv} \cdot \exp(-\alpha \, E_{wv}) \cdot a_{H^+}] \quad (8)$$

where k is the rate constant and $\alpha = F/2RT$. After the wave, the blocking of the active sites (probably kinks) for iron dissolution by adsorbed hydrogen intermediates[64–67] and the decrease of a_{H^+} near the electrode surface must be taken into consideration. At the minimum of the transient, $dE^{cor}/dpH = 0$, and, taking into account that the metal surface is practically free of hydrogen evolution adsorbed intermediates or adsorbed anions, Eq. (8) leads to

$$E_{min}^{cor} = \ln[k_H \, (a_{H^+})^{E_{min}}/2k_{Fe}(\Theta_{Fe})^{E_{min}}]/3\alpha \qquad (9)$$

Therefore, the changes in E^{cor} with time can be explained by the time-dependent opposing effects of the pH near the surface and the degree of coverage of the surface by the active sites for iron dissolution, which in turn decreases due to the adsorption of inhibitory species (hydrogen, anions) but increases with the increase in the number of kinks through the parallel dissolution of atoms in other positions and, at longer exposure times, with surface roughness.

2. The Open-Circuit Steady State

According to ASTM G15-71, the corrosion potential is the potential of a corroding surface in an electrolyte relative to a reference electrode measured under open-circuit conditions. It is customary to denote by the term steady-state corrosion potential the value reached asymptotically for which $|\Delta E^{cor}/\Delta t| < 1$ mV h^{-1}. It will be seen below that a "stationary" corrosion potential according to the above definition does not imply a steady-state interface because substantial changes in some metal surface or solution characteristics do not influence significantly the corrosion potential value but have a major impact on polarization response.

The stationary corrosion potential has been measured directly, under open-circuit conditions, or determined from the intersection of the initial or steady-state Tafel lines. Despite the fact that hundreds of articles concerning iron corrosion in acid media have been published, only a small number include the open-circuit characteristics, probably due to the dynamic nature of the interface and to the difficulties in reproducing the results.

One of the first facts to emerge from numerous investigations was that the pseudo-stationary corrosion potential in acidic solutions shifts, in a fairly reproducible way, in the negative direction with an increase in the pH of the bulk electrolyte. Unfortunately, few authors have reported the exposure time, that is, the time elapsed between the moment of immersion and the measurement of the "steady-state" potential, but the spreadness of the data is evident from the following example. Based on measurements apparently made under the same conditions, the slope $\Delta E_s^{cor}/\Delta$ pH for iron in acidic sulfate solutions in the pH range 0–4 was reported as –40 mV by Bala,[68,69] –45 ± 5 mV by Heusler,[70] and –62 mV by Voigt,[71] thus covering all possible mechanisms. A number of experimental E_s^{cor}/pH points given in the literature for iron in sulfuric and perchloric acid solutions are plotted in Fig. 11. Figure 12 shows the histogram obtaining by plotting relating the numbers of papers reporting values of the slope $\Delta E_s^{cor}/\Delta$pH within certain intervals of values. Obviously, the data do not fit either a normal distribution or a double-maxima one, despite the fact that two midpoints are evident, one in the interval (−45, −50) and the other in the interval (−60, −65) mV. Therefore, we are not faced with random errors that can be handled by statistical analysis alone.

In addition to the rather systematic study of the pH dependence of corrosion potential and corrosion rate, some other influences have also been investigated. Thus, Drazic et al.[60,61] observed that after a strong anodic pulse, the corrosion potential became more negative than after a long waiting time and the corrosion rate increased. Haruyama and co-workers[40] found that the metallic electrode under plastic deformation

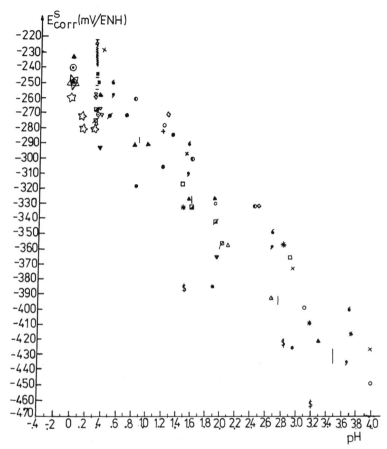

Figure 11. The pseudo-steady-state corrosion potential versus pH for iron in SO_4^{2-} and ClO_4^- solutions, as reported by different authors. (See Refs. 37, 43, 45, 51–53, 68–71, 80, 92, 95, 101, 109, 137, 167, 180, 193, 206, 218, 219, 225, and 232.)

exhibited a negative shift of immersion potential and an increase in corrosion current as well as in anodic flow when the electrode was constrained at the reversible potential.[72–74] It was suggested that the emergence of fresh surface was followed by adsorption of inhibitory species[75] or intermediates.[76]

Despite the fact that the small amount of experimental data available does not afford a firm basis for correlations, it appears that the purity of

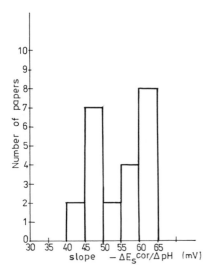

Figure 12. Histogram showing the distribution of values reported in the literature for the dependence of the steady-state corrosion potential on pH, $\Delta E^{cor}/\Delta pH$, for iron in sulfuric and perchloric acid solutions. (The literature values are taken from Refs. 37, 43, 45, 51–53, 68–71, 80, 92, 95, 101, 109, 137, 167, 180, 193, 206, 218, 219, 225, and 232.)

the iron and the hydrodynamics, which influence the value of the corrosion potential, have no significant effect on the slope $\Delta E_s^{cor}/\Delta pH$. Nevertheless, for carbon-steel in Sörensen solutions, this slope increased from –42 mV after 5 s to –55 mV after 30 min from the moment of immersion.[77]

If one defines the corrosion potential at the pseudo-steady state as the mixed potential established when the iron dissolution current approximately balances the hydrogen evolution one, this last reaction occurring randomly at the free surface, the following simple formula can be written:

$$E_s^{cor} = \{\ln (k_H/2k_{Fe}) + \ln[a_{H^+}(t)/\Theta_k(t)]\} \div 3\alpha \tag{10}$$

where $\Theta_k(t)$ and $a_{H^+}(t)$ are functions describing the time dependence of the coverage of the surface by nonblocked sites, that is, active sites for iron dissolution, and of hydronium ion activity near the electrode, respectively. Equation (10) leads to

$$[(E_{s2}^{cor} - E_{s1}^{cor})/(pH_1 - pH_2)]$$

$$= 2.3 \{1 + \log[(\Theta_{k2}/\Theta_{k1})^{(pH_2 - pH_1)}]\}/3\alpha$$

$$= -39 \text{ mV} + \log [(\Theta_{k2}/\Theta_{k1})^{2.3(pH_2 - pH_1)}/3\alpha] \tag{11}$$

where pH_1 and pH_2 are the pH values near the surface after the waiting times t_1 and t_2, respectively, for $t_2 > t_1$. On the basis of this relationship, one can explain the increase in the experimental value of the slope $|\Delta E_s^{cor}/\Delta pH|$ with an increase in the immersion time, as reported in Ref. 77, in terms of the following argument. Because the experimental value of the slope $|\Delta E_s^{cor}/\Delta pH|$ is greater than -39 mV, this means that when $pH_2 > pH_1$, then $\Theta_{k2} > \Theta_{k1}$; that is, the degree of coverage with free kinks is greater at higher pH. If the pH in the bulk is different from the pH near the electrode surface, the relationship between the measured and the corrected slope is

$$[(E_2^{cor} - E_1^{cor})/(pH_1 - pH_2)_{bulk}] =$$

$$[(E_2^{cor} - E_1^{cor}) \exp(pH_1 - pH_2)_{x=0}]/[(pH_1 - pH_2)_{x=0}(pH_1 - pH_2)_{bulk}] \quad (12)$$

It can be seen from this relation that if the difference $pH_2 - pH_1$ is greater near the surface than in the bulk of electrolyte and if this difference increases with time (because the higher the pH, the slower the H^+ diffusion), the slope $\Delta E^{cor}/\Delta pH$ must increase with an increase in the waiting time.

The corrosion rate data published in the literature are generally values averaged over a period of time. They are based on measurements of the loss of material (the weight loss method), of the ferrous ions present in the solution (photocolorimetric method), and of the pH change in the solution and on volumetric methods (collection of evolved hydrogen). Electrochemical procedures have frequently been used. Thus, corrosion current has been determined at the intersection of the initial or steady-state cathodic and anodic Tafel lines, or any of these lines with $E = E^{cor}$. Another way in which information has been acquired is through Stern–Geary linear polarization and impedance analysis.

Reported results on the dependence of corrosion rate on pH in acid sulfate or perchlorate solutions do not coincide. Thus, the reaction order in OH^- has been found to lie between 0[78–80] and -1.[81] Values close to -0.5 have also been reported.[22] The variability in the data does not follow any statistical distribution, leading to the conclusion that unknown, noncontrolled conditions mask the pH effect.

The influence of internal stresses on the corrosion current is also not clear. Thus, Foroulis[82,83] observed that the corrosion rate of zone-refined iron was the same whether it had been cold-worked or annealed.

III. ANODIC POLARIZATION

1. The Transitory Regime of Anodic Polarization

Transient measurements have often been used in electrochemical studies as a means of obtaining deeper insight into reaction mechanisms and estimating quantitatively kinetic parameters. Transient measurements have been obtained by applying a rectangular pulse or sinusoidal polarization to the electrode, which has previously reached a certain open-circuit or polarization steady state. The response of the system to these perturbations, recorded oscillographically, may then be analyzed and interpreted.

(i) The Transitory Regime after Rectangular Pulse Perturbations

Galvanostatic transients cannot be avoided due to the necessity to make corrections for the *ohmic drop*, which is the first kind of information obtained from the pulse polarization. The second kind of information obtained concerns the *double-layer capacity*. An unexpected feature that is observed during anodic dissolution of iron-group metals in the active range is *superpolarization peaks* that appear on the pulse galvanostatic measurement. Under the same conditions, potentiostatic transients exhibit a *minimum value for the initial current* situated immediately after the major charging of the double-layer capacity. (Correspondingly, an inductive behavior of the AC impedance was observed—i.e., inductive loops at low frequencies.) Additional kinetic information provides the *transition time*, that is, the time necessary for the relaxation processes between nonsteady state and steady state (corresponding to the characteristic frequency on the capacitive and inductive loops).

Finally, upon interruption of the pulse galvanostatic polarization, the decay curve exhibits a *minimum*, that is, *underpolarization* with the potential more negative than the open-circuit potential, E_{oc}. Hurlen[84] found that the difference between E_{min} and E_{oc} is approximately the same as between E_{peak} and E_s, but Barnartt[206] noticed that the total potential change on opening the circuit, that is, $|E_{min} - E_{oc}|$, is smaller than $|E_{max} - E_{oc}|$.

(a) The double-layer capacitance

Two kinds of measurements have been mainly used to calculate the double-layer capacitance: the slope at the beginning of the rising part of the galvanostatic transient and the characteristic frequency determined at the top of the first capacitive loop, which is in the kilohertz range. Usually, the double-layer charging–discharging is a very rapid process. Nevertheless, capacitances of up to several hundred microfarads per square centimeter were observed in acid solutions.[22,85,86] Generally, it is considered that such large values cannot be accounted for by simple double-layer capacitances. Moreover, specific measurements revealed that the capacitance resulting from anodically polarized iron depends strongly upon electrode potential and pH value.[87–93] On the other hand, in the region of cathodic polarization the capacitance does not change much with electrode potential and pH.[85]

According to Lorenz and Heusler,[37] the capacitance can be interpreted as the sum of an almost potential-independent double-layer capacitance of 20–30 μF cm^{-2} and an adsorption capacitance C_a, increasing with anodic potential and pH according to $(\delta \ln C_a/\delta E)_{pH} = F/RT$ and $(\delta \ln C_a/\delta pH)_E = 1$, respectively. Nevertheless, this hypothesis is not the only possible one, and some rapid faradaic pseudocapacitance process may also be coupled with the double-layer relaxation.[51,94,118] The main experimental evidence in support of such a coupling is the depressed form of the capacitive loop, which can be interpreted as a consequence of overlapping of time constants. However, one must keep in mind that all loops obtained with polycrystalline metals are slightly depressed semicircles.[96] Accordingly, the depressed form has sometimes been interpreted as resulting from the distribution of the processes occurring on a polycrystalline surface.[94,97]

In fact, the expression relating the double-layer capacity, C_{DL}, to the galvanostatic pulse current, i, that is,

$$C_{DL} = [i - \lim i_F(t \to 0)]/[\lim dE/dt(t \to 0)] \qquad (13)$$

can account by itself for the increase in C_{DL} with anodic current, i, and pH. Apparently, large values of C_{DL} are due to the assumption that, initially, the faradaic current, i_F, can be practically neglected, that is, $\lim i_F(t \to 0) = 0$. In fact, the higher the pH and E, the higher is $\lim i_F(t \to 0)$, so that the numerator in Eq. (13) is in fact much lower than i and leads to reasonable capacity values.

Moreover, the fact that C_{DL} determined from cathodic transients does not change with pH and E suggests that near the corrosion potential the hydrogen evolution mainly occurs on the bare planar atoms so that the active surface for this reaction is practically constant.

On the anodic side, one can notice the similarity between $(\delta \ln C_a/\delta E)_{pH} = F/RT$ and $(\delta \ln C_a/\delta pH)_E = 1$ and the corresponding dependences of the initial current, i_i, namely, $(\delta \ln i_i/\delta E)_{pH} = F/RT$ and $(\delta \ln i_i/\delta pH)_E = 1$. Consequently, it may be the influence of lim $i_i(t \to 0)$ that affects Eq. (13).

(b) The inductive behavior of anodic transients

The first investigators to observe superpolarization peaks in response to different large-signal pulse polarization of iron dissolution in the active range and report some of their characteristics were Roiter *et al.*[98]

Two characteristic times, with kinetic significance, have been observed on the anodic galvanostatic transients: the *peak time*, τ_{peak}, that is, the time elapsed from the moment of switching on the rectangular current pulse until the peak value is reached; and the *transition time*, τ_{ss}, that is, the time necessary for the relaxation processes between the nonsteady state and the steady state. Because the measurement of the transition time is uncertain, due to the asymptotic character of the decay curve, it is preferable to use the time constant, a term borrowed from electrodynamics, which refers to the time necessary to reach 37% of the steady-state value. The following main features characterize the anodic polarization transients:

(i) In the studies by Roiter *et al.*,[98] the peaks appeared larger, the greater the current density.

(ii) The transient or initial polarization curve, η versus log i_i or η_{peak} versus log i, yielded a linear region with a slope $\Delta\eta/\Delta \log i_i$ or $\Delta\eta_{peak}/\Delta \log i$ of 60 ± 5 (mV dec^{-1}) at $T = 298$ K, independent of the iron substrate used, the composition of the solution, the hydrodynamics, or the starting potential, as reported by Heusler and co-workers[37,70,86,99,100,103] and Lorenz and co-workers.[46,47,101, 102,104]

(iii) The concentration of ferrous ions in the electrolyte has no influence on the anodic transient.

(iv) The surface density of kinks is lower than its corresponding value under steady-state polarization.[34]

(v) After the current pulse is switched off, the decay transient is entirely situated at potentials more negative than the open-circuit potential before anodic polarization (E_{oc}). The curve displays a minimum[70] and then slowly increases, but the potential remains 1–2 mV more negative than E_{oc} even after long waiting times, suggesting that a part of the interface was not totally reversible. The difference between the minimum and E_{oc} is approximately the same as between E_{peak} and E_{ss}.

(vi) The hydrodynamics has little effect on the rising transient (a variation of about 2% was reported by Bignold and Fleischmann[54] for an increase in the rotation velocity from 500 to 7000 rpm), but it influences the decay curve. This suggests that the decay transient is partially under the control of some transport phenomenon in solution; partially, it is the consequence of a slow equilibration of a reaction leading to coverage of the surface by some inhibitory compound (i.e., H_{ads} or an iron dissolution product) because in order to have $E_{decay}(t) < E_{oc}$, the electron source local current, $i_{Fe/Fe^{2+}}$ must be greater than the electron sink local current, i_{H^+/H_2}.

(vii) The shape of the transient depends on the polarization program. Roiter *et al.*[108] polarized the iron in acidic $FeSO_4$ solutions by successive galvanostatic pulses. The system response as a function of the interruption time can be seen in Fig. 13, from which it can be noticed that the shorter

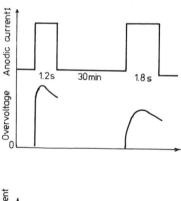

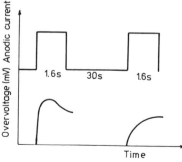

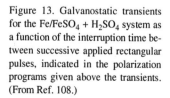

Figure 13. Galvanostatic transients for the Fe/FeSO$_4$ + H$_2$SO$_4$ system as a function of the interruption time between successive applied rectangular pulses, indicated in the polarization programs given above the transients. (From Ref. 108.)

the interruption time, after an anodic pulse as long as 1.2 s, the smaller the second peak, and this peak completely disappeared if the interval between the pulses was as short as 30 s. A similar polarization program was used by Plonski[64] for carbon-steel SA 106 gr B in complexing buffer solutions of $0.1M$ HCl + $0.1M$ disodium citrate, pH 4.2, under stagnant conditions. The response of the system after each of three consecutive 0.1-s rectangular pulses of 4.2×10^{-4} A cm^{-2} with long interruption times of 28–50 min is plotted in Fig. 14. In these experiments the shape of the galvanostatic transients did not change, but the transients moved toward negative values along with the shift of the open-circuit potential. The change in the interface during the polarization was probably small enough and the interruption time long enough that the system returns to its initial state in this particular solution.

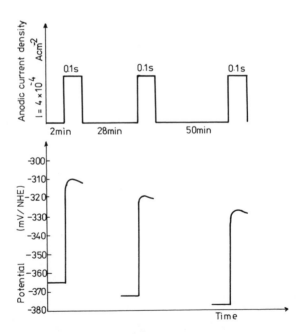

Figure 14. Galvanostatic transients successively recorded during the evolution of the open-circuit potential of carbon-steel (Fe 98.4%) in $0.1M$ HCl + $0.1M$ disodium citrate solutions. The polarization program is given in the upper part of the figure. (From Ref. 64.)

A more detailed study of the phenomenon of superpolarization during the anodic dissolution of iron in acidic $1M$ SO_4^{2-} solutions was carried out by Drazic and Zecevic.[56] The polarization program consisted in two successive anodic galvanostatic pulses, the first one, called the "activation" pulse, of high intensity (300–150 mA cm^{-2}) and 1 min in duration, followed by a second one ranging between 0.3 and 200 mA cm^{-2} and >20 ms in duration. The interruption time between two successive pulses varied between 0.5 s and 15 h. (The waiting time between the immersion and the switching off of the activation pulse as well as the interruption time corresponding to the steady-state polarization measurements was not specified.) All effects examined, namely, the effects of pulse density, pH, interruption time under open circuit between the two pulses (denoted t_{inter}), hydrodynamics, purity, and cold working of iron, referred to the transient response to the second pulse. The following results from this study, confirming old findings or revealing new aspects, are noteworthy:

(i) Their results confirmed that the time needed to achieve a steady-state polarization is lower at higher current density (e.g., several tens of milliseconds at 150 mA cm^{-2} and several hundreds of seconds at 0.3 mA cm^{-2}, for iron in $1M$ H_2SO_4).

(ii) Also, at the higher pH, the peaks on the transients were less sharp, with smaller differences between the maximum value and the steady-state one.

(iii) The new findings are related to the effect of the interruption time between the activation pulse and the second pulse on the response to the latter:

1. The height of the superpolarization peak increases with the interruption time (see Figs. 15 and 16). For example, in $1M$ H_2SO_4 for $i = 300$ mA cm^{-2}, the maximum ratio between the peak overvoltage after an interruption time of 15 h and that after one of 4 min was as high as 2.2. However, it can be seen from Fig. 5 in Ref. 56 and also from Fig. 15 here that there is a region of stability in which the superpolarization peak remains constant, between 1 and 2 h for $i = 15$ mA cm^{-2} and between 20 min and 1 h for $i = 30$ mA cm^{-2} and $1M$ H_2SO_4. (Generally, the corrosion potential is also stable between 20 and 60 min under static conditions and in the range between 10 min and 1 h under hydrodynamic conditions; these regions are usually chosen for polarization programs starting from the corrosion potential.)

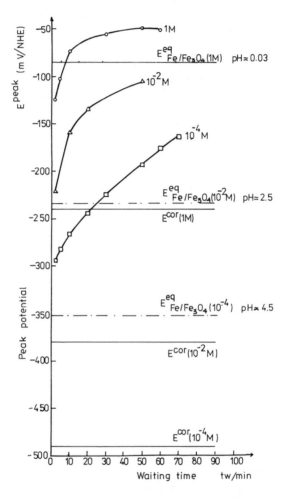

Figure 15. Peak overpotentials as a function of the waiting time for iron in acid sulfate solutions of different pHs (as indicated on the curves). The experimental points are taken from Ref. 56. Equilibrium potential lines for the thermodynamically possible reactions are superimposed.

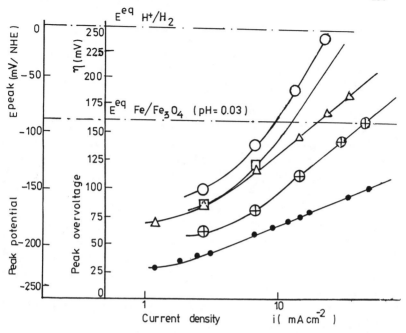

Figure 16. Peak overpotentials as a function of the pulse current density for different waiting times. The steady-state Tafel plots are also shown for comparison. The experimental points are for the following waiting times and initial Tafel slopes, for $60 < \eta < 120$ mV: O, 4 min, 65 mV dec^{-1}; Δ, 9 min, 68 mV dec^{-1}; $\square$, 2 h, 95 mV dec^{-1}; $\oplus$, 15 h, 95 mV dec^{-1}; $\bullet$, steady state, 44 mV dec^{-1}. (From Ref. 56, by courtesy of the authors.)

Nevertheless, it should be recalled that Roiter *et al.*[98] found that if the pulse is applied after a previous one (of the same intensity in their experiment), the superpolarization is lower but increases with increasing interruption time (see Fig. 13).

2. Tafel-like lines in the semilogarithmic plot were obtained with slopes increasing from 40 to 140 mV dec^{-1} for an increase in the length of the interruption time from 0.5 s to 15 h, the slopes being about 60 mV dec^{-1} if the periods between pulses were 1 or 2 min (see Fig. 16).

This remark does not extend to the initial polarization curve corresponding to the steady-state part between 35 and 85 mV, where the net curvature that can be observed exhibits a much lower slope, $d\eta^{peak}/dt$. The Tafel region near the corrosion potential is particularly interesting because it can be more probably related to processes occurring under open circuit.

The following additional comments can be made in regard to these experiments:

(i) Because the galvanostatic pulse under study is applied after different interruption times ranging between 0.5 s and 15 h after the previous activation pulse of $t = 1$ min and $i = 150$ or 300 mA cm^{-2}, the electrode system response is a mixed one composed of the decay of the first pulse and the rising part of the second.

(ii) It was established by Bech-Nielsen[23] for iron (1500 rpm) in $H_2SO_4 + Na_2SO_4$ solution, pH 3.45, that the transition range starts at $i \approx$ 4 mA cm^{-2}, and the prepassive range near $i \approx 30$ mA cm^{-2} (see Fig. 3C). This solution is similar to the solution ($10^{-4}M$ H_2SO_4, $1M$ SO_4^{2-}) considered in Fig. 15. This means that the activation pulse of 150 mA cm^{-2} is situated in the prepassive range, and the second pulse toward the end of the transition range. Therefore, the system response reflects the influence of the decay of the prepassivated surface on the rising part specific to the transition range reactions.

(iii) Initial Tafel slope values of 60–70 mV dec^{-1} have been reported many times (see, e.g., Refs. 119, 171, 194, 202, 222, 224, 251, 261, 362, and 473 in Ref. 38). It is difficult to believe that all the experimentalists waited systematically only 2 min after immersion and never longer in order to reach a more stable region of the corrosion potential. Also, 15 h of contact of the metal with a solution of pH ≈ 0.03 would lead to substantial surface damage and screening of the surface by the hydrogen evolved under open-circuit conditions.

(iv) The hydrodynamic conditions did not influence the height of the transient. Based on this observation, Drazic and Zecevic[56] excluded transport phenomena as a possible reason for the appearance of the superpolarization peak. Nevertheless, one must accept that the time elapsed to reach the peak is not long enough to change significantly the chemistry on the solution side, to say nothing of the fact that a part of the current is consumed by the double-layer charging. The contribution of transport phenomena can be evident at longer times, on the decay part of the transient, and undoubtedly this was found.[51]

(v) Cold working was not found to influence the transients. Based on this behavior, Drazic and Zecevic[56] made the general conclusion that dislocation density and phenomena connected with the surface morphology are not likely to be the cause of superpolarization.

Drazic and Zecevic[56] interpreted their experimental results by advancing the idea that the superpolarization peak was due to the partial blockage of the surface with adsorbed hydroxo particles (at high coverages), which inhibits the main reaction on the surface. Consequently, the peak height and the decay time depend on the coverage and stability (i.e., cross-linking) of this surface layer. It was also postulated that the time effect can be explained in terms of "aging" (i.e., cross-linking) of the aforementioned hydroxo particles, particularly expected in SO_4^{2-} -ion containing solutions, since it is well known that sulfate ions form bridges between the hydroxo complexes of transition metals.

Nevertheless, for the reasons analyzed above, the extrapolation of these interesting results and interpretations to processes occurring near the corrosion potential must be made with caution.

Plonski et al.[109] performed a preliminary investigation concerning the influence of the polarization program on the shape of the galvanostatic anodic transients for Armco iron in stagnant, oxygen-free, acidic sulfate solutions, in the pH range 1.2–4.3. The polarization consisted of three consecutive rectangular current pulses of equal intensity (4.3×10^{-5} or 2.5×10^{-4} A cm^{-2}) applied according to the following program: I. waiting at open circuit = 1 h; II. first pulse length = 2 min; III. interruption time = 5 min; IV. second pulse length = 20–800 s; V. interruption time = 5 min; VI. third pulse length = 5 min. Three typical galvanostatic transients obtained from this program are juxtaposed in Fig. 17. On the basis of these transients, the following qualitative observations can be made:

- The $\eta-t$ curves display two maxima, the first in the range of 100–300 ms, and the second after 1–40 s.
- The peak polarization curves do not rigorously obey a linear relationship. Very approximately, it was found that $\Delta\eta^{peak1}/\Delta \log i = 66$–76 mV dec^{-1}, and $\Delta\eta^{peak2}/\Delta \log i = 72$–79 mV dec^{-1}.
- After 300-s polarization, the pseudo-steady-state Tafel line slope was $\Delta\eta_{300}/\Delta \log i = 38$–57 mV dec^{-1}, where "pseudo" underlines the uncertainty attached to the identification of any steady state.
- The two peaks are separated by a minimum value of the overvoltage which is only 3–10 mV lower than the first peak value, so that it can be easily overlooked experimentally.

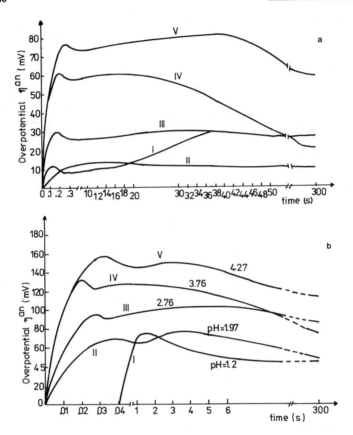

Figure 17. Galvanostatic transients for Armco iron in $H_2SO_4 + K_2SO_4$ (1M SO_4^{2-}) solutions, under stagnant conditions, as a function of pH (indicated on the curves) for two current intensities. (a) $I = 4.3 \times 10^{-5}$ A; (b) $I = 2.5 \times 10^{-4}$ A. (From Ref. 109.)

• Both peak times are lower, the higher the current density. At lower current density, the influence of pH on τ_{peak} increases with increase in pH; at high current density, the influence of pH decreases.

• Both peak overvoltages grow with increase in pH.

Nevertheless, a more refined and sophisticated polarization program, with magnetic recording of the transient during the same long pulse and a large number of replicates, must be conceived in order to establish accurately the origin and characteristics of the two peaks.

The pH dependence of the anodic dissolution rate at the minimum of the i–t potentiostatic transient was also determined. As a rule, the initial reaction order in OH^-, $v_{OH^-}^i$, is lower than the corresponding value at the steady state, $v_{OH^-}^s$. Generally, values of $v_{OH^-}^i$ between 0.8[70,86,110,111] and 1[45–47,54,71,101,112] have been frequently reported; in an isolated case, $v_{OH^-}^i = 0.45$ was found.[46]

The initial Tafel lines for iron anodic dissolution and cathodic deposition have been measured starting from various steady states.[86,70,100] They all intersected at the equilibrium potential of the iron electrode,[100] leading to a set of initial current densities. In view of this behavior, Lorenz and Heusler[37] concluded that "initial reaction orders," with respect to, for example, hydroxyl ions, derived from measurements of initial polarization curves have no meaning. Nevertheless, other opinions can be offered. If, for example, one considers that the equilibrium reaction $Fe^{2+} + 2e^- \Leftrightarrow Fe$ occurs at specific sites "k," the intersection at $E = E_{Fe/Fe^{2+}}^{eq}$ of the initial Tafel lines (anodic or cathodic) leads to a set of "intersection" currents, i_{inter}, given by the relation

$$i_{inter} = 2Fk_{Fe}(\Theta_k^i)_{E,pH} \exp(2\alpha E_{Fe/Fe^{2+}}^{eq})$$
$$= -2Fk_{Fe^{2+}}(\Theta_k^i)_{E,pH} \cdot \exp(-2\alpha E_{Fe/Fe^{2+}}) \cdot a_{Fe^{2+}} \qquad (14)$$

where k_{Fe} and $k_{Fe^{2+}}$ are the electrochemical rate constants with respect to a certain reference potential for iron dissolution and deposition, respectively, and $(\Theta_k^i)_{E,pH}$ is the degree of coverage by specific sites, which depends on the initial steady state. Therefore, the intersection current is different from the exchange current, i_0. In order to obtain the exchange current from the intersection of two initial Tafel lines, one must start the polarization from the equilibrium potential, for which the coverage has its equilibrium value, $(\Theta_k^i)_{E_{pH}^{eq}}$, so that

$$i_{inter} = i_0 = 2Fk_{Fe}(\Theta_k^i)_{E_{pH}^{eq}} \exp(2\alpha E_{Fe/Fe^{2+}}^{eq})$$
$$= -2Fk_{Fe^{2+}}(\Theta_k^i)_{E_{pH}^{eq}} \cdot \exp(-2\alpha E_{Fe/Fe^{2+}}^{eq}) \cdot a_{Fe^{2+}} \qquad (15)$$

In view of this behavior, the initial reaction order with respect to, for example, hydroxyl ions, derived from measurements of this kind, leads to information about the pH dependence of the degree of coverage with free kinks at the starting point of polarization.

The corrosion potential is a special steady state at which the rates of anodic dissolution and cathodic reduction of an oxidized substance happen to be equal. Therefore, it is strongly influenced by the kinetics of iron dissolution and by the nature of this oxidized species and the kinetics of its reduction. Consequently, the initial reaction order with respect to hydroxyl ions is related to the kinetics of both reactions and therefore provides information about their mutual influence at the corrosion potential. This viewpoint is at variance with the opinion[34] that the determination of electrochemical reaction orders in hydroxyl ions from transient measurements yields physically insignificant parameters, because the results are determined by, for example, the pH-dependent initial state of the iron electrode.[37,46,101,113,114]

On the other hand, the measurement of the characteristic times is affected by major difficulties. First, one must establish the coordinates of a point on a curve that evolves more or less asymptotically. Second, various processes such as adsorption/desorption of species, changes in pH, alteration of surface structure and morphology, or/and increase in surface roughness, all characterized by different time constants, overlap. Finally, the interface never reaches a real steady state.

Some authors found transition times of the order of tenths of a second, independent of the starting potential, polarization, and the pH value, but dependent on the temperature.[37,54,70,86,115–122] According to Lorenz and Heusler,[34] longer times of many minutes were correlated with the time necessary to establish a steady-state morphology.

When active iron dissolution in H_2SO_4, pH O, occurred at a potential 60 mV cathodic to the open-circuit potential, at which hydrogen evolution prevailed, the electrode surface remained practically undamaged and retained its bright appearance after immersion for 24 h at this potential, and the steady-state current was reached only after a 20-h polarization. According to Keddam et al.,[123] this means that the interaction between iron and hydrogen, which is a very slow process, controls the electrode kinetics.

The same slow changes in the current was observed also at polarization very slightly anodic to the open-circuit potential. These observations were interpreted in terms of the blocking nature of strongly adsorbed hydrogen.[124] In a solution of pH 0, at a current density of >10 mA cm^{-2}, the stationary state was reached within 10 min.

(ii) Impedance Analysis

Similar and additional information on transients has also been obtained by analysis of the faradaic impedance. A sinusoidal voltage of sufficiently small amplitude (a few millivolts) is applied to the electrode, which is at a certain steady-state potential. As a result, a sinusoidal current, of the same frequency, will pass through the cell. At a given frequency, the electrode acts similarly to a combination of capacitors, resistors, and chokes. The resulting experimental electrode impedance is a function of starting potential, electrode surface, and the composition of the solution, especially pH and anions. The resulting current is analyzed by means of an analyzer of transfer functions. A comparison between $V + \Delta V$ and $I + \Delta I$ leads to the impedance, Z, of the electrode/electrolyte interface. The impedance is a complex quantity which depends on frequency, ω. Usually, it is plotted on Cartesian axes—the real part Re(Z) on the abscissa and the imaginary part Im(Z) on the ordinate, so that $Z(\omega) = \text{Re} + j\,\text{Im}$ (Nyquist plane complex representation). The real part Re is related to the ohmic passive components of the interface, that is, resistances; the imaginary part is related to the reactive components such as capacitances and inductances, so that one can write $Z = \text{Re} + j\,(L\omega - 1/C\omega)$. For historical reasons, the capacity loop is plotted in the first quadrant. Variation of the frequency over a large region (usually 1 mHz–100 kHz) leads to a separation of the elementary phenomena as a function of their time constants. For a rapid perturbation of an electrochemical system, the sluggish phenomena, such as diffusion and anion adsorption, have no time to react. So, at very high frequencies (hundreds of kilohertz), that is, at short times, very rapid phenomena such as charging of the double-layer capacitance in parallel with very rapid electron-transfer steps can be studied. In order to obtain information on the slowest processes, such as diffusion or changes in surface structure and morphology, very low frequencies (millihertz) must be used. In between, electron transfer and chemical reactions occurring at moderate speeds impose their characteristics on the impedance. In addition to this, because the time constants are generally potential-dependent, one explores different potential intervals in order to attain a better separation of the relaxation regions of various processes. The relationship between the time constant, τ, of a process and the characteristic frequency of the representative loop, f_v, is given by $\tau = 1/2\pi f_v$. In cases in which more than one inductive loop appeared, branching mechanisms including self-catalytic and noncatalytic processes have been proposed. Generally, when the characteristic frequency did not depend on the current, a catalytic reaction was considered to be responsible.

AC impedance measurements in the active range of the dissolution of polycrystalline iron showed, after the depressed capacitive semicircle at relatively high frequencies, different inductive loops at low frequencies in the Nyquist representation.[118–120,125,126] The number of inductive loops increased from one at electrode potentials close to the corrosion potential to three at the most positive electrode potential in the active range. The resonance frequencies of the inductive loops are relatively low, being in the range of hertz to millihertz.

Figure 18 shows the impedance diagrams obtained by Keddam *et al.*[123] for a Johnson Matthey iron disk electrode (1600 rpm) in $1M$ acid sulfate solutions, pH 0, at various polarization points: hydrogen evolution domain (K), open-circuit corrosion potential (A_o), and anodic dissolution (B_o–F_o). The steady-state conditions at which the impedance measurements were performed are marked by the corresponding letter on the polarization curve given in the same figure. Diagram K, measured at $E = E^{cor} - 60$ mV, shows only one capacitive loop. The authors' explanation was that the hydrogen which covered the entire surface hindered the metal

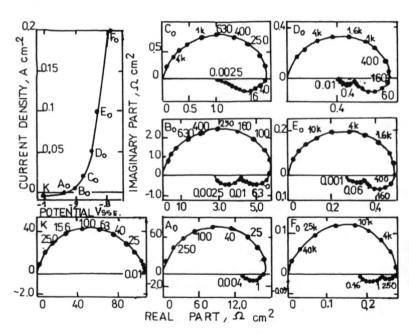

Figure 18. Frequency dependence of complex impedance measured with an iron disk electrode in H_2SO_4 (pH 0) solution at various steady-state polarization points (SSE, saturated sulfate electrode). (From Ref. 123, by courtesy of the authors.)

dissolution; the relaxation of the coverage could not be seen by the impedance.[127] Thus, nothing can be said about the metal dissolution in the region $E^{cor} > E > E^{eq}_{Fe/Fe^{2+}}$, because this process is screened to a considerable extent. In the anodic range $E^{cor} < E < E^{cor} + 140$ mV, the impedance diagrams display two inductive loops, their characteristics being potential-dependent. The impedance measurements on an active iron electrode showing inductive parts were interpreted by Keddam et al.[123,125] in terms of a branching mechanism comprising many dissolution paths as consecutive and catalytic one- and two-electron-transfer reactions. Each inductive loop was associated with the formation of a certain mono- or divalent iron intermediate, at a specific rate constant, being successively the rate-controlling step in the active, transition, and prepassive regions. There is no information on the chemical structure of these intermediates. Arguments against this reaction scheme can be found in Ref. 34.

Recently, impedance measurements over a very wide frequency range have been applied to the study of dissolution/passivation processes of iron in sulfate, sulfate + chloride, and chloride solutions, within the pH range 0–5, by Barcia and Mattos.[94,129] The authors found that the number of inductive loops (one to three) and the characteristic frequencies depended on the pH, DC current flux, and chloride concentration. The effect of chloride addition was especially detected at low pH values (0–2) and mainly consisted in a decrease in the number of inductive loops (from three to one), accompanied by a gradual loss of the self-catalytic properties of the system. In order to explain the effects of anions on the inductive loops, adsorbed intermediates containing SO_4^{2-} and Cl^- in their structure and having different catalytic capabilities were proposed.[94,129]

Similar investigations were carried out by Schweickert et al.[118,119] for a polycrystalline iron disk electrode, with a rotation speed of 70 s^{-1}, in acidic sulfate solutions, $0.3 <$ pH < 6, at 298 K. At pH 0.3, these authors found two inductive loops at $E = -225$ mV vs. NHE and three at more positive potentials. According to these authors, the impedance characteristics are determined not only by charge-transfer steps involving different adsorbed intermediates, but also by the influence of electrocrystallization steps and surface relaxation phenomena, which can be rate-determining steps. The authors postulated a catalytic iron dissolution occurring at the kink sites, the number of these active centers depending exponentially on the electrode potential, and the surface relaxation of the active centers having a time constant close to 1 s. Their model yields a faradaic impedance with two time constants in the inductive region of the transfer locus and one time constant in the capacitive one, thus covering the active and transition ranges.

MacFarlane and Smedley[128] measured the electrochemical impedance for the dissolution of iron in very concentrated chloride solutions (4.5M) of varying pH (0–4). The impedance diagram exhibited a high-frequency capacitive loop, followed by one inductive loop at low overpotentials and a second inductive loop at high overpotentials. This behavior has been interpreted in terms of two parallel paths.

(iii) Possible Explanations of the Origin of the Inductive Behavior of the Transients Suggested by Various Authors

Regions of decreasing voltage on the galvanostatic transients and of increasing current on the potentiostatic ones, as well as inductive loops of an AC circuit, correspond to an enhancement in the metal dissolution rate with an increase in the polarization time.

Many phenomena can lead to such behavior. Over the years, many possible causes have been suggested and analyzed, all of them based on the general principle advanced in 1955 by Gerischer and Mehl[130]: "The origin of a capacitive or inductive electrode impedance would be the relaxation of the surface coverage by adsorbed intermediate." These may be classified as follows:

• An increase with time of some catalytic species, such as mono- or divalent iron hydroxo compounds, directly or indirectly favored by anodic polarization[70,86,123,125,129]

• An increase with time in active surface sites (such as kinks) generated by a parallel anodic dissolution of atoms in edges or steps[37,118,119,135]

• An anodically favored decrease with time of the concentration of some inhibitory species, previously adsorbed at more negative potentials (such as anions and hydrogen intermediates).[66,67,127,131,132]

The indirect influence of local pH change, through one of the above-mentioned effects, has also been analyzed.[54,115]

For cases in which more than one inductive loop appeared, branching mechanisms involving catalytic and noncatalytic processes have been proposed.[123,125] Generally, when the characteristic frequency did not depend on the current, a catalytic reaction was considered to be responsible. Also, in order to explain the effects of anions on the inductive loops, adsorbed intermediates containing anions such as SO_4^{2-} or Cl^- in their structure and having different catalytic capabilities were also postulated.[129]

In order to explain the number of inductive loops, Schweickert and co-workers[118–120] postulated a catalytic iron dissolution mechanism occurring at kink sites, the number of these active surface centers depending exponentially on the electrode potential.

An analysis of the different reaction schemes proposed in the literature in the last 10 years and the mutual criticisms of their authors will be given later in this chapter.

2. Steady-State Anodic Polarization

Steady-state polarization curves have been obtained using two types of procedures:

(i) In dynamic methods, the electrode system is continuously polarized at an optimum potential or current sweep rate, that is, one that is low enough to reach the steady state[33,36,38] but also high enough to avoid major noncontrolled changes of the metal surface roughness and morphology or the interference of concentration polarization. (According to Drazic and Zecevic,[56] if the decay time of the superpolarization peak is about 100 s, the sweep rate should be as low as 0.1 mV s^{-1}.) The polarization curves are continuously recorded.

When the potential changes its direction, a cyclic voltammogram is recorded during the same experiment; useful qualitative information from the possible hysteresis is thus obtained.

(ii) In relaxation methods (as well as in chronopotentiometry or chronoamperometry), the electrode system is abruptly polarized by a current pulse or a potential pulse, starting from the open circuit or from any other stationary state of a previous polarization, and the galvanostatic or potentiostatic transients are recorded and analyzed. The polarization curves are traced point by point after different polarization times, arbitrarily chosen by the experimentalist as characterizing the steady state, which is usually assumed to have been reached when the transients display an asymptotic trend with a slope $|\Delta E/\Delta t| < 1$ mV min^{-1} or $|\Delta I/\Delta t| < 1\%$ per minute.

Despite the difficulties encountered in the interpretation of the stationary polarization curves, due to the interference of different processes occurring with rates that depend in an intricate way on time, potential, pH in the bulk of the solution, anions, and metal surface characteristics, and to the slow rate of change of some supposed parameters, such as pH near the electrode and surface activity, it is generally accepted that the steady-state approach contributes to establishing selective criteria for iron corrosion and anodic dissolution mechanisms. Agreement between experimental and theoretically predicted kinetic quantities characterizing the steady state is a *sine qua non* condition for the validation of any reaction mechanism.

Consequently, different reaction schemes have been discussed based on the shape of the polarization curve, steady-state Tafel slope, b_a^s, and reaction orders with respect to pH, $v_{OH^-}^s$, and anions, $v_{A^-}^s$. Recent research has provided new information based on the dependence of the kinetic quantities on the time elapsed under open-circuit conditions prior to polarizaton,[50,52] on the metal surface structure and morphology, controlled by chemical attack, heat treatment, or mechanical deformation,[15,40,104,136] and also on the polarization range (in the active region) in which the E–log i curve is recorded.

The steady-state polarization curves obtained by various authors for active iron dissolution, in the range of polarization between the corrosion potential and the beginning of the transition to the prepassive range, showed undoubtedly that the dissolution current grows with the anodic potential and with increase in the solution pH. Linear regions over approximately two decades of current density, shifting in parallel with changes in pH, were found by many experimentalists. A typical example is given in Fig. 19[137] for zone-refined polycrystalline iron dissolution in acid sulfate solutions, for which $b_a^s = 40 \pm 2$ mV and $v_{OH^-}^s = 1.0 \pm 0.1$, in the pH range 0.97–3.3. Also, parallel Tafel regions, but characterized by $b_a^s = 31 \pm 2$ mV dec^{-1} and $v_{OH^-}^s = 1.6$, were previously reported by Bonnhoeffer and Heusler[80] for iron in sulfate and perchlorate solutions, pH 0.3–2.5.

Tafel-like regions in which the slope decreased with an increase in the solution pH were also found.[69,71] For example, see Fig. 20,[69] where the Tafel slope increased from 30 mV dec^{-1} for pH 2.1 to 58 mV dec^{-1} for pH 0.0.

Finally, polarization curves displaying a net curvature over a narrow region of current density have also been found, being characterized, as in the example in Fig. 21, taken from Ref. 33, by a derivative $dE^s/d \log i^s$ ranging between 40 and 80 mV dec^{-1} and a reaction order in OH$^-$ concentration close to 0.7 in the low-current-density range and *ca.* 0.3 or 0.0 in the high-current-density range.

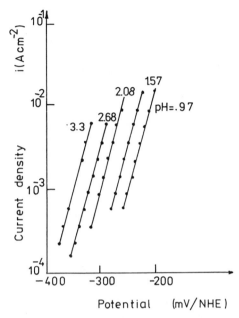

Figure 19. Steady-state polarization curves for zone-refined iron in $0.5M$ SO_4^{2-}, recorded after at least 1 day of immersion. (From Ref. 137, by the courtesy of the authors.)

The histogram in Fig. 4 summarized experimental data from the stationary measurements at $T = 298 \pm 5$ K of iron dissolution in acid sulfate and perchlorate solutions. Although there is no apparent reason to believe that the measurements have not been performed with high accuracy and despite the fact that many of the values reported fall into one of two groups, namely, $b_a^s = 30 \pm 5$ mV dec^{-1}, $v_{OH^-}^s = 1.8 \pm 0.2$ and $b_a^s = 40 \pm 5$ mV dec^{-1}, $v_{OH^-}^s = 1.0 \pm 0.2$, the spreadness of data is evident. Clearly, the stationary polarization curves have not been recorded under identical conditions.

Consequently, a more thorough analysis of experimental and theoretical factors that can influence the steady-state kinetics but which are not evident from the stoichiometric reaction, and sometimes not from the detailed mechanism and the corresponding kinetic equations,[36] has been carried out in the past decade. Experimentally, particular attention has been given to the waiting time and metal surface state effects; theoretically,

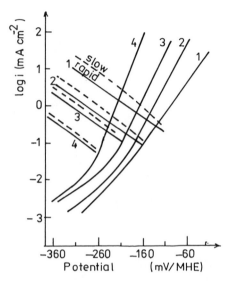

Figure 20. Partial polarization curves for pure Fe in
$2M$ SO_4^{2-} solutions under conditions of slow end of
rapid stirring. The pH values and Tafel slopes, b_a^s, are:
1, pH 0.0, 58 mV dec^{-1}; 2, pH 0.75, 41 mV dec^{-1}; 3,
pH 1.20, 38 mV dec^{-1}; 4, pH 2.1, 30 mV dec^{-1}. (From
Ref. 69, by courtesy of the authors.)

exhaustive analyses concerning the possible inhibitory role of hydrogen
intermediates and of anions have also been performed.

(i) The Polarization Concentration Effect

According to some authors,[54,117] hydrodynamic conditions do not
influence the anodic results. On the other hand, there are experimental
findings attesting to the necessity to carry out steady-state polarization
experiments under conditions of controlled convection, using rotating
disk electrodes or circulating electrolytes. For example, Morel[33] studied
the influence of the rotation speed of a disk electrode on the steady-state
polarization curves recorded with a linear change of the electrode potential
of iron in $0.5M$ H_2SO_4. The results indicated that potentiostatic conditions
are reached only for rotation speeds as high as 1500 rpm and that the
so-called Tafel slope increases with a decrease of the rotation speed below

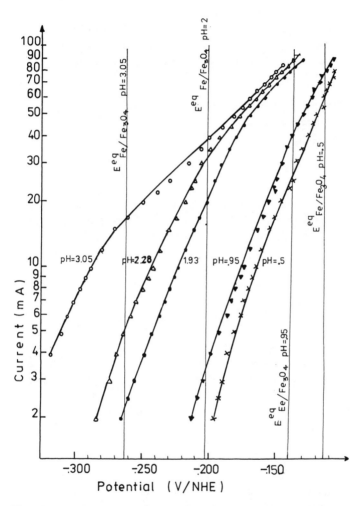

Figure 21. Steady-state polarization curve for a Johnson Matthey iron rotating disk electrode (potentiokinetic sweep rate = 80 mV min^{-1}, rotation speed = 1600 rpm) in 0.05M SO$_4^{2-}$ solutions of different pH values (indicated on the curves). The equilibrium potential lines for the thermodynamically possible reactions are superimposed. (From Ref. 33.)

1000 rpm. Keddam et al.[123,125] noticed that, at higher overvoltage, the current was proportional to the square root of the rotation speed of the disk electrode, which is characteristic of a diffusion-controlled process.[141] Similarly, it was found that the anodic dissolution of a stagnant Fe/H_2SO_4 electrode is limited by convective diffusion and the formation of some layer at the electrode surface.[142-146]

(ii) The Effect of Waiting Time on the Steady-State Anodic Polarization Curves

Generally, it is considered that an asymptotic trend with a slope $|\Delta E^{cor}/\Delta t| < 5$ mV h^{-1} characterizes the open-circuit steady state. This condition is usually reached after 15 min from the moment of electrode immersion.

Despite this, owing to the sluggish nature of phenomena taking place at the interface, changes in the waiting time from 15 min to 24 h lead to various steady-state polarization curves. For example, Akiyama et al.[52] found that the anodic Tafel slope of zone-refined iron in acidic sulfate solutions decreased with the waiting time, τ_w—from 55 mV dec^{-1} after 2 min to 42 mV dec^{-1} after 10 h, considered to be the time necessary to reach a steady-state value.

A detailed study of the influence of the waiting time on the steady-state polarization curves of Armco iron in $0.5M$ H_2SO_4 under stagnant conditions was performed by Drazic and Drazic.[50] Similarly to Akiyama et al.[52] the authors found a decrease of the Tafel slope from 50 mV dec^{-1} after $\tau_w = 15$ min to 35 mV dec^{-1} after 24 h, accompanied by a shift of the polarization curves toward lower current densities (see Fig. 22). The authors suggested that the reason for this inhibitory effect is the slow adsorption of HSO_4^- or competitive adsorption of HSO_4^- and Cl^- ions,[50] but it could also be due to the slow covering of the surface sites active for metal dissolution with hydrogen intermediates.[64,66,67,132,148]

On the basis of these experimental findings, Drazic et al.[147] questioned the validity of the steady-state characteristics as diagnostic criteria for analysis of the reaction mechanism, considering them to be "artifacts."

(iii) Hysteresis Phenomena on the Steady-State Polarization Curves

It is known that a hysteresis appears on the anodic polarization curve for iron in acid solutions when the polarization is scanned first in the direction of increasing polarization and then in the direction of decreasing polarization.[149] Hysteresis is a phenomenon that occurs in the range of

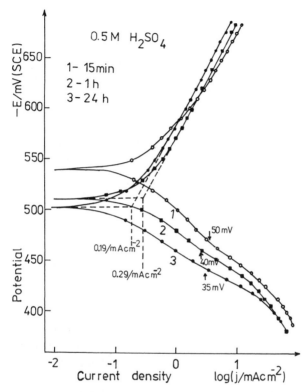

Figure 22. Anodic polarization curves for Armco iron in 0.5M H$_2$SO$_4$ solution, recorded after different waiting times: 1, 15 min; 2, 1 h; 3, 24 h. (From Ref. 50, by courtesy of the authors.)

potential in which the hydrogen evolution reaction coexists with the iron dissolution. The effect is less pronounced in sulfate solutions but is very evident in halogen-ion-containing solutions.[133,150]

The hysteresis phenomenon was extensively studied by Morel.[33] Figure 23 shows the hysteresis displayed on the current–potential curve recorded under potentiostatic conditions, near the corrosion potential, on single-crystal iron in 0.5M H$_2$SO$_4$. As can be seen, for overvoltages higher than 120 mV versus the corrosion potential, the hysteresis disappeared. For polycrystalline iron, under similar conditions, the author did not find any hysteresis. This phenomenon was more pronounced and the reverse

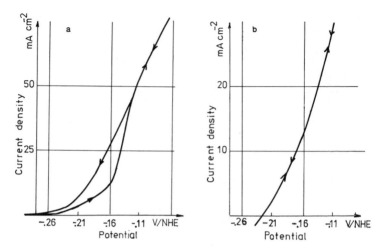

Figure 23. Cyclic polarization curves recorded near the corrosion potential under potentiostatic conditions for a polycrystalline (A) and a single-crystal Johnson Matthey iron rotating disk electrode (B) in $0.5M$ H_2SO_4 solution (pH 0.3), displaying hysteresis only for the single-crystal electrode. (From Ref. 33.)

current was lower, the more negative the starting potential. The influence of inhibitors and catalysts of iron corrosion supplied additional information on the origin of the hysteresis. Thus, the author found an increase in anodic current density (at constant potential) without any influence on the Tafel slope and no hysteresis in the presence of Na_2S, which is an inhibitor of the hydrogen evolution reaction.

On the basis of these experimental findings, Morel[33] arrived at the earlier conclusion of Bonhoeffer and Heusler[80] that the hysteresis is due to chemisorption of atomic hydrogen, so that during the anodic polarization the surface area corresponding to active sites increases due to the slow potential-dependent hydrogen desorption. It is known that pure iron absorbs 15 times less hydrogen and corrodes 10 times slower than Armco iron. According to Morel,[33] hysteresis is observed for the high-purity single-crystal iron because it is homogeneously and extensively covered by H_{ad}. On heterogeneous polycrystalline iron, atomic hydrogen recombines into H_2 and also absorbs into the metal, so that the degree of coverage with H_{ad} decreases. On the other hand, Matsuda and Franklin[151] considered that at low anodic overvoltage, the adsorbed hydrogen is oxidized, followed by absorbed hydrogen which desorbs electrochemically only at high polarization.

However, it was found by Drazic and Drazic[61] that polycrystalline Armco iron in $0.5M$ H_2SO_4 also exhibited hysteresis, the Tafel-like slope being 67 mV dec^{-1} upon reversal of the polarization whereas the value on the first scan in the anodic direction was $b_a^s = 30$–36 mV dec^{-1} (see Fig. 24). The hysteresis effect was greater when the anodic cyclic polarization was applied after a previous cathodic cyclic polarization. The slope of the anodic polarization curve after the anodic "cleaning" of the surface was higher and depended on the highest current density used and the time taken for the measurements, during which the readsorption process proceeds to a certain extent.

Cathodic hydrogen evolution proceeded faster on the anodically cleaned surface, but, during the cathodic polarization, coverage with the adsorbed blocking particles was restored. If the reverse scan was taken at a very slow rate, for instance, allowing 60 min for each point, the reverse curve would tend to coincide with the first one.[38]

Drazic and Drazic[38,61] explained the hysteresis behavior by the blocking effect of adsorbed anions. Thus, at high current densities, due to the fast removal of metal atoms from the surface and the slow adsorption of the inhibiting particles, the surface is virtually free of blocking species.

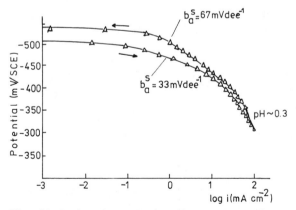

Figure 24. Quasi-steady-state (1-min waiting per point) anodic polarization curves for Armco iron in $0.5M$ H_2SO_4 (pH 0.3). Arrows indicate the direction of the scan. The Tafel slopes are indicated on the curves. (From Ref. 61, by courtesy of the authors.)

Hence, the reversal of the anodic polarization produces much higher anodic currents.

At variance with the hysteresis displayed in highly acidic solutions near the corrosion potential, Worch and Forker[136] found a decrease in the current density (at the same potential), indicating not a cleaning but rather an inhibitory effect produced at high current densities upon reversal of the polarization direction within and beyond the Tafel range. This decrease in the current density was accompanied by changes in the values of b_a^s and $v_{OH^-}^s$ from about 40 mV dec^{-1} and 0.9 to 30 mV dec^{-1} and 1.8, respectively, which were preserved also after anodic reversal (see Fig. 25). According to these authors, during dissolution at high current densities, etch pits were formed so that the kinetics changed from those for the uniform dissolution which occurred with the participation of 1 OH$^-$ per atom ($b_a^s = 40$ mV dec^{-1}, $v_{OH^-}^s = 1$) to those for dissolution via atoms at kink sites with the assistance of 2 OH$^-$, characterized by $b_a^s = 30$ mV dec^{-1} and $v_{OH^-}^s = 2$.

However, it is not clear why this kind of change did not occur also during the first scan in the anodic direction at high current densities and why the dissolution occurred at a lower rate upon reversal of the polarization.

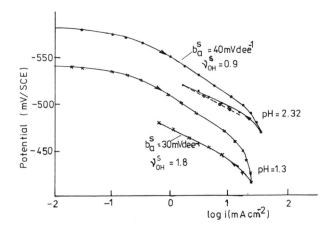

Figure 25. Steady-state current density–potential curves for polycrystalline iron in $H_2SO_4 + Na_2SO_4$ solutions of different pH values (indicated on the curves). Potential changes stepwise upon reversal of its direction. (From Ref. 136.)

In any case, it appears that the hysteresis displayed near the corrosion potential and that exhibited at high current densities have different origins.

(iv) Relation of the Surface Structure and Morphology to the Kinetics of the Steady State

On the basis of some experimental findings from investigations using electrochemical and microscopic techniques, it is commonly accepted that the kinetic data characterizing the stationary anodic dissolution of iron are influenced by the characteristics of the metal, and vice versa. It seems that, among other factors, the state of the iron surface, the type and content of impurities, and the surface structure and defects (dislocation type and density, internal stresses, absorbed hydrogen, etch pit shape, crystallinity, etc.) play an important role. Detailed analyses of this subject were given by Haruyama,[40] Worch and Forker,[136] and Heusler.[152,153]

Nonetheless, at the present state of knowledge, a definite correlation between the initial surface state and the dissolution kinetics is not well supported by experimental evidence so that the effect of surface structure on the kinetics cannot be positively predicted.

(a) The role of impurities

A summary of the experiments performed with iron containing various types and levels of impurities is presented in this section.

The influence of carbon content is not clear. According to Abdul Azim and Sanad[154] and also Kasparova et al.[155] no effect of carbon at concentrations between 10^{-8} and 1.4% on anodic dissolution of iron was observed. Nevertheless, Flitt and Bockris[201] found a cathodic Tafel slope for the hydrogen evolution reaction on corroding heterogeneous iron, containing 2% C, in $0.05M$ H_2SO_4 of about -178 mV dec^{-1} in comparison with the value of -118 mV dec^{-1} usually reported for pure iron. Akiyama et al.[52] found a decrease in metal dissolution and an increase in hydrogen evolution on Armco in comparison with the same reactions on zone-refined iron. Likewise, an inhibition of the anodic rate and an acceleration of the cathodic reaction with an increase in the carbon content of steel from 0.002 to 1.05% were reported by Przewlocka and Bala.[156]

Although accelerated dissolution is usually attributed to the enrichment of iron with less noble components,[40] the presence of Rh and Pt on the iron surface was found to have an accelerating effect on the iron dissolution rate together with a catalytic effect on the hydrogen evolution reaction according to Mikhailovskii and Gaysenko.[157] The effects have been explained by a local increase of pH near the noble-metal inclusion, which leads to a low degree of coverage by inhibiting adsorbed hydrogen on the surrounding iron metal and an increase in the catalytic role presumed to be played by OH^- ions.

Evidently, on the basis of the hitherto published results, it is difficult to predict unambiguously the relationship between iron dissolution kinetics and impurity content.

(b) The role of metallurgical factors

Early papers by Lorenz and co-workers,[46,158,159] in which the authors reported a change in the steady-state values of b_a^s and $v_{OH^-}^s$ from about 40 mV dec^{-1} and 1 for nondeformed and recrystallized (after deformation) iron to about 30 mV dec^{-1} and 2 for iron slowly deformed by cold working, drew attention to a possible explanation of the kinetics and even the mechanism of iron dissolution based on mechano-metallurgical factors. Subsequently, studies were performed on single-crystalline and polycrystalline iron of different purities under well-defined metallurgical and mechanical conditions. Generally, the structural and morphological state of the metal surface was characterized prior to and after electrochemical investigation, especially by light and scanning electron microscopy. Characterization of the surface entailed establishing the roughness factor of the specimen surface, ranging from atomically roughened surfaces to surfaces with distinct pits (determined by means of a roughness testing device), the shape, number, and the orientation of the faces within the etch pits (by means of a microreflective goniometer), and the type and density of dislocations (edge, screw, etc.) emerging on the metal surface (using a Kossel camera).

Thus, the following kinds of single-crystalline and polycrystalline iron surfaces displaying well-defined defects of various natures and densities have been prepared:

(i) Nondeformed,[136] annealed,[158] electropolished,[161] and/or recrystallized specimens,[136,158] characterized by atomically roughened surfaces with a dislocation density as low as 5×10^7 cm^{-1};[136]

(ii) Specimens exhibiting conical or cylindrical etch pits (dislocation etching) characterized by higher or lower index plane faces, obtained by chemical preetching in $0.1M$ HNO_3 or electrochemical preetching by anodic dissolution at high current densities beyond the Tafel range[136]

(iii) Specimens subjected to low-temperature deformation at high strain rate, generating screw dislocations in polycrystalline material; for example, fracture surfaces obtained starting from highly purified iron, crushed at low temperature in a notch-impact bending test[136]

(iv) Single crystal strongly deformed by rolling (with an initial dislocation density of 5×10^7 cm^{-1}), exhibiting after deformation a dislocation density of about 5×10^{10} cm^{-1} [136]

(v) Recrystallized specimen of polycrystalline iron strongly deformed by imposition of a stretching strain, resulting in an increase in dislocation density from 3×10^8 cm^{-1} to 5×10^{10} cm^{-1} [136]

(vi) Single crystal having faces of low crystallographic index, favoring etch pit formation[162]

(vii) Very slowly deformed iron, exhibiting atomically close-packed faces,[104] favoring the formation of textures in which the crystallites are oriented in a specific direction of stress

(viii) Specimens having a mechanically ground,[161] scratched,[163,164] or abraded[112] surface, leading to a Beilby layer characterized by a structural roughening of the surface and an increased dislocation density

(ix) Specimens subjected to *in situ* plastic deformation such as instantaneous elongation,[72,165–168] constant elongation,[74,111,169–172] and continuous scraping,[75,76] yielding fresh surfaces that exhibit loosely bound metal atoms (adatoms) released on the surface by glide slipping in a first stage and fresh active dislocations introduced in a second stage which serve as a long-lasting active sites for both anodic and cathodic reaction. *In situ* mechanical deformations of this type provide information regarding the transient behavior from a fresh to a stationary surface and led to a new transient method based on a stepwise change of surface structure or morphology at constant potential or constant current.

(x) Specimens obtained by loading the metal with hydrogen, produced by a strong cathodic prepolarization. It is known that hydrogen accumulates in the layer near the surface and also diffuses preferentially along dislocation lines into the interior of the metal.[174,175] The dissolved hydrogen produces mechanical stresses that relax by formation of dislocation loops. The lattice elements that are in the neighborhood of these loops become looser as a result, and with subsequent anodic polarization they can be dissolved more easily. This induces an increased formation of etch pits.[175] However, Allgaier[15] found a decrease in the rate of anodic dissolution of iron in $0.5M$ ClO$_4^-$ solution, pH 0.95, due to a prior cathodic polarization; a possible explanation, among others, was the prevailing inhibitory effect of adsorbed hydrogen.

From investigations performed with dislocation-free crystals as well as with specimens prepared as described above, the following conclusions have been drawn concerning the influence of mechano-metallurgical factors on the steady-state iron anodic dissolution in perchlorate and sulfate solutions in the active region. In a crystalline surface there are many classes of sites of different binding energies for a metal atom. For each class there is a certain rate constant of dissolution; thus, adatoms that are less tightly bound dissolve fastest, followed by atoms in kink sites. Dissolution rate constants for planar atoms and atoms in complete steps (creating surface vacancies) are the smallest. The iron transfer reactions do not proceed randomly at arbitrary sites on a crystalline surface. According to the theory of Kossel and Stranski, crystals are dissolved by removal of atoms at kink sites in monatomic steps. This process shifts the position of a kink but does not change the surface concentration of kink sites. Except in the case of kink atoms, the surface concentration of a class of sites decreases during the dissolution of atoms at these sites.

A single crystal can display one-dimensional lattice defects such as edge and screw dislocations and dislocation rings. The probability of forming holes is increased at the site where an edge dislocation ends in the surface. Steps extend from screw dislocations emerging at the surface rise to etch pits.

The relation of structure to kinetics is not one-sided. The structure influences the observed rate, and the electrochemical processes influence the structure.

The occurrence of the two sets of kinetic data—$b_a^s = 30$ mV dec^{-1}, $v_{OH^-}^s = 2$ and $b_a^s = 40$ mV dec^{-1}, $v_{OH^-}^s = 1$—depends (if at all) in an intricate and not decisive way on the different structural and morphological states of the metal surface. Thus, both sets of values have been obtained for the dissolution of the various types of iron specimens examined—nondeformed, recrystallized, and deformed in the various ways described above. Some examples are given below.

The dissolution of *nondeformed* iron is characterized by $b_a^s = 40$ mV dec^{-1} and $v_{OH^-}^s = 1$ according to Lorenz and co-workers,[46,158,159] Bech-Nielsen and Reeve[176,177] (at low pH), Vorkapic and Drazic,[174] and Worch and Forker[136] (for polycrystalline iron) but is also characterized by $b_a^s = 30$ mV dec^{-1} and $v_{OH^-}^s = 2$ according to Worch and Forker[136] (for a highly purified single crystal), Bech-Nielsen and Reeve[176,177] (at high pH), and Heusler and co-workers[37,162] (for a single crystal).

The dissolution of *deformed* iron leads to $b_a^s = 30$ mV dec^{-1} and $v_{OH^-}^s = 2$, as reported by Worch and Forker[136] (for a chemically preetched single crystal exhibiting etch pits and dislocations, for the fracture surface of the single crystal, and also for galvanostatically preetched polycrystalline iron), Lorenz and co-workers[46,158,159] (for cold-worked iron), Bech-Nielsen and Reeve[176,177] (at high pH), and Vorkapic and Drazic[174] (for iron heavily charged with hydrogen). The other set of data, $b_a^s = 40$ mV dec^{-1} and $v_{OH^-}^s = 1$, was also obtained for the dissolution of deformed iron, as reported in Refs. 176 and 177 and in Ref. 136 for single-crystal iron deformed by rolling and for polycrystalline iron deformed by stretching strain.

Also, the extent to which a change in *dislocation density* influences the dissolution kinetics is not clear. For instance, for highly purified polycrystalline iron in H_2SO_4, pH 2–4, an increase in dislocation density from 3×10^8 to 5×10^{10} cm^{-1} (as the result of a stretching strain exerted on the specimen) did not change the shape of the polarization curve; likewise, an unstrained single crystal having a dislocation density of 5×10^7 cm^{-1} and a single crystal deformed by rolling having a dislocation density of up to 3×10^{10} cm^{-1} displayed the same Tafel slope and reaction order in hydroxyl ion.[136]

However, Worch and Forker[136] managed to establish the following correlations between the electrochemical dissolution of iron and the structure and morphology of the surface based on their characterization of the surface after dissolution:

(i) There are metastable surface atoms that are less strongly bound than atoms at kink sites, for example, atoms at crystal corners and edges or at the sites of emergence of dislocations on the metal surface.

(ii) The anodic dissolution of iron in acid media yields a steady-state metal surface on which the number of etch pits depends on the applied potential, the crystallographic orientation of the surface, and deformation structure.

(iii) On the atomically rough surface, a uniform (on the scale of light microscopy) dissolution takes place from the very beginning of the anodic polarization. On the tilted (111) face, large etch pits are formed and subsequently grow together, leading, after an extended time of polarization, to a coarser surface.

(iv) In the course of etch pit formation, as a rule, the faces having the lowest rate of dissolution are formed, so pits will originate whose walls consist of atomically close-packed stepped faces. According to Kossel[237] and Stranski's theory,[239] they are dissolved via kink sites.

(v) At the points of emergency of dislocations, etch pits are preferentially formed.

(vi) The dissolution of metal at the step originating from a *screw* dislocation terminating at the surface results in the formation of pyramidal etch pits. They become steeper with increasing overpotential, but heat-treated iron never displays stable screw deslocations.

(vii) Etch pits on (112) iron surfaces grew along extended *edge* dislocations ending at the surface. The density of these dislocations was very small in the single crystals prepared by dynamic recrystallization. The etch pits were trigonal. Periodically, the etch pits slipped to grow deeper and developed terraces. It was shown that only (112) and (110) surfaces of iron are morphologically stable during dissolution in perchlorate or sulfate solutions. Surfaces with other orientations disintegrate into structures with (112) and (110) surfaces.

The interpretation given by Worch and Forker[136] for the above experimental behavior is that the two sets of kinetic quantities, that is, $b_a^s = 30$ mV dec^{-1} and $v_{OH^-}^s = 2$ on the one hand and $b_a^s = 40$ mV dec^{-1} and $v_{OH^-}^s = 1$, on the other, characterize the dissolution of two kinds of iron atoms, namely, more weakly and more strongly bound surface atoms, respectively. Therefore, these authors considered that the removal of the corner and edge atoms, occurring easily and rapidly, needs the help of only one hydroxyl ion and, as it is followed at longer times by the removal of tightly bound atoms, constitutes a non-steady-state situation. The dissolution of the strongly bound atoms requires the participation of two OH$^-$ ions and controls the steady-state kinetics. In this context, the authors explained the reaction orders with respect to OH$^-$ ions by adopting the Bockris–Drazic–Despic or the Heusler mechanisms, that is, by assuming that the hydroxyl ion is a reactant in iron dissolution.

A lot of important information relating the dissolution morphology to the dissolution kinetics was provided by the investigations of Heusler and Allgaier[15,162,199] and Folleher and Heusler[103] on iron single-crystal faces oriented vicinal to (112) in acid perchlorate solutions. By means of *in situ* and *ex situ* morphological and electrochemical DC measurements, these authors established that the (112) plane exposed to the electrolyte dissolves essentially without interference from dislocations. There were only a few pits, their contribution to the total current being no more than 1%. The metal dissolved from kinks in steps. During anodic dissolution, a steady-state surface was reached that exhibited flat three-sided pyramids enclosed by (112) faces which intersect in $\langle 113 \rangle$ directions. Pairs of monatomic steps in the corresponding $\langle 113 \rangle$ directions were found on each (112) face. The authors measured the angles between pairs of monatomic steps and found that it increased with increasing anodic potential; that is, the mean distance between equivalent kink sites located at the monatomic step decreased. By determining the average distance between kinks in the steps, x_k, and the mean step distance, x_s, as a function of electrode potential and pH, the authors established the relationships $(\delta E / \delta \ln (x_k)^{-1})_{a_i} = RT/F$ and $(\delta \log x_s / \delta \mathrm{pH})_{/a_i} = a_{\mathrm{H}^+, \mathrm{E}} = -0.6$. The distance between such monatomic steps at pH 1 was found to be independent of the electrode potential, and the distance between kinks at constant potential was independent of pH. The surface kink concentration was determined from the formula $C_k = (x_s \cdot x_k)^{-1}$. Anyhow, the independence of the kink distance on pH was assumed to mean that dissolution of an iron (II) ion from a kink site and from the intersection of two steps results in the intermediate formation of $FeOH^+$ according to $Fe \cdot H_2O \rightarrow FeOH^+ + 2e^- + H^+$, while for the dissolution from the apex an additional hydroxyl ion is required.

(v) Influence of Anions on Iron Dissolution in Acid Media

The role of anions in iron dissolution is still a matter of dispute. According to recent interpretations, all anions influence the kinetics and mechanism of metal corrosion and anodic dissolution to some degree and in different ways, depending on their nature and concentration, the solution pH, and the potential range. Thus, near the corrosion potential, anions are chemisorbed more strongly at dissolution sites than on the rest of the surface, thereby reducing the area free for metal dissolution to a larger extent than that for hydrogen evolution.[133] On the basis of their inhibitory effect, as determined by the decrease of the exchange current,[51] of corrosion,[178] and of anodic currents, the shift of the corrosion potential toward positive values, and the increase of the Tafel slope,[51,179,180] the following order of adsorbability was established: $I^- > Cl^- > SO_4^{2-}$, $HSO_4^- > ClO_4^- > PO_4^{3-}$. The anions compete for the metal surface with each other and with water molecules and hydroxyl ions.

The inhibitory effect is especially pronounced at low current densities, that is, at low anodic overvoltage, and sharply disappears at a certain potential, termed the "potential of unpolarizability,"[133] at which in many possible ways (i.e., electrochemical removal, reactions forming intermediate complexes with the atom, or direct dissolution of Fe $\cdot$ A$^-$ $\rightarrow$ Fe^{2+} + A$^-$ + 2e), the anions are removed from the metal surface without any substantial increase in overvoltage. After this cleaning of the surface, the behavior of the iron is the same as in the absence of adsorbable anions in the same range of potential. As the current is raised, that is, at very high ferrous ion flux, especially in stagnant solutions, reaction products such as salts precipitate, blocking again the metal surface.

(a) Electrolytes containing only "inert" oxygenic anions

According to Heusler and co-workers,[70,79,80,86,113] Lorenz and co-workers,[16,17,114,181] and also Vojnovic,[182] alkali perchlorates or sulfates are almost inert electroytes; that is, neither their cations nor their anions influence the dissolution process "significantly." Nevertheless, adsorption of sulfate ion on iron from sodium sulfate solutions at open-circuit potential was measured by Hackerman and Stephens.[59]

On the other hand, these electrolytes were reported to have a stimulatory effect on iron dissolution in early papers by Kolotyrkin, Florianovich, and co-workers[183–186] as well as by Schwabe,[78,211] but according to others, it seems that the ionic strength and pH may not have been constant in these experiments. However, in 1980 Reshetnikov[187] published the results of studies of the polarization curves for Armco iron in sulfate solutions. He also found a positive reaction order in SO$_4^{2-}$ concentration, for example, $v_{SO_4^{2-}}^s = 0.9$, together with $b_a^s = 40$ mV dec^{-1} and $v_{OH^-}^s = 0.9$. Similarly to Kolotyrkin, he proposed a mechanism in which sulfate ion takes part in the electron-transfer step: Fe$^+$(OH)$_{ad}$ + SO$_4^{2-}$ $\rightarrow$ FeSO$_4$ + OH$^-$ + e^-. (Drazic et al.[188] considered that, because FeSO$_4$ is unstable in acidic aqueous solutions, it cannot be a reaction product; however, Heusler[152] pointed out the stability of the neutral FeSO$_4$ complex in sulfuric acid during the transfer of hydrolyzed ferrous ions, despite its instability in the electrolyte.)

The influence of varying amounts of sulfate ions on the corrosion and polarization behavior of iron in acid perchlorate solutions, at constant pH, was investigated by Ström et al.[236] The results obtained indicated an increase of the anodic current density and of the Tafel slope in the first steady-state linear range, the Tafel slope increasing from 30–40 mV dec^{-1}

in pure perchlorate to 30–50 mV dec^{-1} in the solution with added sulfate. They also found a reaction order with respect to sulfate ion, $z_{SO_4^{2-}}$ equal to 0.4 for 0.2M ClO$_4^-$ and 0 for 1M ClO$_4^-$ solution, respectively. The reaction order in OH$^-$ in 0.2M perchlorate was 1.7; upon addition of sulfate, a value of 1.0 was observed.

Recently, Bala[69,192] also reported experimental data attesting that SO$_4^{2-}$ and HSO$_4^-$ ions at high concentration and low pH values are reduced on the iron surface by adsorbed hydrogen, the reduction products SO$_3^{2-}$, S$_2$O$_4^{2-}$, S$_2$O$_3$, and S^{2-} acting as strong stimulators for iron dissolution even at very low concentrations.

On the other hand, some experimental results have been interpreted in terms of an inhibitory action of sulfate ions not too far from the corrosion potential. Among them, the increase in current density upon the reverse of polarization which appears on the steady-state polarization curve for iron in H$_2$SO$_4$ + Na$_2$SO$_4$ solutions[33,60] was attributed by Drazic et al.[60] to the anodic removal of the adsorbed sulfate ions, SO$_4^{2-}$ and HSO$_4^-$, with a corresponding enhancement of the area free for dissolution of the metal. The same explanation was given[50] for the shift of the anodic polarization curve toward positive potentials with the increase in the waiting time, during which the inhibiting sulfate ions adsorb on the metal surface.

Other new experimental findings were explained in a similar manner. Thus, Florianovich et al.[189] and Florianovich and Mikheeva[112] observed a 10-fold increase in the dissolution of Armco iron for a mechanical in situ refreshment of the metal surface in comparison with the dissolution without refreshment (see Fig. 26). This result was interpreted by Florianovich and Mikheeva[112] in terms of a decrease in the catalytic capacity of rapidly adsorbed SO$_4^{2-}$ ions at the "dissolution sites" due to the slow adsorption of the same ions at the remaining sites of the surface, termed "inhibiting sites." A rather similar situation can be found in Ref. 133. The process responsible for anion removal implies the reaction between an atom from the uncovered area, a hydroxyl ion from the solution, and a neighboring atom with an adsorbed anion on top of it. Obviously, if all metallic atoms are covered by adsorbed anions, this reaction becomes improbable, so that bare atoms contiguous to the covered one can be considered as "inhibiting sites" for this type of reaction. Nevertheless, on freshly emerged surfaces, many other phenomena must be taken into account,[40] such as charging of the electrical double layer,[165] formation and dissolution of loosely bound metal atoms (adatoms),[74] emergence and

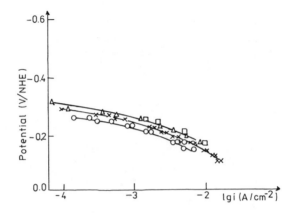

Figure 26. Polarization curves for iron in $0.05M$ H_2SO_4, recorded at different time intervals τ after mechanical *in situ* refreshment of the metal surface. O, nonrefreshed; x, $\tau = 0.9$ s; Δ, $\tau = 0.28$ s; $\square$, $\tau = 0.09$ s. (Experimental points taken from Ref. 112 and 189.)

dissolution of high-index planes[169] and active dislocations,[170,190] increases in temperature due to thermoelectric effects, and so on.

(b) Electrolytes containing halogenide anions

The inhibitory effect of halogenide ions on the active dissolution of iron was first reported in 1930 by Walpert,[178] who found that addition of halides to sulfuric acid solutions decreased the corrosion rate, the decelerating action increasing with the halide concentration. A summary of the experimental results reported by various authors since that time is given here.

Addition of halide yielding a concentration of halogenide ions below a certain value (the higher the value, the lower the anion adsorbability), for instance, $10^{-4}M$ for [I⁻] in $0.5M$ H_2SO_4[133] and $>10^{-2}M$ for [Cl⁻] in $0.25M$ H_2SO_4,[191] produces no measurable effect on the polarization curves, thus suggesting that a critical degree of coverage by adsorbed anions must be reached in order for these anions to exert an inhibitory effect and influence the kinetics. However, after 24-h contact time of Fe with a solution containing 10^{-3}–$10^{-2}M$ [Cl⁻] in $0.5M$ H_2SO_4, Drazic and Drazic[50,61] found a substantial shift of the polarization curves (cf. Figs. 27

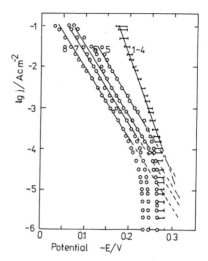

Figure 27. Steady-state polarization curves for iron in 0.25M sulfuric acid containing potassium chloride at concentrations of $10^{-5}M$, $10^{-4}M$, $10^{-3}M$, and $10^{-2}M$ (1–4), $10^{-1}M$ (5), 0.5M (6), 1M (7), and 2M (8) at 25°C. (From Ref. 191, by courtesy of the authors.)

and 28). This disagreement was explained by the fact that the replacement of water by adsorbed anions at the corrosion potential is relatively slow. In fact, on the basis of radiotracer measurements, Heusler and Cartledge[133] found that the time necessary to reach a steady state for adsorption of iodide ion from $1.5 \times 10^{-4}M$ for KI in 0.5M H_2SO_4 was a few minutes.

By recording the polarization curves of iron in 0.1M [Cl$^-$] + 0.25M H_2SO_4 after different waiting times, Drazic and Drazic[50] found that the position and shape of these polarization curves changed when the contact time before polarization was changed and that, in order to establish a constant curve, it was necessary to wait as long as 24 h (see Fig. 29, taken from Ref. 50); they explained this behavior by the long time necessary to reach a stationary coverage of the iron surface with Cl$^-$ ions. As a consequence, they concluded again that the waiting time effect is predominantly due to the slow adsorption of anions. The overall effect is a considerable diminution of the corrosion rate and a shift of the corrosion potential toward positive values; concomitantly, a substantial increase in the overvoltage on the anodic polarization curve and a slight increase in the overvoltage for hydrogen evolution occur. Near the corrosion potential, the halogenide ions are adsorbed, more or less reversibly, on the iron surface, blocking the dissolution sites to a larger extent than the oxygenic anions.

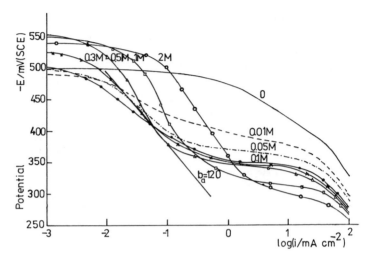

Figure 28. Anodic polarization curves for Armco iron in $0.5M$ H_2SO_4 containing various concentrations of added NaCl (indicated on the curves). (From Ref. 50, by courtesy of the authors.)

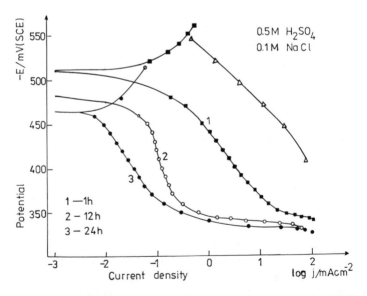

Figure 29. Anodic polarization curves for Armco iron in $0.5M$ $H_2SO_4 + 0.1M$ NaCl solution, recorded after different waiting times (indicated on the figure). (From Ref. 50, by courtesy of the authors.)

Anodic transients displaying overvoltage maxima were also observed in the presence of halides. Initial Tafel lines having slopes of approximately 63 mV dec^{-1}, that is, the same slope as for "inert" electrolytes, were reported in Ref. 133 for Fe in 0.5M $H_2SO_4^+$ [I$^-$] up to 0.2M.

The polarization curves are S-shaped with a linear region at overvoltage values lower than ca. 120 mV. The linear part is narrower, the higher the chloride ion activity. Concomitantly, the steady-state Tafel slope increases from 30 mV dec^{-1} at [Cl$^-$] < $10^{-2}M$ to about 60 mV dec^{-1} for [Cl$^-$] = 2M, as the product $a_{Cl^-} \cdot a_{H^+}$ increases[191] (see Fig. 27). At higher overvoltage, the potential of unpolarizability (or the current of unpolarizability) is reached, and the current increases sharply[50] (Fig. 28).

The characteristic kinetic quantities reported by various authors for iron corrosion and active dissolution in aqueous chloride solutions in the region of potential in which chloride ions exert an inhibitory effect (i.e., negative to the potential of unpolarizability) are presented in Fig. 30. At open circuit, $\Delta E_s^{cor}/\Delta pH$ ranges between –40 (reported by Bala[192]) and –63 mV dec^{-1} (reported by McCafferty and Hackerman[13]), and $\Delta \log i_s^{cor}/\Delta pH = -0.6 \pm 0.1$. Steady-state anodic Tafel slopes near 60 mV dec^{-1}, $v_{OH^-}^s = 1$, and values of $v_{Cl^-}^s$ between –1 and –0.5 have been more frequently reported, but the scatter in the of data is evident.

Earlier experiments conducted by McCafferty and Hackerman[13] for iron dissolution in solutions of high chloride concentration (1–6M) and [H$^+$] between 0.1M and 6M revealed an accelerating effect of OH$^-$, ($v_{OH^-}^s = 1$) at [H$^+$] < 1M and an accelerating effect of H$^+$ ($v_{OH^-}^s = -2$) at higher acidity. Recently, Drazic and Drazic[50] studied anodic dissolution of Armco iron in 0.5M H_2SO_4 with addition of NaCl (Cl$^-$ concentration ranging from $10^{-3}M$ to 2M). In their experiments, Cl$^-$ ions at concentrations up to ca. 0.1M acted as anodic inhibitors; when present in concentrations higher than 0.1M, they reversed their action, with a reaction order in Cl$^-$, $v_{Cl^-}^s$, of 1 and a Tafel slope close to 120 mV dec^{-1}. The conclusion drawn from these experiments was that the polarization curves obtained for solutions of low chloride activity correspond to iron dissolution inhibited by adsorbed anions; for electrolytes very concentrated in halide ions, the kinetics is controlled by halogenide desorption via complex formation.

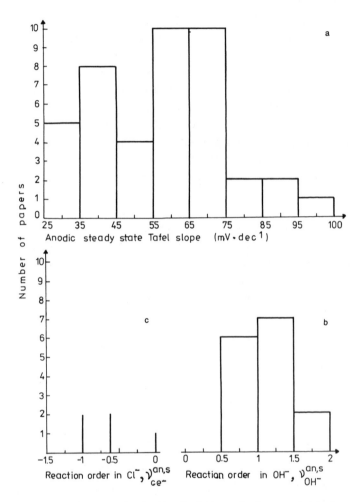

Figure 30. Histograms showing the distributions of values reported in the literature of the anodic steady-state Tafel slope (a), the reaction order in OH⁻ (b), and the reaction order in Cl⁻ (c), for iron corrosion and active dissolution in aqueous chloride solutions at potentials negative to the potential of unpolarizability. The literature values are taken from Refs. 10, 13, 44–49, 51, 79, 82, 84, 101, 112, 161, 191, 209, 210, and 228–236.

IV. MECHANISMS OF ACTIVE DISSOLUTION OF IRON IN ACID MEDIA

The history of corrosion science is marked by the tireless efforts of researchers to find a mechanism able to account for the complex behavior of dissolving iron. However, despite the large number of possible mechanistic interpretations of the processes involved that have been reported in the literature, a complete understanding, on a molecular level, of the behavior of the iron electrode is still lacking.

There are two principal reasons for this lack of understanding. First, iron corroding in acid media represents a complex system in which a large number of independent and interdependent processes occur. Thus, there are reasons to suppose that the metal dissolution and hydrogen evolution reactions proceed in a stepwise manner, via the formation of intermediate species, occurring at preferential sites on the metal surface. Along with these two main reactions, the decomposition of adsorbed water molecules and the adsorption and possible participation of hydroxyl ions and anions as reactants, catalysts, or inhibitor species should also be taken into consideration. Second, the very narrow range of polarization (not exceeding 70 mV for pH > 2 and 120 mV for pH < 1.5) in which additional branching reactions do not interfere, the gradual changes in the metal structure and morphology, as a result of both the adsorption of various species and the lattice destruction by electrochemical dissolution and their mutual influence on each other, as well as the continuous change of concentrations in the reaction layer have led to highly variable and sometimes contradictory data. Consequently, the experimentalist is faced with the difficult task of more or less arbitrarily selecting from the large spectrum of values those that should be considered to have mechanistic significance.

Faced with this disheartening state of affairs, scientists proposed and analyzed a great number of possible reaction models, using analytical and, beginning in 1969,[195] numerical methods, which unfortunately, even today, describe quantitatively and even qualitatively only a part of the experimental results. The procedure was to work out the theoretical predictions for several hypothetical reaction sequences, under particular simplifying conditions concerning the ratio between the partial standard currents, and compare them with experiment. Most of these mechanisms have already been discussed many times, and recently they have been commented on in a number of review articles (see Refs. 34, 36, 37, 83, and 188).

In this section, the new reaction schemes proposed by various authors in the past 10 years will be presented and analyzed, and the consistency of each of these schemes with the experimental data will be assessed.

As will be seen, at present both the concept of the transfer of two electrons in one step as proposed by Heusler[86,70] and that of the transfer of one electron per step proposed by Bockris et $al.$[51] are accepted by various authors, sometimes in the framework of the same mechanism.

The following arguments may be made in favor of the transfer of two electrons in one step:

(i) There are mechanisms for which it is not necessary to introduce single-electron transfer in order to explain all the experimental results. Moreover, it is difficult to interpret the experimental data without invoking a two-electron-transfer step.

(ii) The metal/solution interface is characterized by an intermediate state involving electron transfer between two specific adsorbed species such as Fe OH_{ad}^{+} /Fe OH_{ad}^{-} or Fe_{ad}^{2+} /Fe; because the two species are in intimate contact, the concept of simultaneous transfer of two electrons is not unreasonable.

(iii) Moreover, the thermodynamic properties of the adsorbed species will change to the extent that its chemical potential is changed by adsorption[32] so that the objections raised against the simultaneous transfer of two electrons based on the solvent reorganization energy required must be reconsidered, taking into account the particular situation of two adsorbed contiguous species.

(iv) A two-electron transfer (irrespective of the specific site on the metal surface where the reaction occurs) also satisfies the principle of microscopic reversibility, the mechanism near equilibrium being the same in both directions of the process. Thus, it is often assumed that the adsorbed hydroxo compounds of iron $FeOH_{ad}^{+}$ and $FeOH_{ad}^{-}$ play an important role in the reaction mechanism even in acidic solutions. For the reaction sequence

$$Fe \cdot H_2O_{ad} = H^+ + FeOH_{ad}^- \overset{-2e^-}{=} FeOH_{ad}^+ + H^+$$
$$\overset{+2e^-}{=} Fe^{2+} \cdot H_2O_{ad} = Fe^{2+} \cdot H_2O_{sol} \qquad (16)$$

calculation of the stoichiometric number, ν, from Parsons' equation, $\nu = nFb_ab_c/[2.303\ RT(b_c - b_a)]$, leads for $b_a = -b_c = 60$ mV dec^{-1} to the value $\nu = 1$.

(v) Furthermore, according to Heusler,[37,100]

Because of the greater potential dependence of the direct charge transfer than of the monoelectron transfer, the first one must always predominate at high overvoltage even if at the equilibrium potential the monoelectron transfer is energetically favored, so that it cannot be considered as barred. Moreover, if the intermediates occurring in multistep mechanisms are relatively unstable, direct charge transfer may become the fastest of the parallel paths even at equilibrium.

According to the other point of view, the energy required for the simultaneous transfer of two electrons in a single step is prohibitive, and therefore all mechanisms involving it are ruled out from the very beginning. The following arguments can be made in favor of this point of view:

(i) The leading idea is that the electron transfer proceeds mainly through the tunnel effect; that is, the two electrons have to move out of the metal to some distance by tunneling across the thickness of the double layer, at least for metal deposition.

(ii) Arguments offered to support the concept that there is a higher probability for consecutive single-electron transfer on the grounds that it is energetically much more advantageous are based on a consideration of solvent reorganization energies for the differently charged particles.[193]

(iii) In favor of single-electron transfer is also the experimentally proven fact that, under particular conditions, iron can form complexes [(such as the carbonyl complexes $Fe(CO)_5$, $Fe_2(CO)_9$, etc.] in which the metal is in an unusual oxidation state $(0, +1)$ and also that it forms stable nonstoichiometric oxides.[194] Therefore, the concept of monovalent iron compounds as intermediates in the adsorbed state seems also reasonable and is adopted by many authors.

Both reaction schemes involving only single-electron transfer[195–197] and those involving a two-electron-transfer step predict a purely capacitive behavior of the transients in the region of low steady-state Tafel slope, in contradiction to the experimental findings, but this difficulty can be surmounted in several ways, all having some experimental support. The first is based on the concept that blocking species such as adsorbed anions[38,133] or hydrogen[33,66,132] are electrochemically desorbed under anodic polarization (by an electrochemical reaction[133] or by an undermining effect[61,147]; see Section IV.3 and Fig. 33). Another possibility is an increase with time in the surface concentration of kink atoms, supposed to be both preferential dissolution sites and building sites of the lattice, by the formation of kinks through the parallel anodic dissolution of atoms in edges or steps.[34,153] Finally, catalyst formation and accumulation by anodic reactions have been invoked by some authors.[100,118,119]

There are also some questionable explanations for the inconsistency between a low Tafel slope and the inductive behavior that characterizes the consecutive charge transfer. One of them is to accept that the symmetry factor has a totally "asymmetric" value, such as the limiting values 0 or 1, and that it is strongly potential dependent[123,125]; the other is to deny the mechanistic significance of steady-state characteristics, that is, of the Tafel slope and reaction orders, because they are influenced by the sweep rate and waiting time arbitrarily chosen by the experimentalists.[188]

As expected, it is practically impossible and futile to attempt to take into account, in the framework of one theoretical approach, all the possible processes occurring at the metal surface. For this reason, elucidation of the mechanism of iron dissolution must be an iterative process. Therefore, authors who have developed and analyzed various reaction schemes have taken into consideration only a certain phenomena and overlooked the others. In this way, one can establish the capability of a certain process to account for a part of the experimental behavior. The next step is then to assess the contribution of each process to the overall mechanism.

1. Mechanisms Involving Surface Blockage by Adsorbed Hydrogen

The coexistence of hydrogen evolution with iron dissolution at the corrosion potential justifies an exhaustive analysis of the influence of coverage of the metal by atomic and molecular hydrogen on the kinetic behavior of the iron electrode.

A recent theoretical treatment was performed by Plonski[64,132,148,198] for a mechanism entailing electrochemical desorption of hydrogen coupled with ferrous ion formation in one step, both reactions occurring randomly at the metal surface. The author found that, for certain relations between the standard partial current densities[65,148,197] corresponding to the reaction steps in both directions and characterized by a high rate of iron dissolution and a low rate of hydrogen electrochemical desorption, the electrode system displays the following kinetic quantities, which are in agreement with some experimental data:

- At the corrosion potential: $\delta E^{cor}/\delta pH = -47$ mV, $\delta \log i^{cor}/\delta pH = -0.6$.
- For the anodic dissolution: $\delta \eta /\delta \log i_i = 59$ mV dec^{-1}, $\delta \eta /\delta \log i_s = 29$ mV dec^{-1}, $\delta \log i_i/\delta pH = 0.5$, and $\delta \log i_s/\delta pH = 1$ at lower overvoltage, and $\delta \eta /\delta \log i_s = 59$ mV dec^{-1} and $\delta \log i_s/\delta pH = 0$ at higher overvoltage values.

This analysis showed that the above mechanism cannot account for higher dependences on the pH such as $\delta E^{cor}/\delta pH = -60$ mV and $\delta \log i_s/\delta pH = 1.8 \pm 0.2$, values that have often been reported in the literature. In addition, this mechanism entails a rather complete coverage of all the surface metal atoms with adsorbed hydrogen in the anodic Tafel range of polarization instead of a low relative coverage as calculated by some authors. The assumption of a homogeneous iron surface from a chemical reactivity standpoint is a poor one because the metal consists of molecularly flat areas bounded by edges, the atoms at kink sites in the edges being the most energetically favored for dissolution. Moreover, this model cannot explain the specific action of anodic inhibitors or halogenide ions, whose adsorption influences the iron dissolution to a greater extent than the hydrogen evolution.

The next iterative analysis of the influence of adsorbed hydrogen on the iron dissolution kinetics included the possible correlation between the surface structure and the reactions occurring at specific sites, as revealed by experimental and theoretical studies by Allgaier and Heusler,[14,15,162,199] Lorenz and co-workers,[17,102,105,117–121] and Keddam and co-workers.[125,140] All these authors agreed that the dissolution rate constant should be proportional to the weakness of the binding of the surface atoms to the bulk metal, thus decreasing in the order kink > step > plane. The rate of metal dissolution is proportional to the rate constant and to the number of atoms in the position concerned, which decreases in the order plane > step > kink.

On the other hand, a novel suggestion concerning the hydrogen evolution on corroding heterogeneous iron (containing 2% C) in 0.05M H_2SO_4 has been adopted by Flitt and Bockris[201] and by Bockris et al.[202] These authors measured a high cathodic Tafel slope of about -178 mV dec^{-1} and (like Thomas[203] and Cleary and Greene[204]) explained this result on the basis of the hypothesis that the hydrogen evolution kinetics varies locally as a function of the surface nonuniformity. Thus, they advanced the idea that the hydrogen evolution reaction takes place preferentially on carbide regions of the surface so that the local degree of coverage with adsorbed hydrogen could be as high as 0.7, the relative degree of coverage averaged over the whole surface being only between 0.01 and 0.1.

Taking into account the above-mentioned possible influence of the metal structure on iron dissolution and hydrogen evolution, Plonski[66,67] examined its simultaneous effect on the iron corrosion and dissolution kinetics. The treatment was confined to the conditions under which a roughly constant surface structure can be assumed (such an assumption can be valid in the anodic region near the corrosion potential and for short transients), so that slow processes such as anion adsorption, hydrogen diffusion into the metal, or substantial changes of pH or in metal surface activity could be neglected. Theoretical relationships have been derived assuming a Volmer–Tafel hydrogen evolution at the planar atoms "p" and a high rate of iron dissolution and a Volmer–Heyrovsky hydrogen reaction at the reactive sites "k", according to the following sequence of reactions:

$$Fe_p \xrightarrow{A_0} Fe_{sol}^{2+} + 2e^-$$

$$Fe_p + H^+ + e^- \underset{A_1}{\overset{C_1}{\rightleftarrows}} Fe_pH_{ad}$$

$$2Fe_pH_{ad} \xrightarrow{B_2} Fe_pH_{2ad} + Fe_p$$

$$Fe_pH_{2ad} \xrightarrow{B_1} Fe_p + H_{2sol}$$

$$FeFe_k \xrightarrow{A_{0'}} Fe_{sol}^{2+} + 2e^- + Fe_k$$

$$Fe_{k+}H^+ + e^- \underset{A_{1'}}{\overset{C_{1'}}{\rightleftarrows}} Fe_kH_{ad}$$

$$Fe_kH_{ad} + H^+ + e^- \xrightarrow{C_2'} Fe_pH_{2ad} + Fe_p$$

$$Fe_kH_{2ad} \xrightarrow{B_1'} Fe_k + H_{2sol} \tag{17}$$

where A_0, A_1, B_1, B_2, C_1 and $A_{0'}$, $A_{1'}$, $B_{1'}$, $C_{1'}$, $C_{2'}$ are rate constants of the partial reactions at the planar atoms and at the kink sites, respectively. The relative thicknesses of the arrows signify the ratio between the different standard partial currents that must be satisfied in order to have a low relative degree of coverage with adsorbed hydrogen at planar atoms and a high one at kink sites, as well as a preferential dissolution of kink atoms in comparison with planar atoms.

The partial current density is defined as the value of the current divided by the geometrical area of the electrode, in either the cathodic or anodic direction, corresponding to a degree of coverage with the reactant surface species equal to unity. Therefore, for a given potential, it represents the maximum value of the partial current. At the corrosion potential, the partial current densities are $i_{a_{0'}} = 2FA_0'z^2$, $i_{a_1} = FA_1z$, $i_{C_{1'}} = FC_1'a_{H^+}z^{-1}$, $i_{C_{2'}} = FC_2's_{H^+}z^{-1}$, and $i_{b_{1'}} = FB_1'$ flowing through the kink atom surface, and $i_{a_0} = 2FA_0z^2$, $i_{a_1} = FA_1z$, $i_{c_1} = FC_1a_{H}z^{-1}$, $i_{b_1} = FB_1$, and $i_{b_2} = FB_2$ flowing through the planar atom surface, where $z = \exp(FE^{cor}/2RT)$. As will be seen below, the ratios between these standard currents decide the electrode kinetic characteristics.

The kinetic equations describing the relationships between the current density, i, averaged over the geometrical surface area, on the one hand, and the coverage with the intermediates Fe_kH_{ad}, Fe_kH_2ad, Fe_pHad, and Fe_pH_2ad, denoted by T_1, T_2, Θ_1, and Θ_2, respectively (assuming Langmuir-type adsorption), on the other hand, and the electrode potential have been established for three characteristic situations, namely, the steady-state corrosion potential, the initial moment of polarization, and the steady-state polarization.

As a result of a thorough analysis, four limiting cases consistent with the experimental data in the low region of polarization and one limiting case at higher anodic overvoltage have been distinguished. For all these cases, the relationships $i_{a_{0'}} \gg i_{a_0}$ and $i_{b_1}, i_{b_2} \gg i_{c_1} \gg i_{a_1}$ are valid, which leads to negligible dissolution of planar atoms and a high degree of coverage of the kink sites "k" by adsorbed hydrogen, together with rather bare planar sites. In addition to this, in order for hydrogen evolution on planar atoms to be the predominant cathodic reaction, the inequalities $i_{c_1} \gg i_{a_{1'}} \gg i_{c_{2'}}$ and $i_{c_1} \gg i_{c_1'}$ must be satisfied. In this case, the steady-state cathodic polarization curve displays a Tafel slope $\delta\eta^H/\delta \log i_s = -118$ mV dec^{-1} and a reaction order in hydronium ion $v_{H^+}^s = 1$, as found experimentally.

All the conditions arising in the limiting cases are summarized in Table 1. The kinetic characteristics of the limiting cases at the corrosion

Table 1
Relationships between the Standard Partial Current Densities Arising in the Four Limiting Cases

	Relationships between the partial standard currents	
Case	At the corrosion potential	Under anodic polarization

A. Relationships for the four limiting cases displaying a low steady-state Tafel slope in the linear region of polarization near the corrosion potential and a high coverage of the dissolution sites with hydrogen, concomitantly with rather bare planar atoms.

I	$i_{b_1'}, i_{c_1'} \gg i_{c_2'} \gg i_{a_1'}$	$i_{b_1'}, i_{c'}x^{-1} \gg i_{c_2}x^{-1} \gg i_{a_1'}x$
II	$i_{b_1'}, i_{c_1'} \gg i_{a_1'} \gg i_{c_2'}$	$i_{b_1'}, i_{c_1}x^{-1} \gg i_{a_1'}x \gg i_{c_2}x^{-1}$
III	$i_{c_1'}, i_{c_2'} > i_{a_1'}, i_{b_1'}$	$i_{c_1}x^{-1}, i_{c_2}x^{-1} > i_{a_1'}x, i_{b_1'}$
IV	$i_{a_1'} \gg i_{c_2'} \gg i_{b_1'}$	$i_{c_1}x \gg i_{c_2}x^{-1} \gg i_{b_1'}$
	$i_{c_1'} \gg i_{b_1'}$	$i_{c_1'}x^{-1} \gg i_{b_1'}'$
	$i_{c_1'}i_{c_2'} \gg i_{a_1'}i_{b_1}'$	$i_{c_1}x^{-1}i_{c_2}x^{-1} \gg i_{a_1'}x, i_{b_1'}$

B. Common relationships leading, at high anodic polarization, for the fifth limiting case, characterized by a low coverage of the dissolution sites by adsorbed hydrogen, $b_s^{an} = 59$ mV dec^{-1}, and $v_{OH}^{an,s} = 0$

V	Same as in the limiting case from which it results	$(i_{a_1}x + i_{c_2}x^{-1})/i_{c_1}x^{-1} \gg (1 + i_{c_2}x^{-1}/i_{b_1'})$

C. Common relationships leading to a low coverage by adsorbed hydrogen averaged over the entire surface, a cathodic Tafel slope for the hydrogen evolution reaction of -118 mV dec^{-1}, and $v_{H^+}^{cat,s} = 1$, concomitantly with a preferential dissolution of kink atoms

I–V	$i_{b_1}, i_{b_2} \gg i_{c_1} \gg i_{a_1}$	$i_{b_1}, i_{b_2} \gg i_{c_1}x^{-1} \gg i_{a_1}x$
	$i_{c_1} \gg i_{c_1'}, i_{a_1'}, i_{c_2'}; i_{a_0'} \gg i_{a_0}$	
	$i_{c_1}\Theta_{CF}{}^a > i_{a_1}\Theta_{C1} + i_{c_1'}T_{CF} - i_{a_1'}T_{C1} + i_{c_2'}T_{C1}$	

---

[a] where $\Theta_{CF} = 1 - \Theta_{C1} - \Theta_{C2}$ and $T_{CF} = 1 - T_1 T_2$ is the degree of coverage by free metallic atoms, at the corrosion potential, for the planar atoms (Θ) and the kink atoms (T), respectively.

potential and under anodic polarization are presented in Tables 2 and 3, respectively.

Several important conclusions may be drawn from these tables. First of all, these results demonstrate that a corrosion mechanism in which the atoms at active sites preferentially dissolve and also adsorb hydrogen accounts for all the kinetic data reported in the literature for active corrosion and dissolution of iron in the narrow Tafel region of anodic polarization. A number of experimental facts can thus be explained by the

Table 2
Summary of the Open Circuit Steady-State Characteristics for the Limiting Cases

Case	Predominant species at kink sites	Degree of coverage by bare kink atoms, T_{CF}	Corrosion current, I^{cor}/F	$\Delta \log I^{cor}/\Delta pH$	Corrosion potential, E^{cor}	$\Delta E^{cor}/\Delta pH$ (mV)
I	$Fe_k \cdot H_{ads}$	$\dfrac{C_2'}{C_1'}$	$\left(\dfrac{2A_0'C_2'}{C_1'}\right)^{1/3}(C_1 a_{H^+})^{2/3}$	-0.67	$\dfrac{2RT}{3F}\ln\dfrac{C_1 C_1'}{2A_0'C_2'}a_{H^+}$	-39
II	$Fe_k \cdot H_{ads}$	$\left(\dfrac{C_1}{2A_0'}\right)^{2/5}\left(\dfrac{A_1'}{C_1'}\right)^{3/5}a_{H^+}^{-1/5}$	$\left(\dfrac{2A_0'A_1'}{C_1'}\right)^{1/5}C_1^{4/5}a_{H^+}^{3/5}$	-0.60	$\dfrac{2RT}{5F}\ln\dfrac{C_1 C_1'}{2A_0'A_1'}a_{H^+}^2$	-47
III	$Fe_k \cdot H_{2ads}$	$\left(\dfrac{B_1'}{C_1'}\right)^{3/4}\left(\dfrac{C_1}{2A_0}\right)^{1/4}a_{H^+}^{-1/2}$	$\left(\dfrac{2A_0'B_1'}{C_1'}\right)^{1/4}C_1^{3/4}a_{H^+}^{1/2}$	-0.50	$\dfrac{RT}{2F}\ln\dfrac{C_1 C_1'}{2A_0 B_1'}a_{H^+}^2$	-59
IV	$Fe_k \cdot H_{2ads}$	$\left(\dfrac{A_1'B_1'}{C_1'}\right)^{2/3}\left(\dfrac{C_1}{2A_0'}\right)^{1/3}a_{H^+}^{-1/3}$	$\left(\dfrac{2A_0'A_1'B_1'}{C_1'C_2'}\right)^{1/6}C_1^{5/6}a_{H^+}^{1/2}$	-0.50	$\dfrac{RT}{3F}\ln\dfrac{C_1 C_1'C_2'}{2A_0'A_1'B_1'}a_{H^+}^3$	-59
V	Fe_{Pe}	1	$(2A_0)^{1/3}C_1^{2/3}a_{H^+}^{2/3}$	-0.67	$\dfrac{2RT}{3F}\ln\dfrac{C_1}{2A_0}a_{H^+}$	-39

Table 3
Summary of the Initial and Steady-State Anodic Polarization Characteristics for the Limiting Cases Described in Table 2, When Only the Preponderant Terms $i'_{a0}T_{CF}x^2$ for i_i and $i'_{a0}T_{SF}x^2$ for i_s Are Taken into Account^a

Case	Initial current $i'_{a0}T_{CF}x^2$	$\Delta E/\Delta \log i_i$ (pH = const) (mV dec^{-1})	$\Delta \log i_i/\Delta pH$ (E = const)	Steady-state current $i'_{a0}T_{SF}x^2$	$\Delta E/\Delta \log i_s$ (pH = const) (mV dec^{-1})	$\Delta \log i_s/\Delta pH$ (E = const)	Shape of anodic transient
I	$\dfrac{2A'_0C'_2}{C'_1}y^2$	59	0.0	$\dfrac{2A'_0C'_2}{C_1}y^2$	59	0	Capacitive
II	$\left(\dfrac{2A'_0A'_1}{C'_1}\right)^{3/5} C_1^{2/5} a_{H^+}^{-1/5} y^2$	59	0.2	$\dfrac{2A'_0A'_1}{C'_1} a_{H^+}^{-1} y^4$	29.5	1	Inductive
III	$\left(\dfrac{2A'_0B'_1}{C'_1}\right)^{3/4} C_1^{1/4} a_{H^+}^{-1/2} y^2$	59	0.5	$\dfrac{2A'_0B'_1}{C'_1} a_{H^+}^{-1} y^3$	39.5	1	Inductive
IV	$\left(\dfrac{2A'_0A'_1B'_1C_1}{C'_1C'_2}\right)^{1/2} a_{H^+}^{-1/2} y^2$	59	0.5	$\dfrac{2A'_0A'_1B'_1}{C'_1C'_2} a_{H^+}^{-2} y^5$	23.5	2	Inductive
V	Corresponds to the limiting case at the corrosion potential from which it results	59	—	$\lim_{y\to\infty} i_s/F = 2A'_0 y^2$	59	0	Inductive

$^a x = \exp[F(E_s - E_{cor})/(2RT)]$, $y = \exp[(FE_s)/(2RT)]$, $T = 25°C$.

electrochemical desorption of hydrogen intermediates which, by blocking the metal dissolution sites, act as anodic inhibitors. Thus, the two sets of values frequently found experimentally, namely,

$$\Delta E^{\text{cor}}/\Delta \text{pH} = -47 \text{ mV}, \Delta \log i^{\text{cor}}/\Delta \text{pH} = -0.65, b_a^i = 60 \text{ mV dec}^{-1}, b_a^s = 30 \text{ mV dec}^{-1}, v_{\text{OH}^-}^i = 1, v_{\text{OH}^-}^s = 2$$

and

$$\Delta E^{\text{cor}}/\Delta \text{pH} = -60 \text{ mV}, \Delta \log i^{\text{cor}}/\Delta \text{pH} = -0.5, b_a^i = 60 \text{ mV dec}^{-1}, b_a^s = 40 \text{ mV dec}^{-1}, v_{\text{OH}^-}^i = 0.5, v_{\text{OH}^-}^s = 1,$$

concomitantly with anodic inductive transients (i.e., $i_i < i_s$), are both predicted by this mechanism. The first group of values essentially corresponds to case II and can be explained by the following relative magnitudes of the standard partial currents at kinks: very rapid electrochemical discharge, i_{c_1}, and molecular hydrogen physical desorption, i_{b_1}, both much faster than the anodic desorption, i_{a_1}, the slowest process being the cathodic desorption of atomic hydrogen, i_{c_2}. It is important to notice that with an increase in H^+ activity, i_{c_2} can become greater than i_{a_1}, so that case II turns into case I. Consequently, this could be an explanation for the experimental increase in the steady-state Tafel slope with decreasing pH. The second set of experimental data corresponds to case III, that is, a very high standard rate of Volmer–Heyrovsky cathodic hydrogen discharge in comparison with the rates of oxidation of atomic hydrogen and of physical desorption of molecular hydrogen, at kinks. Theoretical polarization curves computed on the basis of kinetic formulas describing cases II and III are plotted in Figs. 31 and 32, respectively.

It may be noted that, with a decrease in hydrogen ion activity, case III turns into case IV, which means a decrease of the steady-state Tafel slope from 40 to 24 mV dec^{-1} concomitantly with an increase in $v_{\text{OH}^-}^s$ from 1 to 2. This provides a possible explanation for the finding of Bala[68] for an iron rotating disk electrode in acidic sulfate solutions that the Tafel slope increased from 30 to 40 mV dec^{-1} with a decrease in pH from 2 to 1.

Other experimental findings that can be explained by the determining role of hydrogen codeposition on the kinetic behavior of corroding iron are:

(i) The decrease of the dissolution rate, at open circuit and under anodic polarization, with an increase in waiting time, which can be

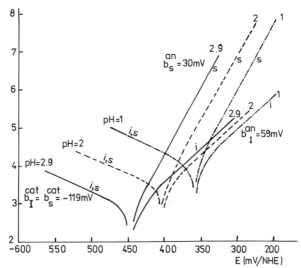

Figure 31. Theoretical polarization curves, computed on the basis of kinetic formulas taken from Tables 1 and 2 describing a preferential iron dissolution and hydrogen adsorption at kink sites concomitantly with a Volmer–Tafel hydrogen evolution reaction at the planar atoms, for ratios between the electrochemical rates approaching the limiting case II at the corrosion potential. Approximate theoretical characteristics at the corrosion potential and in the linear region at moderate polarization: $E^{cor} = (2RT/5F)\ln[(C_1 C_1' a_H^{2+})/(2A_0' A_1')]$, $\Delta E^{cor}/\Delta pH = -47$ mV, $v_{OH^-}^{cor} = -0.60$, $b_s^{an} = 30$ mV dec^{-1}, $v_{OH^-}^s = 1$; $b_i^{an} = 58$ mV dec^{-1}, $v_{OH^-}^i = 0.5$, concomitantly with inductive anodic transients ($i_i < i_s$). The pH values are labeled on the curves.

explained as being due to the time necessary to reach a certain degree of coverage by adsorbed hydrogen

(ii) The decrease or disappearance of galvanostatic peaks after anodic prepolarization, during which the kink coverage by hydrogen is lowered

(iii) The finding that there are promoters of hydrogen embrittlement which increase the hydrogen coverage without acting as anodic inhibitors. This means that in the competition between the adsorption of promoters and of hydrogen on the preferentially dissolving sites, the hydrogen

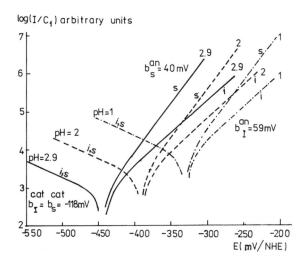

Figure 32. Theoretical initial (---) and steady state (—) polarization curves, for the same mechanism as in Fig. 31, calculated for the ratios between the electrochemical rate constants approaching the limiting case III at the corrosion potential and fitting the Bockris–Drazic–Despic set of data.[81] Approximate theoretical characteristics: at the corrosion potential, $E^{cor} = (RT/2F) \ln [(C_1 C'_1 a_{H^+}^2)/(2A'_0 B'_1)]$, $\Delta E^{cor}/\Delta pH = -59$ mV, $v_{OH^-}^{cor} = -0.50$; in the linear region at moderate polarization, $b_s^{an} = 40$ mV dec^{-1}, $v_{OH^-}^s = 1$, $b_{i_{an}} = 58$ mV dec^{-1}, $v_{OH^-}^i = 0.5$, inductive transients ($i_i < i_s$); and at cathodic polarization (hydrogen evolution reaction), $b_s^{cat} = b_i^{cat} = -118$ mV dec^{-1}, $v_{H^+}^s = v_{H^+}^i = 1$. The pH values are labeled on the curves. (Reproduced from Refs. 66 and 67.)

adsorbs more strongly and rapidly. In this case, the increase in the degree of coverage of the planar atoms with the weakly bound hydrogen would increase only the diffusion of hydrogen into the metal, thus promoting embrittlement.

(iv) The shift of the corrosion potential toward negative values after a strong anodic pulse.

(v) The first inductive loop at low frequencies displayed during the faradaic impedance measurements.

(vi) The increase of the differential capacity, C_{DL}, with immersion time and the high value of C_{DL} measured using an anodic DC pulse after a cathodic one.

This treatment also provides a theoretical support for the earlier ideas of Epelboin and co-workers[238] concerning the role of hydrogen as responsible for the hysteresis observed near the corrosion potential, as well as those of Allgaier[15] concerning the shift of the anodic polarization curves toward lower current densities for cathodically prepolarized iron.

Consequently, the result of the analysis of a mechanism based on iron dissolution at kink sites partially inhibited by adsorbed hydrogen is that the anodic hydrogen desorption can be responsible for the kinetic behavior of the iron–acid system, in a low range of overpotential.

Nevertheless, one must be aware of the fact that once concordance between any reaction scheme and the experimental data has been established, it is not a foregone conclusion that all other concepts about the iron mechanism are invalidated. First of all, any reaction scheme represents a simplified model. *Inter alia*, one must take into consideration that the inductive behavior exhibited by the transient response of the electrode during the step-pulse polarization as well as the first inductive loop displayed in AC measurements could be the result of superposition of hydrogen anodic desorption and some other unknown effect produced by the iron reaction itself. Finally, artifacts are always possible as long as our models are based on the existence of hypothetical intermediates.

2. Autocatalytic Mechanisms

In the context of present-day concepts, branching autocatalytic mechanisms are frequently proposed, especially in attempts to explain the number of inductive loops found at low frequencies[216–218] and sometimes the effect of anions[64,198] on those loops displayed at polarization values encompassing the active and transition ranges. Generally, many new models have been incorporated, in slightly different ways, the first autocatalytic mechanism proposed by Heusler.[70,86]

(i) The Initial Version of the Catalytic Mechanism

The initial version of the catalytic mechanism is represented by the reaction sequence

$$Fe + OH^- \underset{C_0'}{\overset{A_0'}{\Leftrightarrow}} (FeOH)_{ad} + e^-$$

$$\overset{A_0}{Fe + OH^- + (FeOH)_{ad} \rightarrow FeOH^+ + (FeOH)_{ad} + 2e^-}$$
$$FeOH^+ + H^+ \Leftrightarrow Fe^{2+} \cdot H_2O \tag{18}$$

where the monovalent iron hydroxo species $(FeOH)_{ad}$ is the catalyst.

The kinetic quantities characterizing the limiting cases of this reaction model coupled, near the corrosion potential, with the hydrogen ion discharge

$$H^+ + e^- \rightarrow H \tag{19}$$

are summarized below, using the following notation for the standard partial corrosion currents (where A_0', C_0', A_0, and C_1 are electrochemical rate constants):

$$i_{a_0}^{cor} = 2FA_0'a_{OH^-}\exp(2\alpha E^{cor})$$

$$i_{a_0'}^{cor} = FA_0'a_{OH^-}\exp(\alpha E^{cor})$$

$$i_{c_1}^{cor} = FC_1 a_{H^+}\exp(-\alpha E^{cor})$$

$$i_{c_0'}^{cor} = FC_0'\exp(-\alpha E^{cor}) \tag{20}$$

and also

$$x = \exp(\alpha\eta) \tag{21}$$

Case I. When the following inequalities are obeyed,

$$i_{a_0}^{cor} \gg i_{c_0'}^{cor} \gg i_{a_0'}^{cor} \gg i_{c_1}^{cor}; \qquad i_{c_0'}^{cor}x^{-1} \gg i_{a_0'}^{cor}x \tag{22}$$

the kinetic relationships become

$$E^{cor} = [\ln(C_0'a_{H^+}/2A_0)]/3\alpha; \quad \Delta E^{cor}/\Delta pH = -39\ mV \tag{23}$$

$$i^{cor}/F = 2^{-1/3}A_0^{-1/3}A_0'C_0^{1/3}(a_{H^+})^{-2/3}; \quad v_{OH^-}^{cor} = -0.67 \tag{24}$$

$$i_s = i_{a_0}^{cor}i_{a_0'}^{cor}x^4/i_{c_0'}^{cor}; \quad b_a^s = 30\ mV\ dec^{-1} \quad and \quad v_{OH^-}^s = 2 \tag{25}$$

Case II. The conditions

$$i_{a_0}^{cor}, i_{c_1}^{cor} > i_{c_0'}^{cor}x^{-1} \gg i_{a_0'}^{cor}x \tag{26}$$

lead to

$$E^{\text{cor}} = \{\ln[C_1 a_{\text{H}^+}^2/2A_0]\}/3\alpha; \quad \Delta E^{\text{cor}}/\Delta\text{pH} = -78 \text{ mV} \tag{27}$$

$$i^{\text{cor}}/F = 2^{-1/3}A_0^{-1/3}A_0'C_0'^{-1}C_1^{4/3}(a_{\text{H}^+})^{2/3}; \quad v_{\text{OH}^-}^{\text{cor}} = 0.6 \tag{28}$$

$$i_s = i_{a_0}^{\text{cor}}i_{a_{0'}}^{\text{cor}}x^4/i_{c_{0'}}^{\text{cor}}; \quad b_a^s = 30 \text{ mV dec}^{-1} \quad \text{and} \quad v_{\text{OH}^-}^s = 2 \tag{29}$$

Case III. When the following inequalities are obeyed,

$$i_{a_0}^{\text{cor}} \gg i_{c_{0'}}^{\text{cor}}x^{-1} \gg i_{a_{0'}}^{\text{cor}}x; \qquad i_{c_1}^{\text{cor}} \gg i_{a_{0'}}^{\text{cor}} \tag{30}$$

the kinetic relationships become

$$E^{\text{cor}} = 3[\ln(C_1 C_0' a_{\text{H}^+}/2A_0 A_0')]/5\alpha;$$
$$\Delta E^{\text{cor}}/\Delta\text{pH} = -71 \text{ mV}; \quad v_{\text{OH}^-}^{\text{cor}} = 0.6 \tag{31}$$

$$i_s = i_{a_0}^{\text{cor}}i_{a_{0'}}^{\text{cor}}x^4/i_{c_{0'}}^{\text{cor}}; \quad b_a^s = 30 \text{ mV dec}^{-1} \quad \text{and} \quad v_{\text{OH}^-}^s = 2 \tag{32}$$

In all the above cases, the steady-state degree of coverage by the catalyst, Θ_{cat}^s, is

$$\Theta_{\text{cat}}^s = i_{a_{0'}}^{\text{cor}}/i_{c_{0'}}^{\text{cor}}x^{-2} \ll 1 \tag{33}$$

Case IV. The conditions

$$i_{a_0}^{\text{cor}} \gg i_{c_{0'}}^{\text{cor}} \gg i_{c_1}^{\text{cor}}; \quad i_{a_{0'}}^{\text{cor}} \gg i_{c_1}^{\text{cor}} \tag{34}$$

lead to

$$E^{\text{cor}} = [\ln(C_0' a_{\text{H}^+}/2A_0)]/3\alpha; \quad \Delta E^{\text{cor}}/\Delta\text{pH} = -39 \text{ mV} \tag{35}$$

$$i^{\text{cor}}/F = 2A_0 C_0'/A_0'; \quad v_{\text{OH}^-}^{\text{cor}} = 0 \tag{36}$$

$$i_s/F = i_{\text{lim}}^s = 2A_0 C_0'/A_0; \quad b_a^s \to \infty; \quad v_{\text{OH}^-}^s = 0 \tag{37}$$

and

$$\Theta_{\text{cat}}^s = 1/[1 = (i_{c_{0'}}^{\text{cor}}x^{-2}/i_{a_{0'}}^{\text{cor}})] \simeq 1 \tag{38}$$

It is easy to demonstrate that in the first three cases, the electrode system displays an inductive behavior, that is, $i_i < i_s$. Based on the above-mentioned limiting conditions,

$$i_s/F = 2A_0\Theta_{\text{cat}}^s(1 - \Theta_{\text{cat}}^s)a_{\text{OH}^-}\exp(2\alpha E) \tag{39}$$

and

$$i_i/F = 2A_0\Theta_{cat}^{cor}(1 - \Theta_{cat}^{cor})a_{OH^-}\exp(2\alpha E) \tag{40}$$

which yields

$$i_i/i_s = [\Theta_{cat}^{cor}(1 - \Theta_{cat}^{cor})]/[\Theta_{cat}^s(1 - \Theta_{cat}^s)]$$

$$\simeq \Theta_{cat}^{cor}/\Theta_{cat}^s \tag{41}$$

where

$$\Theta_{cat}^s = 1/[1 + (i_{c_{0'}}^{cor}x^{-2}/i_{a_{0'}}^{cor})] \tag{42}$$

and

$$\Theta_{cat}^{cor} = 1/[1 + (i_{c_{0'}}^{cor}/i_{a_{0'}}^{cor})] \tag{43}$$

so that $\Theta_{cat}^s > \Theta_{cat}^{cor}$, and consequently $i_i < i_s$, and also $b_a^i = 60$ mV dec^{-1}.

The following remarks emerge from the analysis of these cases. Cases II and III satisfy the principal experimental characteristics of iron dissolution kinetics—initial and steady-state Tafel slopes, reaction orders in OH$^-$, and corrosion current–pH and corrosion potential–pH relationships, together with inductive transients in the region of low Tafel slopes. In addition to this, if in case III the condition $i_{c_1} \gg i_{a_{0'}}$ is not strictly respected, case IV can exercise some influence, so that the slopes $\delta E^{cor}/\delta$pH, $\delta \log i^{cor}/\delta$pH, and $\delta \log i_s/\delta$pH slightly decrease and b_a^s slightly increases, thus covering the whole spectrum of experimentally reported values.

Thus far, the catalytic mechanism, in its initial form, then appears to be satisfactory. However, a problem arises when the new experimental findings are considered, especially concerning the inhibitory effect of the waiting time, as this reaction scheme predicts an accelerating effect because the fractional coverage by catalysts must increase with time under open-circuit conditions. This means that, even if the mechanism of iron corrosion includes catalytic reactions, they are screened by some other parallel processes leading the accumulation of inhibiting species.

(ii) Kink Atoms as Catalytic Sites

Many authors have agreed that the dissolution reaction mainly starts from the kinks sites. Among the various surface defects that characterize the structure of the singular faces [adatoms, vacancies, adatom and cluster vacancies, half-crystal atoms (kinks), steps], the kink position has a salient feature: if one atom is removed from such a position, the next atom takes its place so that the initial configuration remains intact. Consequently, any change in the surface concentration of kink atoms as a result of changes in overvoltage or pH must be due to some special mechanism other than by consecutive removal of atoms from kinks.

There are two concepts concerning the role of kink atoms in the iron dissolution kinetics and mechanism. One of them is that kink atoms are simply the atoms that dissolve according to the reaction

$$FeFe_kH_2O \xrightarrow{k^+} Fe^{2+}OH^- + 2e^- + Fe_k + H^+ \tag{44}$$

The rate of this reaction is given by

$$i_a = 2Fk^+\Gamma_k \exp(FE/RT) \tag{45}$$

where Γ_k denotes the surface concentration of kink sites, which is proportional to the surface activity. Concomitantly, a pH and potential dependent parallel process results in a change of kink surface concentration, Γ_k.

Based on electrochemical and microscopical investigations, it was established that the average distance of kinks in the steps, x_k, depends on the potential in both directions, according to the formula $\Delta|E|/\Delta \ln (x_k)^{-1} = RT/F$, and the mean step distance x_s, is function of pH according to $\Delta \log x_s/\Delta pH = -0.6$.[15] Using these empirical relationships one can write

$$x_k = x_k^0 + (x_k^\infty - x_k^0)(1 - \exp(-2\alpha|\eta|)) \tag{46}$$

and

$$x_s = x_s^0 + k_s(a_{H^+})^{-0.6} \tag{47}$$

where x_k^0 is the average distance of kinks in steps at the equilibrium potential, x_s^0 is the mean step distance at $a_{H^+} = 0$, k_s is a constant of proportionality, and $|\eta|$ is the absolute value of the overvoltage vs. the equilibrium potential of iron electrode. The surface kink concentration was determined using the equation

$$\Gamma_k = 1 \div (x_k \cdot x_s) \tag{48}$$

so that it becomes

$$\Gamma_k = \Gamma_k^0 \div [x_k^\infty/x_k^0 + (1 - x_k^\infty/x_k^0) \exp(-2\alpha|\eta|)] \tag{49}$$

where $\Gamma_k^0 = 1 \div (x_s^0 + k_s(a_{H^+})^{-0.6})x_k^0$. The kink concentration–overvoltage dependency displays an inversed bell shape, satisfies the conditions $\lim_{\eta \to 0} \Gamma_k = \Gamma_k^0$, $\lim_{(\eta \to \pm \infty)} \Gamma_k = \Gamma_k^\infty$, and yields an increase in Γ_k with increasing η in both directions, the minimum value being the equilibrium one. After the introduction of (49) into (45) one obtains the current–overvoltage relationship for this model. If one takes arbitrarily $x_k^\infty / x_k^0 = 0.01$, it results the steady-state Tafel line with $b_a^s = 30$ mV dec^{-1} (for a symmetry factor 0.5), the initial Tafel line with $b_a^i = 60$ mV dec^{-1} and also the inductive transients. As long as the OH$^-$ ions do not participate directly in the reaction (44), the pH influence on iron dissolution rate is smaller than the experimental one. In addition to this, the total current must include the partial current associated with the kink formation, so that the Tafel slope will be a little higher.

A model accounting for changes in the surface concentration of kinks was developed by Heusler[214] from quantitative observations of the steady-state morphology for iron surface vicinal to (112) during anodic dissolution.[15] He described the model as follows:

> The pyramid formed by complete atomic layers on top of three (112) planes in a bbc lattice displays a group of three atoms at the apex more weakly bound to the metal than the atoms in the $\langle 113 \rangle$ edge and on the complete layers (112). The dissolution starts with the atoms in the apex and then proceeds along the edges. As a result, a monoatomic step in the $\langle 112 \rangle$ direction is formed on the (112) surface. Dissolution of an atom from the intersection of the steps results in the appearance of a kink in a step.

According to Worch and Forker,[136] two oxydryls are necessary to dissolve a kink atom, therefore, Γ_{cat} would be proportional to $\Gamma_k(a_{OH^-})^2$.

The second concept concerning the role of kink atoms in the dissolution kinetics and mechanism is that they form catalytic species of different chemical structure and origin such as:

• monovalent iron hydroxo intermediate like for instances $Fe_k^I (OH)_{ad}$, as a result of the reaction

$$Fe_k H_2O \Leftrightarrow Fe_k^I (OH)_{ad} + H^+ + e^- \tag{50}$$

• A hydroxyl radical, OH, adsorbed at a kink atom, $Fe_k^0 (OH)_{ad}$, as proposed by Schweickert et al.,[118,120] the radical in the adsorbed state being previously formed randomly at the metal surface according to the reaction

$$H_2O \Leftrightarrow (OH)_{ad} + H^+ + e^- \tag{51}$$

followed by

$$Fe_k + (OH)_{ad} \rightarrow Fe_k^0 \, (OH)_{ad} \tag{52}$$

• Simply a kink atom with an OH^- ion preferentially adsorbed on it, $Fe_k(OH^-)_{ad}$, or resulting from the reaction

$$Fe_kH_2O \Leftrightarrow Fe_k(OH^-)_{ad} + H^+ \tag{53}$$

In the latter case, the Heusler mechanism discussed above includes the pH effect, because the coverage by the kink catalyst, Γ_{cat}, is proportional to $\Gamma_k a_{OH^-}$. If the subsequent reaction is

$$FeFe_k(OH^-)_{ad} + OH^- \rightarrow FeOH^+ + Fe_k(OH^-)_{ad} + 2e^- \tag{54}$$

the steady-state reaction order in OH^- becomes 2.

Another possibility is to take into account the pH dependence for the process generating kink positions, so that a reaction order $v_{OH^-}^s$ between 1 and 2 is predicted. In any event, the kink-catalyst theory remains to be worked out.

3. Mechanisms Involving Anion Interference

(i) Mechanisms of Iron Dissolution Involving Oxygenic Anions

There are two viewpoints concerning the role of oxygenic anions in the kinetics of active dissolution of iron in acid media in the Tafel range of polarization.

According to Russian researchers,[183,184,186,187] there is a consecutive charge transfer via the intermediate $Fe^I(OH)_{ad}$, the anions HSO_4^- and SO_4^{2-} being *reactant species* taking part in the second step of the electron transfer, i.e.,

$$Fe^I(OH)_{ad} + SO_4^{2-} \rightarrow FeSO_4 + OH^- + e^- \tag{55}$$

This kind of mechanism accounts for the positive reaction order in SO_4^{2-} found experimentally[183,184,186,187] and exhibits the incompatibility between the low steady-state Tafel slope (30–40 mV dec^{-1}) and the inductive behavior of the electrode.[195,197]

On the contrary, according to the second viewpoint, near the corrosion potential oxygenic anions are slowly chemisorbed,[61,147] more strongly at the dissolution sites than at the remaining atoms, thereby reducing the area free for metal dissolution and thus exerting an *inhibitory effect*. Thus, the

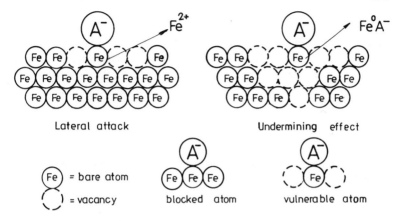

Figure 33. Schematic representation of iron atoms superficially blocked by adsorbed anions removable by lateral attack (left) or by undermining effect (right).

mechanism of dissolution of bare iron atoms or iron atoms covered by adsorbed hydroxyl ions or water molecules, that is, Fe, $Fe \cdot OH_{ad}^-$ or FeH_2O_{ad}, having a degree of coverage $(1 - \Theta_A)$, is still controversial. The iron atoms with adsorbed oxygenic anions atop of them can be electromechanically[61,147] removed from the metal lattice.

In recent papers,[61,147] Drazic and co-workers suggested that, upon dissolution of the free neighboring atoms, iron atoms superficially blocked by adsorbed anions can leave the surface by "electromechanical" processes. When the blocked atom is surrounded by surface vacancies, it becomes *laterally vulnerable* to attack by the solution and leaves the surface by lateral electrochemical dissolution. Furthermore, it can also become vulnerable to attack due to the removal of the underlying atoms, a process termed the *undermining effect* (see Fig. 33). In this second case, the initially blocked atoms chemically dissolve in the solution, outside the electrode interface. At very high current densities, clusters of metallic atoms are detached together with the adsorbed species. As recently shown by Horvath and Kalman,[207] a periodic change in metal surface area must be considered; during the cluster formation, the area increases, and it then sharply decreases at the moment of removal of the cluster.

In any event, any inhibiting effect must cease at the potential of unpolarizability, E_u,[133] where the rate at which the adsorbed species leave the metal surface is higher than the rate of adsorption.[37]

In a first approximation, Drazic and Drazic[61] assumed a Langmuir adsorption–desorption of anions together with a linear relationship between the electromechanical desorption rate, v_3, on the one hand, and the product of the anodic current density associated with the dissolution of the bare metal atoms, $i^{an}_{(1-\Theta_A)}$, and the relative coverage, Θ_A, on the other hand; i.e.,

$$v_3 = K_3 \Theta_A \, i^{an}_{(1-\Theta_A)} \tag{56}$$

where $i^{an}_{(1-\Theta_A)}$ is, of course, proportional to $(1 - \Theta_A)$. The authors developed the kinetic equations for iron electrochemical dissolution on the basis of the Bockris–Drazic–Despic mechanism and, taking into account only the change in surface coverage by anions (i.e., neglecting the contribution of the current associated with the dissolution of metal atoms underneath the anion), performed numerical computations for different values of the kinetic parameters. Like Heusler and Cartledge[133] (see Fig. 34), they obtained S-shaped steady-state polarization curves; over a narrow range of anodic overvoltage, these display different (more or less) linear regions with Tafel slopes as high as 120 mV dec^{-1}, as expected from the assumption of single-electron-transfer steps (see Fig. 35).

Recently, Plonski[179] analyzed the ability of the theory of electromechanical removal of adsorbed anions to fit the experimental initial and steady-state polarization curves of active iron in acid media. Unlike Drazic and Drazic,[61] the author applied the concept of electrochemical dissolution of vulnerable atoms to a mechanism consisting in a slow-discharge hydrogen evolution and a one-step ferrous ion formation. In order to distinguish the possible role of anions from any other interferences, the

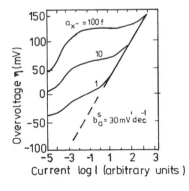

Figure 34. Anodic polarization curves (in arbitrary units) calculated for iron dissolution according to the catalytic mechanism, including the inhibitory effect of adsorbed anions and their potential-dependent desorption via complex formation. (From Ref. 133, by courtesy of authors.)

Figure 35. Theoretical steady-state
anodic polarization curves for iron dissolution in acid media, according to
the consecutive mechanism, in the
presence of different inhibitor concentrations, computed on the basis of kinetic equations including the
electromechanical desorption. (From
Ref. 61, by courtesy of the authors.)

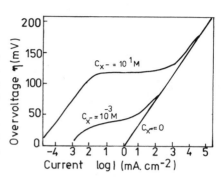

metal surface structure and the blocking effect of adsorbed hydrogen were
ignored.

Polarization curves, computed on the basis of the equation

$$i/i^{cor}_{\Theta_A=0} = [\exp(2\beta_a F\eta/RT) - \exp(-\beta_c F\eta/RT)](1 - \Theta_A)(1 + K_3\Theta_A) \quad (57)$$

where $i^{cor}_{\Theta_A=0}$ is the corrosion current corresponding to the totally bare metal
surface, are plotted in Fig. 36. For certain values of the ratios between the
standard partial rates of anion adsorption and desorption and of iron
dissolution and hydrogen ion discharge, the inductive transients can be
obtained together with straight-line polarization curves, displaying
steady-state Tafel slopes between 29 and 53 mV dec^{-1} and initial Tafel
slopes between 52 and 59 mV dec^{-1}, over an overvoltage range of at least
50 mV, as frequently reported in the literature. In spite of this agreement,
one cannot know whether the electromechanical removal of anions is the
main cause of the iron electrode characteristics as well as of the loss of
surface activity over time or whether it must be taken into account only
as a corrective factor along with more important effects such as that of
hydrogen adsorption or of surface structure.

(ii) Mechanisms of the Dissolution of Iron Involving Halogenide Ions

(a) Reaction mechanisms as a function of specific dissolution sites

A schematic representation of the suggested mechanisms for iron
dissolution from iron surfaces in acid solutions containing halogenide
ions, X^-, is presented in Fig. 37. As may be seen from this figure, there
are three kinds of dissolution sites at which branching reactions can occur.

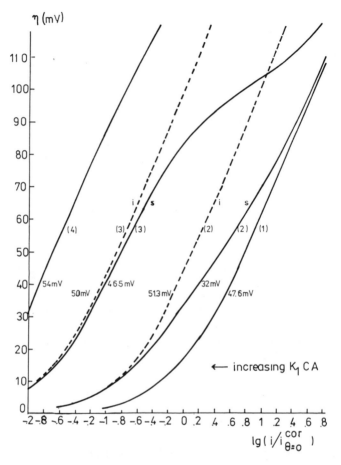

Figure 36. Anodic polarization curves for metal dissolution in one step concomitantly with the electromechanical desorption of the inhibiting adsorbed anions, fitting the initial and steady-state experimental curves for iron in acidic sulfate solutions. (From Ref. 179.)

Bare metallic atoms or metallic atoms covered by hydroxyl ions or water molecules, having a relative coverage $(1 - \Theta_A)$, constitute the first kind of dissolution site. Two reaction routes have been proposed as being followed on this part of the surface:

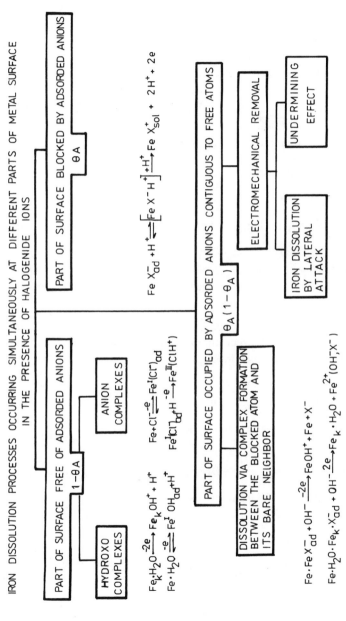

Figure 37. Schematic representation of the iron dissolution reactions in electrolytes containing halogenide ions, occurring simultaneously at different parts of the metal surface, as proposed by various authors.

(i) At low X^- concentration, at shorter waiting times, iron dissolves according to the same mechanism as in the absence of halogenide ions.

(ii) At high X^- concentration, hydroxochloro (at high pH) or hydrochloro (at low pH) iron(I) complexes are formed as intermediates by reactions of iron with the halogenide ion in solution, such as [50]

$$Fe^IOH_{ad} + X^-_{sol} \Leftrightarrow Fe^IOHX^-_{ad} \overset{-e^-}{\Leftrightarrow} Fe^{II}OHCl_{ad} \overset{+H^+}{\Leftrightarrow} Fe^{2+} + X^- + H_2O \quad (58)$$

or, as proposed in Refs. 187, 208, and 209,

$$Fe + X^-_{sol} + H_2O \Leftrightarrow Fe^IX\,OH^-_{ad} + H^+ + e^- \quad (59)$$

or, as proposed in Refs. 208 and 210,

$$FeH_2O_{ad} + X^-_{sol} + H^+ \Leftrightarrow Fe^I(XH^+)_{ad} + e^- + H_2O \quad (60)$$

Metallic atoms covered with adsorbed anions contiguous to bare atoms, having a relative coverage $\Theta_A(1 - \Theta_A)$, constitute the second kind of dissolution site. These atoms can leave the surface according to one of the following possible processes:

(i) *Complex formation* between the covered atom and its bare neighbor followed by an electrochemical desorption. For instance, the following reaction was proposed in Ref. 104:

$$[FeH_2O_{ad}\,Fe_kX^-_{ad}\,] + OH^- \overset{-2e^-}{\rightarrow} Fe_kH_2O + Fe^{II}\,(OH^-,X^-) \quad (61)$$

Alternatively, according to Ref. 191,

$$[Fe(X^-_{ad}Fe)] + OH^- \overset{-2e^-}{\rightarrow} Fe + FeOH^+ + X^-_{sol} \quad (62)$$

In Ref. 211, the following reactions were proposed:

$$FeX^-_{ad}FeOH \overset{-2e^-}{\rightarrow} [FeFe^{II}(X^-OH^-)] + OH^- \overset{+H^+}{\rightarrow} Fe^{2+} + X^- + H_2O \quad (63)$$

The following reaction was suggested in Ref. 212:

$$[Fe(OH)_{ad}Fe^0(H_2O)^+_{ad}A^-] \overset{-e^-}{\rightarrow} Fe(OH)_2 + H^+A^- + Fe \quad (64)$$

(ii) *Electromechanical removal*[61] of adsorbed species, that is, dissolution of metallic atoms, superficially blocked by anions, by undermining or lateral attack, after the surrounding atoms have been dissolved.

Atoms blocked by adsorbed anions, having a relative coverage Θ_A, constitute the third kind of dissolution site. These atoms leave the electrode surface by direct reaction of $Fe \cdot A_{ad}^-$ with other species in the solution. Examples of such reactions are as follows:

$$FeX_{ad}^- + H_2O \Leftrightarrow FeOH_{ad}^- + X_{sol}^- + H^+ \quad (\text{Ref. 213}) \quad (65)$$

$$Fe^ICl_{ad} + H^+ \rightarrow Fe^IClH_{sol}^+ \quad (\text{at high pH; Ref. 208}) \quad (66)$$

$$FeX_{ad}^- + H^+ \Leftrightarrow [FeX^-H^+] \rightarrow FeX_{sol^-}^+ \quad (\text{Ref. 13}) \quad (67)$$

It may be noted that, according to Drazic and Drazic,[61] the principal way in which the adsorbed inhibiting anions leave the surface is via electromechanical desorption; the accelerating effect of halogenide ions at high X^- and H^+ concentrations is attributed to the formation of weak complexes between Fe and X^- in the solution at a critical concentration of halogenide and hydrogen ions.

The first kinetic relationships describing the inhibitory effect of halide ions at low anodic overvoltage followed by the sudden increase of current at the potential of unpolarizability were given by Heusler and Cartledge.[133] The mechanism consists of two charge-transfer reactions occurring in parallel at the half-crystal position[34,214]; that is, at the kinks that are not covered by adsorbed anions, the main dissolution reaction occurs:

$$FeH_2O \cdot Fe_k^I OH_{ad} \Leftrightarrow Fe_k^IOH + FeOH_{sol}^+ + H^+ + 2e^- \quad (68)$$

while, at the kink atoms covered by adsorbed anions but having bare neighboring atoms, a parallel charge-transfer process occurs, yielding an intermediate complex of ferrous ion with halide ion:

$$FeH_2O \cdot Fe_kX_{ad}^- + OH^- \rightarrow Fe_kH_2O_{ad} + [Fe^{2+}(X^-,OH^-)] + 2e^- \quad (69)$$

The overall steady-state rate of the two parallel reactions is

$$i = k'_+ a_{OH^-}\Theta_A(1 - \Theta_A) \exp(2\beta FE/RT)$$
$$+ k_+(a_{OH^-})^2(1 - \Theta_A)^2 \exp[(1 + 2\alpha)FE/RT] \quad (70)$$

It can be seen that, at low current densities, for high values of the coverage when Θ_A is approximately potential- and pH-independent, the first term becomes dominant, and the polarization curve displays a linear part with a slope close to 60 mV dec^{-1}; at high current densities, when $\Theta_A \to 0$, the second term prevails, and the polarization curve coincides with that found in the absence of inhibitor ($b_a^s = 30$ mV dec^{-1} and $v_{OH^-}^s = 2$), a prediction verified experimentally by Heidemayer.[215] The initial polarization curves, calculated by assuming that changes in the adsorption on the iron surface are slow, lead to $b_a^i = 60$ mV dec^{-1} and $i_i < i_s$. The S-shaped polarization curves calculated on the basis of this formula and assuming Langmuir adsorption–desorption have been given in Fig. 34.

As already stated, the potential of unpolarizability, E_u, is given by

$$E_u = (2RT/3F) \ln[(k_1 a_{X^-})/(2k_+' a_{OH^-}\Theta_A)] \tag{71}$$

Consequently, it represents the potential at which the rate of chemical adsorption of anions, $k_1 a_{X^-}(1 - \Theta_A)$, and the rate of their electrochemical desorption, $2k_+' a_{OH^-}\Theta_A(1 - \Theta_A) \exp(2\beta FE/RT)$, are equal. This means that at potentials more positive than E_u, the adsorbed anions are removed too quickly for adsorption to occur over time.[37]

(b) Coverage by anions and the potential of unpolarizability

By comparing the rate of anion removal by electrochemical lateral attack, $v_3 = K_3 i_{(1-\Theta_A)}^{an}$, taken from Ref. 50 as equal to $K_3 i_{\Theta_A=0}^{cor}\Theta_A(1 - \Theta_A) \times \exp(F\eta/RT)$ (mol cm^{-2} s^{-1}), with the current associated with this process, which is proportional to the degree of coverage by laterally vulnerable atoms, Θ_V, that is, $i_V = i_{\Theta_A=0}^{cor}\Theta_V \exp(F\eta/RT)$ (A cm^{-2}), one obtains the parabolic function

$$\Theta_V = FK_3\Theta_A(1 - \Theta_A) \tag{72}$$

where $0 < FK_3 < 1$. According to Ref. 50, K_3 is a constant describing the anodic current efficiency for electromechanical desorption. The maximum value of Θ_V, that is, its largest contribution to the total current, and the maximum rate of removal of adsorbed anions occur at $\Theta_A = 0.5$, when, statistically, every blocked metallic atom has a bare neighbor. This leads to the result that $\Theta_V < 0.25$. Figure 38 shows plots of Θ_A and Θ_V/FK_3 as a function of anodic overvoltage, calculated for particular values of the rate constants of the reactions involved, and illustrates the effect of an increase in the standard adsorption rate $K_1 a_{A^-}$. As can be seen, the

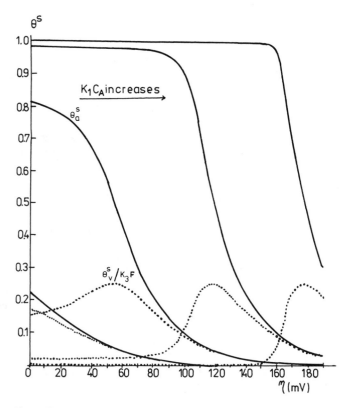

Figure 38. The coverage by adsorbed anions (——) and by superficially blocked, laterally vulnerable metallic atoms (······) as a function of anodic overvoltage, calculated for particular values of the rate constants of the processes involved. (From Ref. 180.)

steady-state degree of coverage by superficially blocked, laterally vulnerable atoms, Θ_V^s, depends on η in a nonmonotonous way, displaying an asymmetric maximum at an overvoltage value corresponding to the inflection point of the Θ_A^s–η curve, for which $d\Theta_A^s/d\eta$ is maximum. Like the potential of unpolarizability,[50,133] the overvoltage value corresponding to the maximum of the Θ_V^s–η curve shifts toward more positive values with an increase in the standard rate of anion adsorption, $K_1 a_{A^-}$, as well as with an increase in the maximum value of the corrosion current, $i_{\Theta_A=0}^{cor}$. Consequently, another interpretation of the potential of unpolarizability

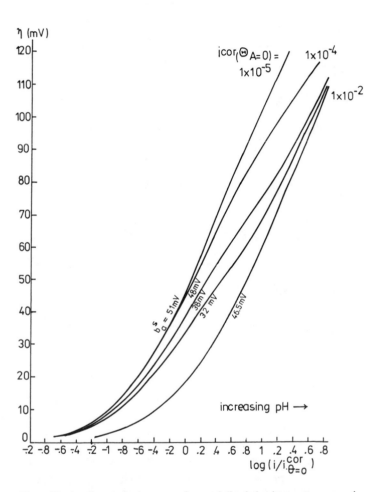

Figure 39. Anodic polarization curves for metal dissolution in one step concomitantly with the lateral attack of laterally vulnerable, superficial blocked atoms, calculated for particular values of rate constants, fitting the oscillatory behavior of the Tafel slope for iron dissolution in chloride electrolytes. (From Ref. 180.)

may be put forward: it is the potential corresponding to the maximum degree of coverage by metallic atoms superficially blocked by adsorbed inhibiting species but contiguous with vacancies.[134,179]

Referring back to Fig. 36, several additional remarks can be made about the polarization curves:

(i) In a certain range of overvoltage, with an increase in adsorbable anion concentration, the anodic polarization curves can display increasing but also decreasing Tafel slopes, in the range between 30 and 50 mV dec^{-1}.

(ii) In the same region of overvoltage, b_a^s can increase both with an increase in pH (i.e., decreasing $i_{\Theta_A=0}^{cor}$) and also with a decrease in pH (i.e., increasing $i_{\Theta_A=0}^{cor}$) from 32 to about 50 mV dec^{-1} (see Fig. 39).

(iii) With an increase in current density, b_a^s first increases, then decreases, and then increases again.

From the above analysis, it may be seen that another reason for the variability of the data in the literature, which Drazic and Drazic[50] and also Bockris and Khan[36] have attributed mainly to the different waiting times and/or polarization times used by the various authors, these times being insufficient to establish the steady state, in some cases is the oscillatory influence of overvoltage, pH, and anion concentration on the Tafel slope. This possibility is supported by such apparently contradictory results as, for example, the values of b_a^s reported by Lorenz and co-workers[45–47] for Fe in HCl–NaCl solutions of 60 mV dec^{-1} for pH 0–1.5 and 30 mV dec^{-1} for pH 1.5–3, in comparison with 30 mV dec^{-1} for pH < 1.5 and 60–70 mV dec^{-1} for pH > 1.5, found by Oftedal.[48]

4. Branching Mechanisms

Around 1980, some branching mechanisms were proposed with the intention of describing the processes occurring in the active, transition, and prepassive ranges of the overall active state, and explaining the different values of the experimental kinetic data obtained by their authors. In addition to this, the supporters of the consecutive electron-transfer concept offered an explanation for the disagreement between the experimental low steady-state Tafel slope and the inductive behavior of the electrode, on the one hand, and the theoretical predictions, on the other, as demonstrated by Plonski[95,195–197] since 1969.

The common concept of these models was the existence of at least one single electron-transfer stage and of special centers on the metal surface which are "active" in catalyst formation and/or adsorption of intermediates. The chemical activity and concentration of these particular sites depended in a complex way on the surface treatments prior to the electrode immersion, the potential, and time. In all these reaction schemes, some adsorbed ferrous compounds were formed in the transition range.

The electron-transfer reactions proposed to occur in the active range were either consecutive or catalytic ones, or both, and will be indicated in the reaction schemes by solid lines. Dashed lines will indicate to the processes which are supposed to occur in the other range of polarization (which will not be discussed in this chapter).

(i) The Drazic-Vorkapic Branching Mechanism (DV)

In 1981, the following variant of the B. D. D. mechanism, called the branching mechanism, was proposed by Drazic and Vorkapic[38,126]:

$$\overset{A_1}{\underset{C_1}{Fe + H_2O \Leftrightarrow FeOH_{ad} + H^+ + e^-}} \qquad (73)$$

$$\overset{A_2}{FeOH_{ad} \rightarrow FeOH_{ad}^+ + e^-}$$

(I)   (II)

$$\overset{B}{FeOH_{ad}^+ \rightarrow FeOH^+} \qquad\qquad \overset{B_1}{\underset{B_2}{FeOH_{ad}^+ + H_2O \Leftrightarrow FeOH_{2ad} + H^+}}$$

$$FeOH^+ + H^+ \Leftrightarrow Fe^{2+} + H_2O \qquad \overset{B_3}{Fe(OH)_{2ad} \rightarrow Fe(OH)_2}$$

$$Fe(OH)_2 + 2H^+ \Leftrightarrow Fe^2 + 2H_2O$$

where A_1, A_2, B, B_1, B_2, B_3, and C_1 denote the rate constants. In the framework of this reaction scheme the paths I and II run in parallel, the switch from one to the other route depends on the surface properties, the second electron transfer is irreversible, and (unlike El Miligy et al.,[24] who propose that ferrous hydroxide is formed only in the transition range) $Fe^{II}(OH)_2$ exists in both adsorbed and desorbed states in the active range of iron dissolution in acid solutions (corrosion potential included). According to the authors, any one of the three intermediates desorption rates, A_2, B, or B_3, can become the rate-determining step, depending on the adsorption properties of the metal (which can be changed, for example, by mechanical or electrochemical pre-treatments). As a result of their kinetic analysis, the authors[38] provided the following diagnostic criteria for the branching mechanism, where Θ is the surface coverage:

Rate-determining step	$\Theta \ll 1$			$\Theta \to 1$		
	α_a	b_a (mV dec^{-1})	ν_{OH^-}	α_a	b_a (mV dec^{-1})	ν_{OH^-}
A_2	1.5	40	1	0.5	120	0
B	2	30	1	0	$\to \infty$	0
B_3	2	30	2	0	$\to \infty$	0

Furthermore, the authors predict that for annealed iron, at lower current densities there is a low Θ and a Tafel slope of 40 mV dec^{-1}, while at higher current densities there is an increase to 120 mV dec^{-1}. They also include the possible appearance of a limiting reaction current when the rate of the step A_2 cannot be increased further with increase of potential due to the subsequent chemical desorption B or B_3 becoming rate-controlling.

For an "activated" surface, i.e., for a metal surface displaying a large number of sites active for the adsorption of intermediates, a slope of 30 mV dec^{-1} and a limiting reaction current at higher polarization are predicted.

Finally, if the rate constants of the two desorption reactions B and B_3 are approximately the same, the two branches proceed with similar rates, and a ν_{OH^-} between 1 and 2 characterizes the kinetics.

Below, an analysis of this reaction scheme is presented, in which it is assumed a constant surface activity for intermediates' adsorption–desorption and symmetry factors for the electron transfer equal to 0.5.

Branching I is characterized by the general kinetic equations

$$i/F = A_{1y}\Theta_F - C_1 H^+ y^{-1}\Theta_1 + A_{2y}\Theta_1 - C_1' H^+ y^{-1}\Theta_F \qquad (74)$$

$$C_M(d\Theta_1/dt) = A_{1y}\Theta_F - C_1 H^+ y^{-1}\Theta_1 - A_{2y}\Theta_1 \qquad (75)$$

where

$y = \exp(FE/2RT)$, C_M = the concentration of metal surface sites,

$$\Theta_1 = FeOH_{ad}/C_M, \quad \Theta_2 = FeOH_{ad}^+/C_M, \quad \text{and} \quad \Theta_F = 1 - \Theta_1 - \Theta_2 \qquad (76)$$

and

$$C_M(d\Theta_2/dt) = A_{2y}\Theta_1 - B\Theta_2 \qquad (77)$$

In order to analyze the branching mechanism near the corrosion potential, the cathodic current associated with the hydrogen evolution reaction, $FC_1'H^+y^{-1}\Theta_F$, is included.

At anodic polarization steady-state, denoted by the index "s," we have

$$\Theta_1^s = 1 \div \text{DENOMS1} \qquad (78)$$

$$\Theta_2^s = (A_2y/B) \div \text{DENOMS1} \qquad (79)$$

where

$$\text{DENOMS1} = 1 + A_2/A_1 + A_2y/B + C_1H^+y^{-2}/A_1 \qquad (80)$$

so that

$$\Theta_F^s = (A_2/A_1 + C_1H^+y^{-2}/A_1) \div \text{DENOMS1} \qquad (81)$$

and

$$i_S/F = [2A_2y - (A_2/A_1 + C_1H^+y^{-2}/A_1)C_1'H^+y^{-1}] \div \text{DENOMS1} \quad (82)$$

At the corrosion potential, denoted by the index c, the relative coverage by the intermediates Θ_1^C and Θ_2^C are given by expressions similar to (77) and (78) with the exception that y is replaced by $z = \exp(FE^{cor}/2RT)$.

As usual, it will be assumed that during the step increase of the potential, which lasts $\ll 1\mu s$, the degree of coverage by adsorbed intermediates has no time to change significantly from the value previously established at the corrosion potential, and the polarization program consists in applying every pulse of potential starting every time from the same open circuit state.

Consequently, the initial current–potential relationship is given by

$$i_i/F = A_1y\Theta_{CF} - C_1H^+y^{-1}\Theta_{C1} + A_2y\Theta_{C1} - C_1'H^+y^{-1}\Theta_{CF} \qquad (83a)$$

or

$$i_i/F = \{[A_2 + A_1(A_2/A_1 + C_1H^+z^{-2}/A_1)]y - [C_1 + C_1'(A_2/A_1$$
$$+ C_1H^+z^{-2}/A_1)]H^+y^{-1}\} \div \text{DENOMC1} \qquad (83b)$$

where

$$DENOMC1 = 1 + A_2/A_1 + A_2z/B + C_1H^+z^{-2}/A_1 \qquad (84)$$

Limiting cases: The usual procedure to identify the limit cases is to establish such inequalities between the terms of Eqs. (81)–(83b) so that only one term is predominant in the numerator and only one term in the denominator.

Limit case I: The relationships between the standard partial currents arising in the first limit case are as follows:

$$C_1H^+y^{-1} \gg A_{1y}, A_{2y}, A_1A_2y^2/B \qquad (85)$$

$$C_1'H^+y^{-1} \ll 2A_1A_2y^2/(C_1H^+y^{-1}) \qquad (86)$$

The introduction of (73) and (74) into (69) and (71) gives

$$i_S = 2FA_1A_2y^3/(C_1H^+) \qquad (87)$$

so that $b_s^{an} = 40$ mV dec^{-1} and $v_{OH^-(E=const.)}^s = 1$, concomitantly with a very high degree of free surface, Θ_F^s, in agreement with the first case in the table above. But, we also have

$$i_i = FA_1y \qquad (88)$$

from which $b_i^{an} = 120$ mV dec^{-1} and $v_{OH^-(E=const.)}^i = 0$.

Because according to (85), (87), and (88)

$$i_s/i_i = 2A_2y/(C_1H^+y^{-1}) \ll 1 \qquad (89)$$

for this limit case, two disagreements with the experimental findings result: an initial Tafel slope of 120 instead of 60 mV dec^{-1}, and especially capacitive transients ($i_s < i_i$) instead of the always-reported inductive ones.

From a careful analysis of Eqs. (81) and (82)in conjunction with (71), one can notice that there is no other possible correlation between the terms of these equations leading to low Tafel slopes. Thus, if according to the inequalities in Eq. (86) the contribution of the hydrogen discharge to the total current is negligible in the linear region of the polarization curve, the predominance of any other term, except $C_1H^+y^{-2}/A_1$ in the denominator, yields Tafel slopes between 120 and ∞. Concomitantly, the coverage by bare metal atoms drastically decreases.

Limit case II: For example, if the physico–chemical desorption of FeOH$_{ad}^+$, (reaction B), controls the kinetics, that implies

$$A_2z/B \gg 1, A_2/A_1, C_1H^+z^{-2}/A_1$$

$$B \ll A_1 z, A_2 z, A_1 A_2 z^3 / C_1 H^+ \tag{90}$$

For this limit case

$$\Theta_F^s = (A_2/A_1 + C_1 H^+ y^{-2}/A_1) \div (A_2 y/B) \ll 1 \tag{91}$$

Also,

$$i^s/F = 2B \tag{92}$$

so that $b_s^{an} \to \infty$, and we cannot have simultaneously the $FeOH_{ad}^+$ chemical desorption as the r.d.s., a high degree of free metal surface, and a low Tafel slope, as predicted by the authors.

Proceeding in the same way, one arrives to similar conclusions for path II.

In addition to this, the mechanism does not work in solutions containing 1M Fe^{2+} at a pH of approx. 3.5 when (see Fig. 10) the equilibrium potential of Fe/Fe^{2+} reaction is close to the corrosion potential, because the step A_2 is completely irreversible and cannot account for the cathodic ferrous ion discharge.

(ii) Keddam–Mattos–Takenouti Branching Mechanism (KMT)

The electrode impedances, recorded starting from different points on the steady-state polarization curve during the dissolution of Fe in 1 M SO_4^{2-} solutions within the ranges: pH, 0–5; d.c., 10^{-4}–5.10^{-2} A cm^{-2}; frequency, 10^{-2}–10^5 Hz, were qualitatively interpreted by computer simulation on the basis of the following reaction model including three dissolution paths and three adsorbed intermediates.[123,125]

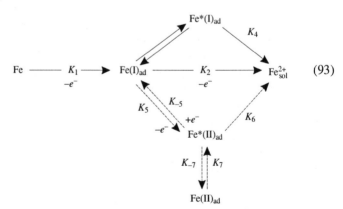

$$\tag{93}$$

where the authors make a distinction between an ordinary adsorption site and an active (*) adsorption site.

The irreversible consecutive paths K_1, K_2 were introduced in earlier papers by Epelboin et al.[140,238] to explain impedance measurements on active iron electrodes showing inductive parts. Their fundamental considerations were intended to provide the theoretical support for the experimental steady-state Tafel slope $b_s^a = 63$ mV dec^{-1} at low overvoltages, and 220 mV dec^{-1} at high overvoltages with the corresponding electrochemical reaction orders of $v_{OH^-}^s = 1$ and $v_{OH^-}^s = 0$, respectively. In order to fit the experimental results, the authors used very asymmetric charge transfer coefficients (symmetry factors) of $\beta_1 = 0.96$–0.99 for the first and $\beta_2 = 0.18$–0.27 for the second single charge transfer step. Also, the so-called chemical step, implying no electron transfer, might be potential dependent.

As in previous papers,[140,238] the basic reactions operating in the active range at low current densities are via K_1, K_2, i.e., the consecutive mechanism is perceived as practically irreversible and controlling the overall rate.

The self-catalytic path via $Fe^*(II)_{ad}$ formation was introduced to account for the increased number of inductive loops when the polarization encompassed the transition and the prepassive ranges.

Neither the role of OH^- ions nor the chemical nature of the intermediate species are considered a priori, so that any stoichiometric reaction cannot be established; hence, theoretical kinetic quantities with mechanistic meaning such as dE_{cor}/d pH, d log i_{cor}/d pH, $v_{OH^-}^s$, and v_{OH}^i are not predictable.

However, some electronic reactions can be written as follows:[123,125,129]

$$Fe \xrightarrow{K_1} Fe(I)_{ad} + e^-$$

$$Fe(I)_{ad} \xrightarrow{K_2} Fe(II)_{sol} + e^-$$

$$Fe(I)_{ad} \underset{K_{-3}}{\overset{K_3}{\Longleftrightarrow}} Fe^*(I)_{ad}$$

$$Fe + Fe^*(I)_{ad} \xrightarrow{K_4} Fe(FeII)_{sol} + 2e^- \tag{94}$$

A lot of drawbacks have been described for the above model. These are:[118]

- electrochemical charge transfer coefficients very close to zero or unity
- non-electrochemical reactions assumed to be potential-dependent
- steady-state Tafel slopes ≥ 60 mV dec^{-1}, in contrast with common experimental findings
- the fact that, phenomenologically, the authors attributed the high degree of coverage with adsorbed hydrogen as resulting from h. e. r. at the corrosion potential
- the responsibility for the lower frequency inductive loop which can be observed at this potential as well as at low anodic polarization, and while the authors overlooked h.e.r. in the numerical computation.

However, the fact that chemical reactions like the catalyst $Fe^*(I)_{ad}$ formation from $Fe(I)_{ad}$ can be potential-dependent is not surprising in view of the later experimental finding that the kink site concentration increases with both anodic and cathodic polarization.[15,37,199] Also, the authors stated from the beginning that at the time they ignored h.e.r., so consequently their analysis was related only to the possible iron reactions, and underlying that the Tafel slope without the influence of hydrogen coverage was close to 0.06 V dec^{-1}. This is a customary practice in physics in order to separate the effects of one variable from the others, and is commonly used in the iterative modelling of processes.

In addition to this, it was demonstrated in section IV.1 that if the hydrogen reaction occurs differentially, i.e., on planar atoms according to Volmer–Tafel and on kink sites obeying the Volmer–Heyrowsky mechanism, the electrode response is inductive and is characterized by a low ss-Tafel slope, with a low coverage by adsorbed hydrogen as averaged over the entire surface, but a high degree of coverage at kinks.

Consequently, it appears that the theoretical analysis performed by the authors is an intermediate one; taking into account the hydrogen reaction will help to surpass partially the present drawbacks. Moreover, as positively stated in Ref. 217, an orthodox treatment of catalytic reactions implies taking into consideration that the catalyst must have among its immediate neighbors at least one bare iron atom on which the catalytic site is shifted. In other words, the relationship between the free neighbors, catalysts, pH, and potential must be established.

The KMT model is still present in the literature and used as basis for related mechanisms including the role of anions on anodic iron dissolution.[129]

(iii) Schweickert–Lorenz–Friedburg Branching Mechanism (SLF)

The kinetic data resulting from the pulse and frequency response measurements on polycrystalline iron (r.d.e) in acidic sulfate solutions $0.3 \leq pH \leq 6$, i $(10^2–5.10^2)$ mA cm^{-2} (after prior anodic polarization by a current density of 70 mA cm^{-2} for 5 min in the measuring cell) were interpreted on the basis of the following branching reaction scheme:

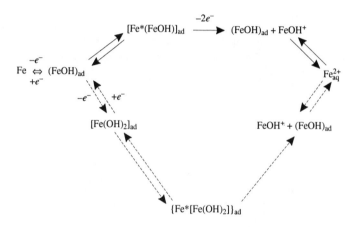

where the asterisk stands for an adsorbed intermediate at an active surface site Fe*. According to this model the active range of iron dissolution is characterized by the sequence of reactions

$$Fe + H_2O \Leftrightarrow Fe(OH)_{ad} + H^+ + e^- \tag{95}$$

followed by

$$Fe^* + (FeOH)_{ad} \Leftrightarrow [Fe^*(FeOH)]_{ad}$$

$$[Fe^*(FeOH)]_{ad} + OH^- \xrightarrow{rds} FeOH^+ + (FeOH)_{ad} + 2e^- \tag{96}$$

that is, a variant of the conventional catalytic mechanism.

The first step is also written[37] as the oxidation of an hydroxyl ion leading to an adsorbed radical

$$H_2O \Leftrightarrow (OH)_{ad} + H^+ + e^-$$ (97)

after which the radical $OH^{\bullet}$ can move toward an active site and form the catalyst. (Such neutral radicals are usually formed during the gamma - radiolysis of water).

It is assumed that the number and the chemical activity of the surface active centers Fe^* and of the adsorption sites depend on the surface substructure of the electrode, the potential, and time. The surface relaxation phenomena are thus responsible for the transient behavior of the electrode.

On the basis of this concept, the authors demonstrated the agreement between the experimental data $b_s^a = 40$ mV dec^{-1} and $v_{OH^-}^s \approx 1$ and the impedance loops in the active range of dissolution, for reasonable symmetry factors $\beta_1 = \beta_2 = 0.4$, $\beta\psi = \beta\phi = 0.5$, where β_1, β_2 refer to the first and second step of charge transfer and $\beta\psi$, $\beta\phi$ to the hypothetical surface exponential functions assumed to correlate the potential with the surface roughness and active centers, respectively.

V. CONCLUDING REMARKS

The present state of knowledge and the unresolved problems concerning active dissolution of iron in acid media may be summarized as follows:

(a) Workers in this field are faced with an abundance of qualitative and quantitative experimental results characterized by a high variability in the data, some of the data being irreproducible or even self-contradictory. Two consequences emerged from this confused situation: (a) a variety of interpretations of the data have been proposed, leading to a large number of reaction schemes, involving consecutive, parallel, and branching paths; and (b) a high degree of uncertainty is attached to the experimental kinetic quantities with mechanistic significance and to the theories developed on the basis of these quantities.

(b) There is a scarcity of data for the solid/fluid interface, mainly due to the difficulty in reproducing and controlling the metal surface at the molecular level and to the dynamic nature of the interface.

(c) In the last decade, surface analysis by *in situ* light microscopy and *ex situ* electron microscopy has provided some information concerning the change of certain surface characteristics as a result of attack by acid solution, such as an increase in the step and kink density, etch pits, and roughness, as a function of pH, polarization, and pretreatment of the electrode surface. It has been found that the changes in the surface structure, morphology, and chemistry during the contact with the electrolyte influences the dissolution kinetics to a larger extent than differences in the initial state of the metal surface as the result of various mechano-metallurgical treatments. On the basis of these experimental findings, catalytic mechanisms involving the kink atoms as specific dissolution sites, and, when an OH⁻ is adsorbed at these sites, as catalytic species, have been suggested by German scientists. Nevertheless, only a small number of studies on the relation between the structure and texture of the crystalline surface and corrosion kinetics have been reported, and this work should be expanded in order to improve the accuracy of the data and to separate the important factors from the irrelevant ones; theoretical models well supported by experiments must then be developed, and their predictions investigated.

(d) The specific adsorption of species from the solution, particularly anions, has been extensively studied in recent years by Serbian scientists. The experimental findings revealed the inhibitory effect of adsorbed anions near the corrosion potential and their possible removal via the dissolution of the underlying metal atoms. Another important aspect is the possible correlation between the time dependence of the polarization curves and of the kinetic characteristics at the corrosion potential and the time-dependent degree of coverage by adsorbed anions.

(e) The role of adsorbed hydrogen, resulting from the hydrogen evolution reaction, on the iron dissolution kinetics has also been exhaustively analyzed in recent years. Reaction schemes based on the concept that the kinks, being preferential sites for metal dissolution and for hydrogen adsorption, could explain the iron electrode behavior in the anodic region of polarization not too far from the corrosion potential have yielded good agreement with experimental data. Thus, these models are able to predict the experimental values for the dependence of the corrosion potential and the corrosion current on pH, the initial and steady-state Tafel slopes, reaction orders in OH⁻ for both the hydrogen evolution and iron dissolution reactions, and inductive anodic transients; in addition, these mechanisms explain the hysteresis near the corrosion potential, the high degree of coverage by hydrogen at kinks, and the low coverage averaged over the whole surface.

(f) There is a general agreement on the multitude of possible phenomena occurring at the metal/solution interface: electric charging, intermediary formation of hydroxo species, adsorption of hydrogen and of anions, and increases in kink and step density as well as in surface roughness. However, there is no consensus on the relative contribution of each of these processes to the dissolution kinetics, or on their mechanism and interdependence.

(g) Real progress in the field will be achieved when the resolution of microscopes or the sensitivity of other instruments, such as the quartz crystal microbalance under an AC regime that has recently been developed by French scientists [233,234] for the investigation of the anodic behavior of iron in sulfuric acid medium, becomes sufficient to reveal, *in situ*, the detailed chemical structure of iron atoms complexed with or covered by adsorbed species.

(h) It is worth emphasizing that there are many initial and also steady-state polarization curves, as a function of waiting and polarization time. Variables which change slowly, such as the surface roughness, can be considered as constant over a comparatively short interval of polarization time so that although they determine the initial state, which must be well defined, they have no significant influence on the measurement. The same is true for the sluggish anion adsorption.

(i) The corrosion potential is a special steady state at which the rates of anodic dissolution and of cathodic reduction of an oxidized substance are equal. The initial reaction order with respect to hydroxyl ions is related to the kinetics of both reactions and therefore provides information about their mutual influence and about the surface structure and chemistry at the corrosion potential.

(j) With respect to the opposing viewpoints regarding the mechanism of iron dissolution, there are insufficient grounds for the rejection of either stepwise electron transfer or two-electron transfer in one step.

(k) Finally, it will be useful to perform a systematic analysis of new models in parallel with a reevaluation of the old reaction schemes, taking into account recently discovered effects, such as by incorporating the preferential dissolution of the kink atoms in the framework of the Bockris–Drazic–Despic (BDD) mechanism or the influence of hydrogen adsorption on the catalytic mechanism.

REFERENCES

[1] F. A. Micheeva, G. M. Florianovich, Ya. M. Kolotyrkin, and F. Ya. Florov, *Zashch. Met.* **23** (1987) 915.

[2] A. M. Kolotyrkin and G. M. Florianovich, *Zashch. Met.* **20** (1984) 1.

[3] M. Smialowski, *Neue Hütte* **2** (1957) 621.

[4] W. Raszynski and M. Smialowski, *Bull. Acad. Pol. Sci., Ser. Sci. Chim.* **8** (1960) 209.

[5] M. A. Machuba and Z. A. Iofa, *Zashch. Met.* **4** (1968) 444.

[6] S. P. Kuznetsova and N. P. Zhuk, *Zashch. Met.* **11** (1975) 726.

[7] D. Vdovenko and P. S. Anikina, *Zashch. Met.* **10** (1974) 157.

[8] L. E. Tsygankova, V. I. Vigdorovich, T. V. Korneeva, and N. V. Osipova, *Zh. Prikl. Khim.* **49** (1976) 1323.

[9] G. G. Penov, A. P. Botneva, and N. P. Zhuk, *Zashch. Met.* **10** (1974) 322.

[10] G. G. Penov, Z. Ya. Kosakovskaia, A. N. Botneva, L. A. Andreeva, and N. P. Zhuk, *Zashch. Met.* **6** (1970) 544.

[11] V. I. Grigo'ev, G. I. Ekilik, and V. V. Ekilik, *Zashch. Met.* **10** (1974) 716.

[12] L. Z. Vorkapic and D. M. Drazic, *Corros. Sci.* **19** (1979) 643.

[13] E. McCafferty and N. Hackerman, *J. Electrochem. Soc.* **119** (1972) 999.

[14] W. Allgaier and K. E. Heusler, *J. Appl. Electrochem.* **9** (1979) 155.

[15] W. Allgaier, Doctoral Thesis, University of Clausthal, 1975.

[16] J. Bessone, L. Karakaya, P. Lorbeer, and W. J. Lorenz, *Electrochim. Acta* **22** (1977) 1147.

[17] P. Lorbeer, Doctoral Thesis, University of Karlsruhe, 1978.

[18] G. Bech-Nielsen, *Electrochim. Acta* **23** (1978) 425.

[19] G. Bech-Nielsen, *Electrochim. Acta* **19** (1974) 821.

[20] G. Bech-Nielsen, M. Mogansen, and J. C. Reeve, *Werkst. Korros.* **31** (1980) 340.

[21] G. Bech-Nielsen, *Electrochim. Acta* **21** (1976) 627.

[22] G. Bech-Nielsen, *Electrochim. Acta* **18** (1973) 671.

[23] G. Bech-Nielsen, *Electrochim. Acta* **27** (1982) 1383.

[24] A. A. El Miligy, D. Geana, and W. J. Lorenz, *Electrochim. Acta* **20** (1975) 273.

[25] D. Geana, A. A. El Miligy, and W. J. Lorenz, *Corros. Sci.* **13** (1973) 505.

[26] D. Geana, A. A. El Miligy, and W. J. Lorenz, *Corros. Sci.* **14** (1974) 657.

[27] D. Geana, A. A. El Miligy, and W. J. Lorenz, *J. Appl. Electrochem.* **14** (1974) 337.

[28] P. Lorbeer and W. J. Lorenz, *Electrochim. Acta* **25** (1980) 375.

[29] P. Lorbeer and W. J. Lorenz, *Corros. Sci.* **20** (1980) 405.

[30] P. Lorbeer and W. J. Lorenz, *Electrochim. Acta* **25** (1980) 849.

[31] F. M. Mikheeva, G. M. Florianovich, and O. S. Sadunishvili, *Zashch. Met.* **13** (1977) 68.

[32] D. M. Drazic and Chen Shen Hao, *Electrochim. Acta* **27** (1982) 149.

[33] Ph. Morel, Thesis, University of Paris, 1968.

[34] W. J. Lorenz and F. Mansfeld, in *Proceedings of the 8th International Congress on Metallic Corrosion*, Mainz, Dechema, Frankfurt September, 1981.

[35] H. Bala, *Electrochim. Acta* **29** (1984) 119.

[36] J. O'M. Bockris and U. M. Khan, *Surface Electrochemistry*, Plenum Press, New York, 1993.

[37] W. J. Lorenz and K. E. Heusler, in *Corrosion Mechanisms*, Ed. by F. Mansfeld, Marcel Dekker, New York, 1987.

[38] D. M. Drazic, in *Modern Aspects of Electrochemistry*, No. 19, Ed. by J. O'M. Bockris and B. E. Conway, Plenum Press, New York, 1989.

[39] G. M. Florianovich, *Zashch. Met.* **27** (1991) 581.

[40] S. Haruyama, in *Proceedings of the Second Japan–USSR Corrosion Seminar*, Japan Society of Corrosion Engineering, Tokyo, 1980.

[41] N. Sato, in *Proceedings of the Second Japan–USSR Corrosion Seminar*, Society of Corrosion Engineering, Tokyo, 1980.

[42]M. Keddam, in *Anodic Dissolution of Pure Metals and Alloys*, Ed. by P. Marcus and J. Outar, in press.

[43]F. Hilbert, Y. Miyoshi, G. Eichkorn, and W. J. Lorenz, *J. Electrochem. Soc.* **118** (1971) 1919.

[44]W. J. Lorenz, H. Yamaoka, and H. Fischer, *Z. Elektrochem.* **67** (1963) 932.

[45]W. J. Lorenz and J. R. Vilche, *Corros. Sci.* **12** (1972) 785.

[46]W. J. Lorenz and G. Eichkorn, *Ber. Bunsenges. Phys. Chem.* **70** (1966) 99.

[47]F. Hilbert, Y. Miyoshi, G. Eichkorn, and W. J. Lorenz, *J. Electrochem. Soc.* **118** (1971) 1927.

[48]T. A. Oftedal, *Electrochim. Acta* **18** (1973) 401.

[49]A. Atkinson and A. Marshall, *Corros. Sci.* **18** (1978) 427.

[50]V. J. Drazic and D. M. Drazic, *J. Serb. Chem. Soc.* **57** (1992) 917.

[51]J. O'M. Bockris, D. M. Drazic, and A. R. Despic, *Electrochim. Acta* **4** (1961) 325.

[52]A. Akiyama, R. E. Patterson, and K. Nobe, *Corrosion* **26** (1970) 51.

[53]M. Keddam and I. H. Plonski, to be published.

[54]G. J. Bignold and M. Fleischmann, *Electrochim. Acta* **19** (1974) 363.

[55]F. Mansfeld, *Corrosion* **29** (1973) 397.

[56]D. M. Drazic and S. K. Zecevic, *Corros. Sci.* **25** (1985) 209.

[57]O. J. Murphy, T. E. Pou, and J. O'M. Bockris, *J. Electrochem. Soc.* **131** (1984) 2787.

[58]F. von Frenzel, *Werkst. Korros.* **26** (1975) 443.

[59]N. Hackerman and S. J. Stephens, *J. Phys. Chem.* **58** (1954) 904.

[60]D. M. Drazic, V. J. Drazic, and V. Jevtic, *Electrochim. Acta* **34** (1989) 1251.

[61]V. J. Drazic and D. M. Drazic, in *Proceedings of the 7th European Symposium on Corrosion Inhibitors* (7 SEIC), *Ann. Univ. Ferrara, N. S., Sez. V*, Ferrara, 1990.

[62]I. H. Plonski, *Rom. J. Phys.* **38** (1993) 589.

[63]I. H. Plonski, in *Proceedings of the 7th International Symposium on Passivity, Passivation of Metals and Semiconductors*, Claushal, Germany, 1994, *Material Science Forum.* Ed. by K. E. Heusler, Trans. Tech., Switzerland, **185–188** (1995) 649.

[64]I. H. Plonski, *Corrosion* **47** (1991) 840.

[65]I. H. Plonski, *Rev. Roum. Phys.* **37** (1992) 601.

[66]I. H. Plonski, *Z. Phys. Chem.* **97** (1993) 1020.

[67]I. H. Plonski, in *Abstracts, 44th Meeting of the International Society of Electrochemistry*, Berlin, Germany, 1993.

[68]H. Bala, *Electrochim. Acta* **30** (1985) 1043.

[69]H. Bala, *Metalurgica i oglewnictwo* **16** (1990) 97.

[70]K. E. Heusler, *Z. Elektrochem.* **62** (1958) 582.

[71]C. Voigt, *Electrochim. Acta* **13** (1968) 2037.

[72]S. Haruyama, S. Asawa, and K. Nagasaki, *Denki Kagaku* **39** (1971) 564.

[73]S. Asawa, S. Haruyama, and K. Nagasaki, *J. Jpn. Inst. Met.* **36**, (1972) 440.

[74]S. Haruyama and S. Asawa, *Corros. Sci.* **13** (1973) 395.

[75]N. D. Tomashov and L. P. Vershinia, *Electrochim. Acta* **15** (1970) 501.

[76]R. T. Foley and P. P. Trzaskoma, in *Passivity of Metals*, Ed. by R. R. Frankenthal and J. Kruyer, The Electrochemical Society, Princeton, New Jersey, 1978.

[77]I. H. Plonski and A. Rusi, *Roum. J. Phys.* **38** (1993) 851.

[78]K. Schwabe and C. Voigt, *J. Electrochem. Soc.* **113** (1966) 886.

[79]K. F. Bonhoeffer and K. E. Heusler, *Z. Elektrochem.* **61** (1957) 122.

[80]K. F. Bonhoeffer and K. E. Heusler, *Z. Phys. Chem. N. F.* **8** (1956) 390.

[81]S. A. Balezin and T. Krasovitskaya, *Zh. Prikl. Khim.* **24** (1951) 197.

[82]Z. A. Foroulis, *J. Electrochem. Soc.* **113** (1966) 532.

[83]Z. A. Foroulis, *Corros. Sci.* **5** (1965) 39.

[84]T. Hurlen, *Electrochim. Acta* **7** (1962) 653.

[85]V. V. Batrakov, G. H. Avad, E. I. Mikhailova, and Z. A. Iofa, *Elektrokhimiya* **4** (1968) 601.

[86]K. E. Heusler, Doctoral Thesis, University of Gottingen, 1957.

[87]E. O. Ayazyan, *Dokl. Akad. Nauk SSSR* **100** (1955) 473.

[88]L. I. Antropov, in *Electrochimie Théorique*, Ed. by A. Assimov, MIR, 1979.

[89]Z. A. Iofa, V. V. Batrakov, and C. N. Ba, *Electrochim. Acta* **9** (1964) 1645.

[90]W. J. Lorenz and H. Fischer, *Electrochim. Acta* **11** (1966) 1597.

[91]W. J. Lorenz, K. Sarropoulos, and H. Fischer, *Electrochim. Acta* **14** (1969) 179.

[92]A. B. Sheinin, M. B. Rytoinskaya, and V. L. Cheifoz, *Zh. Fiz. Khim.* **38** (1964) 2562.

[93]T. N. Voropaeva, B. V. Deryagin, and B. N. Kabanov, *Izv. Akad. Nauk SSSR* **1** (1963) 257.

[94]O. E. Barcia and O. R. Mattos, *Electrochim. Acta* **35** (1990) 1003.

[95]I. H. Plonski, *Rev. Roum. Phys.* **16** (1971) 449.

[96]S. H. Glarum and J. H. Marshall, *J. Electrochem. Soc.* **126** (1979) 424.

[97]O. R. Mattos, Thesis, University of Paris, 1981.

[98]V. A. Roiter, V. A. Yuza, and E. S. Poluyan, *Zh. Fiz. Khim.* **13** (1939) 377.

[99]K. E. Heusler, *Z. Elektrochem.* **66** (1962) 177.

[100]K. E. Heusler, Habilitationsschrift, University of Stuttgart, 1966.

[101]G. Eichkorn, W. J. Lorenz, L. Albert, and H. Fischer *Electrochim. Acta* **13** (1968) 183.

[102]G. Eichkorn and W. J. Lorenz, *Metalloberfläche* **22** (1968) 102.

[103]B. Folleher and K. E. Heusler, *J. Electroanal. Chem.* **180** (1984) 77.

[104]H. Rosswag, G. Eichkorn, and W. J. Lorenz, *Werkst. Korros.* **25** (1974) 86.

[105]J. R. Vilche and W. J. Lorenz, *Corros. Sci.* **12** (1972) 785.

[106]D. Posadas, A. J. Arvia, and J. J. Podesta, *Electrochim. Acta* **16** (1971) 1025.

[107]R. V. Moshtev and N. I. Khristova, *Corros. Sci.* **7** (1967) 255.

[108]V. A. Roiter, V. A. Yuza, and E. S. Poluyan, *Zh. Fiz. Khim.* **13** (1939) 605.

[109]I. H. Plonski, A. Cristescu, G. Dumitru, and A. Andrei, *R. I. IRNE Pitesti* **525** (1979) 49; *Rom. J. Phys.* **40** (1995).

[110]V. S. Muralidharan and K. S. Rajagopalan, *Corros. Sci.* **19** (1979) 199.

[111]V. S. Muralidharan, K. Thangavel, K. Balakrishnan, and K. S. Rajagopalan, *Electrochim. Acta* **29** (1984) 1003.

[112]G. M. Florianovich and F. M. Mikheeva, Paper presented at the 38th Meeting of the International Society of Electrochemistry, Vilnius, 1986.

[113]K. E. Heusler, Habilitationsschrift, TH Stuttgart, 1966.

[114]W. J. Lorenz, Habilitationsschrift, University of Karlsruhe, 1968.

[115]J. O'M. Bockris and H. Kita, *J. Electrochem. Soc.* **108** (1961) 676.

[116]H. Takenouti, Thesis, University of Paris, 1971.

[117]J. A. Harrison and W. J. Lorenz, *Electrochim. Acta* **22** (1977) 205.

[118]H. Schweickert, W. J. Lorenz, and H. Friedburg, *J. Electrochem. Soc.* **127** (1980) 1693.

[119]H. Schweickert, W. J. Lorenz, and H. Friedburg, *J. Electrochem. Soc.* **128** (1981) 1295.

[120]H. Schweickert, Doctoral Thesis, University of Karlsruhe, 1978.

[121]K. Juttner and W. J. Lorenz, in *Proceedings of the 8th International Congress on Metallic Corrosion, Mainz, 1981*, Vol. I, DECHEMA, Frankfurt, 1981.

[122]W. J. Lorenz and A. A. El Miligy, *J. Electrochem. Soc.* **120** (1973) 1698.

[123]M. Keddam, O. R. Mattos, and H. Takenouti, *J. Electrochem. Soc.* **128** (1981) 257.

[124]A. Caprani, I. Epelboin, P. Morel, and H. Takenouti, in *Proceedings, 4th European Symposium on Corrosion Inhibitors*, Ferrara, 1975, Vol III. Ann. University of Ferrara, N.S.

[125]M. Keddam, O. R. Mattos, and H. Takenouti, *J. Electrochem. Soc.* **128** (1981) 266.

[126]D. M. Drazic and L. Z. Vorkapic, *Glas. Hem. Drus. Beograd* **46** (1981) 595.

[127]I. Epelboin, P. Morel, and H. Takenouti, *J. Electrochem. Soc.* **118** (1971) 1282.

[128]D. MacFarlane and S. I. Smedley, *J. Electrochem. Soc.* **133** (1986) 2240.

[129]O. E. Barcia and O. R. Mattos, *Electrochim. Acta* **35** (1990) 1601.

[130]H. Gerischer and W. Mehl, *Z. Elektrochem.* **59** (1955) 1049.

[131]I. H. Plonski, *Rom. J. Phys.* **38** (1993) 259.

[132]I. H. Plonski, *Corrosion* **46** (1990) 581.

[133]K. E. Heusler and G. H. Cartledge, *J. Electrochem. Soc.* **108** (1961) 732.

[134]I. H. Plonski, Paper presented at the 45th Annual Meeting of the International Society of Electrochemistry, Porto, Portugal, 1994.

[135]M. Fleischmann, S. K. Rangarajan, and H. R. Thirsk, *Trans. Faraday Soc.* **63** (1967) 1240.

[136]H. Worch and W. Forker, *Electrochim. Acta* **35**, (1990) 163.

[137]E. J. Kelly, *J. Electrochem. Soc.* **112** (1965) 124.

[138]A. M. Florianovich and F. M. Mikheeva, *Elektrokhimiya* **23** (1987) 1414.

[139]I. Epelboin and M. Keddam, *J. Electrochem. Soc.* **117** (1970) 1052.

[140]B. Bechet, I. Epelboin, and M. Keddam, *J. Electroanal. Chem.* **76** (1977) 129.

[141]E. Fujii, T. Tsuru, and S. Haruyama, in press.

[142]W. Lorenz, *Z. Elektrochem.* **57** (1953) 382.

[143]W. Lorenz, *Z. Phys. Chem.* **202** (1953) 275.

[144]M. Eyring, S. Glasstone, and K. J. Laidler, *J. Chem. Phys.* **7** (1939) 1053.

[145]K. J. Vetter, in Elektrochemische Kinetik, Ed. by Ya. M. Kolotyrkin, Moscow, 1967.

[146]B. E. Conway, in *Theory and Principles of Electrode Processes*, Ronald Press, New York, 1965.

[147]V. J. Drazic, D. M. Drazic, and S. Mitrovski, Paper presented at the 5th International Fischer Symposium on Adsorbates and Inhibitors, Karlsruhe, Germany, June 16–20, 1991.

[148]I. H. Plonski, Paper presented at the 5th International Symposium on Adsorbates, Intermediates and Inhibitors, Karlsruhe, Germany, June 16–20, 1991.

[149]B. Hakanson, N. G. Vannerberger, and G. Bech-Nielsen, *Electrochim. Acta* **28** (1983) 451.

[150]D. M. Drazic and V. Vascic, in *Proceedings of the 9th Yugoslav Symposium on Electrochemistry*, Dubrovnik, 1985.

[151]F. Matsuda and T. C. Franklin, *J. Electrochem. Soc.* **112** (1965) 767.

[152]K. E. Heusler, *Corros. Sci.* **31** (1990) 753.

[153]K. E. Heusler, *Electrochemistry at Well Defined Metal Surfaces, J. Chim Phys.* (1991).

[154]A. A. Abdul Azim and S. H. Sanad, *Electrochim. Acta* **17** (1972) 1699.

[155]O. V. Kasparova, A. V. Plaskaev, Ya. M. Kolotyrkin, E. F. Koloskova, D. S. Kamenetskaya, Yu. V. Maish, I. B. Piletskaya, and V. I. Shiryaev, *Zashch. Met.* **21** (1985) 339.

[156]H. Przewlocka and H. Bala, *Werkst. Korros.* **32** (1981) 443.

[157]Yu. N. Mikhailovskii and L. V. Gaysenko, *Zashch. Met.* **22** (1968) 11.

[158]W. J. Lorenz and G. Lorenz, *Naturwissenschaften* **52** (1965) 618.

[159]W. J. Lorenz, Habilitationsschrift, TH Karlsruhe, 1967.

[160]O. V. Kasparova and A. V. Plaskaev, *Zashch. Met.* **19** (1983) 541.

[161]C. Fabian, M. R. Kazemi, and A. Neckel, *Ber. Bunsenges. Phys. Chem.* **84** (1980) 1026.

[162]W. Allgaier and K. E. Heusler, *Z. Metallkd.* **67** (1976) 766.

[163]G. T. Burstein and D. H. Davies, *J. Electrochem. Soc.* **128** (1981) 33.

[164]G. T. Burstein and M. A. Kearns, *J. Electrochem. Soc.* **131** (1984) 991.

[165]K. Nobe, E. Baum, and F. Seyer, *J. Electrochem. Soc.* **108** (1961) 97.

[166]J. C. Gidding, A. G. Funk, C. J. Christensen, and H. Eyring, *J. Electrochem. Soc.* **106** (1959) 91.

[167]A. Akiyama and Ken Nobe, *J. Electrochem. Soc.* **117** (1970) 999.

[168]T. Shibata, T. Takeyama, and G. Okamoto, in *Passivity and Its Breakdown on Iron and Iron Base Alloys*, Ed. by R. W. Starhle and H. Okada, Japan–USA Seminar, National Association of Corrosion Engineers, Houston, 1976.

[169]R. G. Raicheff, A. Damjanovic, and J. O'M. Bockris, *J. Chem. Phys.* **47** (1967) 2198.

[170]M. A. Devanathan and J. Fernando, *Electrochim. Acta* **15** (1970) 1623.

[171]T. P. Hoar, in *Proceedings of the Conference on Fundamental Aspects of Stress Corrosion Cracking*, National Association of Corrosion Engineers, Houston, 1969.

[172]T. P. Hoar and T. Hurlen, in *CITCE 1956*, Butterworths, London, 1958.

[173] T. P. Hoar and T. Hurlen, in *Proceedings of the VIIIth Meeting of CITCE, Madrid, 1956*, Butterworths, London, 1959.

[174] L. Z. Vorkapic and D. M. Drazic, *Glas. Hem. Drus. Beograd* **42** (1977) 545.

[175] H. S. Wroblowa, Yen Chi Pann, and J. Razumney, *J. Electroanal. Chem.* **69** (1976) 195.

[176] G. Bech-Nielsen and J. C. Reeve, in *Proceedings of the 6th Scandinavian Corrosion Congress*, Gothenburg, 1971.

[177] G. Bech-Nielsen, *Electrochim. Acta* **20** (1975) 619.

[178] G. Walpert, *Z. Phys. Chem.* **A151** (1930) 219.

[179] I. H. Plonski, *Rom. J. Phys.* **40** (1995) 547.

[180] H. J. Podesta and A. J. Arvia, *Electrochim. Acta* **10** (1965) 159, 171.

[181] W. J. Lorenz, G. Eichkorn, and C. Mayer, *Corros. Sci.* **7** (1967) 357.

[182] M. Vojnovic, Dissertation, University of Belgrade, 1981.

[183] Ya. M. Kolotyrkin, *Zashch. Met.* **3** (1967) 131.

[184] G. M. Florianovich, L. A. Sokolova, and Ya. M. Kolotyrkin, *Elektrokhimiya* **3** (1967) 1027.

[185] G. M. Florianovich, L. A. Sokolova, and Ya. M. Kolotyrkin, *Elektrokhimiya* **3** (1967) 1359.

[186] G. M. Florianovich and M. Kolotyrkin, in *Korroziya i Zashchita ot Korrozii*, Vol. 6, Itogi Nauki i Tekhniki, VINITI, Moscow, 1978.

[187] S. M. Reshetnikov, *Zh. Prikl. Khim.* **53** (1980) 572.

[188] V. J. Drazic, D. M. Drazic, and V. Jevtic, *J. Serb. Chem. Soc.* **52** (1987) 711.

[189] G. M. Florianovich, F. M. Mikheeva, and V. A. Goriachkin, in *Proceedings of the Symposium on Dissolution of Metals and Alloys*, Sofia, 1977, p. 1.

[190] S. Haruyama and S. Masaki, unpublished results.

[191] W. J. Lorenz, *Corros. Sci.* **5** (1965) 121.

[192] H. Bala, *Br. Corros. J.* **23** (1988) 29.

[194] C. D. Nenitescu, in *General Chemistry*, Editura Didactica si Pedagogica, Bucharest, 1985.

[193] B. Andrzejaczek and Z. Szklarska-Smialowska, *Bull. Acad. Pol. Sci. Ser. Chem.* **24** (1976) 199.

[195] I. H. Plonski, *J. Electrochem. Soc.* **116** (1969) 944.

[196] I. H. Plonski, *J. Electrochem. Soc.* **116** (1969) 1068.

[197] I. H. Plonski, *Z. Phys. Chem., Ber. Phys. Chem.* **95** (1991) 23.

[198] I. H. Plonski, *Stud. Cercet. Fiz.* **43** (1991) 395.

[199] W. Allgaier and K. E. Heusler, *Z. Phys. Chem. N. F.* **98** (1961) 1975.

[200] K. E. Heusler, *DECHEMA Monogr.* **93** (1983) 193.

[201] H. J. Flitt and J. O'M. Bockris, *Int J. Hydrogen Energy* **7** (1982) 411.

[202] J. O'M. Bockris, J. L. Carbajal, B. R. Scharifker, and K. Chandrasekaram, *J. Electrochem. Soc.* **134** (1987) 1957.

[203] J. G. N. Thomas, *Trans. Faraday Soc.* **57** (1961) 160.

[204] H. G. Cleary and N. D. Greene, *Corros. Sci.* **7** (1969) 821.

[205] G. M. Florianovich, A. Sokolova, and Ya. M. Kolotyrkin, *Electrochim. Acta* **12** (1967) 879.

[206] S. Barnartt, *J. Electrochem. Soc.* **119** (1972) 812.

[207] T. Horvath and E. Kalman, Paper presented at the 45th Annual Meeting of the International Society of Electrochemistry, Porto, Portugal, 1994.

[208] H. C. Kuo and K. Nobe, *J. Electrochem. Soc.* **125** (1978) 853.

[209] P. J. Chin and K. Nobe, *J. Electrochem. Soc.* **119** (1972) 1471.

[210] N. A. Darwish, F. Hilbert, W. J. Lorenz, and H. Roswag, *Electrochim. Acta* **18** (1973) 421.

[211] K. Schwabe and C. Voigt, *Electrochim. Acta* **14** (1969) 853.

[212] G. M. Florianovich, P. M. Lazorenko-Manevich, and Ya. M. Kolotyrkin, in *Korozia i Zaschita Koroziia*, Vol. 16, Itogi Nauki, Tekhniki, VINITI, Moscow, 1990.

[213] A. J. Arvia and J. J. Podesta, *Corros. Sci.* **8** (1968) 203.

[214]K. E. Heusler, in *Encyclopedia of Electrochemistry of the Elements*, Vol. 9, Ed. by A. J. Bard, Marcel Dekker, New York, 1982.

[215]J. Heidemayer, Dissertation, Technische University, Berlin, 1965.

[216]M. Keddam, O. R. Mattos, and H. Takenouti, *J. Electrochem. Soc.* **128** (1981) 1294.

[217]M. Keddam, O. R. Mattos, and H. Takenouti, *Electrochim. Acta* **31** (1986) 1159.

[218]L. Cavallaro, L. Felloni, F. Pulidori, and F. Trabanelli, *Corrosion* **18** (1962) 3971.

[219]G. P. Cammarota, L. Felloni, G. Palombarini, and S. Sosterotraverso, *Corrosion* **26** (1970) 129.

[220]A. A. El Miligy, F. Hilbert, and W. J. Lorenz, *J. Electrochem. Soc.* **120** (1973) 247.

[221]T. Hurlen, *Acta Chem. Scand.* **14** (1960) 1555.

[222]T. Hurlen, *Acta Chem. Scand.* **14** (1960) 1564.

[223]Ya. Zytner and A. L. Rotinyan, *Elektrokhimiya* **2** (1966) 1371.

[224]V. M. Nagirny, R. U. Bondar', and V. V. Stender, *Zh. Prikl. Khim.* **42** (1969) 2236.

[225]G. Pinard-Legry and G. Plante, *Mater. Chem.* **1** (1976) 321.

[226]V. A. Kuznetsov and Z. A. Iofa, *Zh. Fiz. Khim.* **31** (1947) 201.

[227]A. C. Makrides, N. M. Komodromos, and N. Hackerman, *J. Electrochem. Soc.* **102** (1955) 363.

[228]L. Cavallaro, L. Felloni, G. Trabanelli, and F. Pulidori, *Electrochim. Acta* **9** (1964) 485.

[229]L. Felloni, *Corros. Sci.* **8** (1968) 133.

[230]M. Moegensen, G. Bech-Nielsen, and E. Maahn, *Electrochim. Acta* **25** (1980) 919.

[231]J. J. Podesta and A. J. Arvia, *Electrochim. Acta* **10** (1965) 171.

[232]N. Eldakar and K. Nobe, *Corrosion* **32** (1976) 238.

[233]C. Gabrielli, M. Keddam, F. Minouflet, and H. Perrot, Paper presented at the 7th International Symposium on Passivity, Passivation of Metals and Semiconductors, Claushal, Germany, 1994.

[234]M. Keddam, Paper presented at the 45th Annual Meeting of the International Society of Electrochemistry, Porto, Portugal, 1994.

[235]H. Nord and G. Bech-Nielsen, *Electrochim. Acta* **19** (1971) 849.

[236]G. Ström, B. Hâkansson, and N. G. Vannerberg, in: *Proceedings of the 10th Scandinavian Corrosion Congress*, Stockholm, 1986, No. 39, p. 201.

[237]V. J. Drazic, D. M. Drazic, and V. Jevtic, *J. Serb. Chem. Soc.* **52** (1987) 711.

[238]I. Epelboin, C. Gabrielli, M. Keddam, and H. Takenouti, *Electrochim. Acta* **20** (1975) 913.

[239]E. Mc Cafferty and C. Zettlemoyer, *J. Phys. Chem.* **71** (1967) 2452.

[240]A. J. Appleby, H. Kita, M. Chemla, and G. Bronoel, in *Encyclopedia of the Electrochemistry of Elements*, Vol. 9, Ed. A. Bard, Marcel Dekker, New York, 1982.

[241]P. Ruetschi and P. Delahay, *J. Chem. Phys.* **26** (1955) 195.

[242]B. G. Conway and J. O'M. Bockris, *J. Chem. Phys.* **26** (1957) 532.

[243]S. Trasatti, *J. Electroanalyt. Chem.* **39** (1972) 163.

[244]J. O'M. Bockris and D. F. A. Koch, *J. Phys. Chem.* **65** (1961) 1941.

4

Electrochemical Investigations of the Interfacial Behavior of Proteins

Sharon G. Roscoe

Department of Chemistry, Acadia University, Wolfville, Nova Scotia, Canada B0P 1X0

I. INTRODUCTION

A number of studies have been made on the interaction of proteins with solid surfaces in an attempt to determine the molecular conformation or orientation of the adsorbed molecules. This interest in the interfacial behavior of proteins at solid surfaces originates from the need to better understand the mechanisms of processes associated with their use in advanced technical applications and industrial problems. The use of immobilized enzymes in analytical techniques in biotechnology and in chromatography requires knowledge of the interfacial behavior of proteins. Adsorption of proteins and enzymes on various kinds of adsorbents is widely used for the purification, identification, fixation, and separation of these materials.[1] The interaction of proteins with solid surfaces causes major problems in many industrial and medical areas, such as the "fouling" of surfaces in the food processing industry and in medical implant devices and biosensors, as well as microbial growth due to protein adhesion to surfaces. The mechanism of protein interactions with surfaces of artificial materials as well as the ability to control these processes are of great interest for such areas as molecular electronics, biometrics, biocompatible materials, drug release systems, and immunoassays.[2]

The problem of "fouling" of metal surfaces in the food processing industry has been particularly acute in the dairy industry.[3,4] The heat

Modern Aspects of Electrochemistry, Number 29, edited by John O'M. Bockris *et al.*
Plenum Press, New York, 1996.

processing of milk poses a major problem because of deposition of thermally unstable materials on heat-transfer surfaces. The practical problems arising from this are the hydraulic and thermal disturbances that occur during the process and the cleaning operations that have to be carried out in order to bring the exchange surface back to its original state.[5] For the food industry, long-term membrane fouling is one of the drawbacks of techniques based on the application of membrane filtration, such as reverse osmosis for product concentration, ultrafiltration for protein separation, and microfiltration for the separation of microorganisms and product components with large dimensions. For example, these techniques are used in the dairy industry for concentration of milk and cheese whey, in the meat industry for the recovery of protein from blood serum, in the egg industry for preconcentration of dried egg white, and in the beverage industry for clarification of wine, "cold" sterilization of beer, concentration of tomato juice, and clarification of apple juice. However, as a result of membrane fouling, the production capacity decreases and the production costs increase. Film accumulation and microbial colonization in milk storage tanks and spore adhesion on packaging surfaces are all examples of surface phenomena of critical importance to the dairy industry that are initiated by fundamentally similar events.[6] The formation of a biofilm at an aqueous/solid interface is preceded by the adsorption of proteins or glycoproteins which can enhance or mediate adhesion of bacteria. Proteins have been shown to be involved in cellular attachment by the use of proteases to remove an adhering *Pseudomonas* strain of bacteria adsorbed on solid substrates.[7,8] Bacterial adhesion and subsequent metabolism is a major feature of biofouling and serves as a potential source of contamination.[9,10] In many industrial settings, tubular flow systems invariably become fouled because of biofilm development. Thus, in heat exchangers, wastewater transport systems, and secondary oil recovery operations and in many aquatic systems, bacterial biofilms are the cause of energy losses and material deterioration. There is indication that this adhesion occurs only when a macromolecular conditioning film is preadsorbed onto the contact surface. All subsequent adsorptive events have been observed to be dependent upon the conformational state of the preadsorbed film, which in turn depends upon properties of the original contact surface.

   The interfacial behavior of proteins has been particularly important in the design and practical medical introduction of an artificial pancreas and other insulin-releasing systems, because it is extremely important to preserve the native conformation of the protein.[11] One of the reasons for insulin inactivation is the interaction of the protein with the walls of pumps and reservoirs, which are made of corrosion-resistant metals.[12] In another medical area, *in vivo* spoilage of contact lenses has been shown to result from deposition of lysozyme, which is present in human tears.[13] When the enzyme papain was used in an attempt to clean the contact lenses, the result was incomplete removal of lysozyme, accompanied by adsorption of papain itself. Studies of the behavior of proteins at model interfaces can be of great help for the understanding not only of biological problems connected with *in vivo* surface mechanisms, but also of the interactions of plasma proteins with artificial surfaces.[2] Surface-induced coagulation of blood is a critical factor in the design and application of most devices for use with cardiovascular systems.[14] A complex series of events occurs when blood comes into contact with foreign surfaces and may ultimately lead to thrombus formation.[15] The initial event is the adsorption of a layer of plasma proteins, which conditions the surface and subsequently results in platelet adhesion and coagulation. However, not all plasma proteins behave in the same manner. Fibrinogen has been found to greatly enhance platelet adhesion, whereas albumin reduced it, and γ-globulin activated the release reaction.[16] It has been shown that platelets begin to adhere to foreign surfaces after about 30–60 seconds of contact time such as with a glass surface, which is one of the strongest procoagulants.[17] As proteins become adsorbed they may possibly undergo changes so that they become adhesive to normal platelets. There have been suggestions that the dynamic character of protein adsorption, resulting in compositional changes, could create time-dependent thrombosis effects, or protein released from the surface may be altered and become available for activating one of the steps in the coagulation process.[15] Because of the increasing use of medical implant devices, significant effort has been applied to the characterization of the thrombogenic properties of different polymer surfaces in order to provide a sound basis for the choice of materials.[18] This research is making fundamental contributions to the understanding of protein adsorption at solid interfaces.

The complex structure of proteins, coupled with their sensitivity to small changes in environmental conditions, has made elucidation of their behavior on surfaces a real challenge. The solutions to these problems will rely on scientific information acquired through interdisciplinary studies and from a variety of experimental techniques. Relevant background information will be provided in this chapter, and the role of electrochemistry will be discussed.

II. TECHNIQUES USED FOR THE STUDY OF THE INTERFACIAL BEHAVIOR OF PROTEINS

Although a variety of techniques have been used to study protein adsorption at solid/liquid interfaces, not all of the many techniques available have been applied to the measurements of proteins. Recent reviews on the numerous techniques used to study adsorption of a variety of chemical adsorbates at surfaces may be found in publications by Bockris and Khan[19] and Lipkowski and Ross.[20] The techniques that will be focused on in this chapter are those which have been successfully used for measurements of the surface adsorption of proteins. However, the techniques described are not meant to be all-inclusive, nor has a complete literature review been attempted. In many cases, more than one technique has been employed by an author in an attempt to elucidate the interfacial behavior of these complex molecules. Although numerous studies have also been carried out on the interfacial behavior of proteins at air/liquid and liquid/liquid interfaces, these will not be discussed. The methods that have been used for solid/liquid interfacial studies include various types of spectroscopic measurements, including ellipsometry, surface-enhanced resonance Raman scattering, infrared reflection–absorption spectroscopy, multiple internal reflection/attenuated total reflection infrared spectrometry, total internal reflection fluorescence spectroscopy, optically transparent thin-layer spectroelectrochemistry, and photon correlation spectroscopy. Other techniques include small-angle X-ray scattering, radioactive labeling, solution depletion methods, and scanning tunneling microscopy and atomic force microscopy.

1. Ellipsometry

Ellipsometry was one of the first techniques of surface spectroscopy to be applied to *in situ* examinations of interfacial phenomena and was first introduced into electrochemistry by Reddy *et al.*[21] in 1963. The technique is based on the production of linearly polarized light 45° to the plane of incidence by passing light from a source through a collimator, monchromator, and polarizer. It is then elliptically polarized by a compensator and reflected off the surface. The reflected light beam is then converted back into linearly polarized light and passed through an analyzer. The new angle of linear polarization is measured and gives rise to two parameters, Δ and Ψ. Variations on the original techniques have been developed which include measuring the reflectivity of the sample.[22] This has allowed the thickness of the film (t), refractive index (η), and extinction coefficient (k) to be determined from a single experimental measurement.

Ellipsometry has been a popular technique for investigation of the adsorption properties of proteins and has allowed investigations to be carried out on a variety of surfaces. Arnebrant and Nylander[3] measured the amount of the milk proteins κ-casein and β-lactoglobulin adsorbed on metal surfaces *in situ* by ellipsometry as a function of time using phosphate-buffered solutions. Glass slides with vacuum-deposited chromium were treated in different ways to obtain hydrophilic as well as hydrophobic surfaces. The effect of a monolayer of one protein on the subsequent adsorption of the other protein was studied. Competitive adsorption of the two proteins was followed using ellipsometry and ^{14}C-labeled β-lactoglobulin. Their results from sequential adsorption experiments indicated that κ-casein adsorbs after the plateau value of adsorption of β-lactoglobulin has been reached. However, conversely, no adsorption of β-lactoglobulin was found to occur after the plateau value of κ-casein had been reached. When the two proteins were added simultaneously, the surface energy of the substrate was found to influence both the total adsorbed amount and the composition of the adsorbed layer.

Films of β-lactoglobulin adsorbed onto hydrophilic gold surfaces from aqueous solution at pH 4.5–10.0 were analyzed by ellipsometric measurements by Liedberg *et al.*[23] Both the infrared and ellipsometric results indicated a decreasing surface concentration with increasing pH, which corresponded to an increasing net charge of the protein.

The adsorption of β-lactoglobulin onto silica, methylated silica, and polysulfone surfaces was also studied by *in situ* ellipsometry using concentrations of 0.1, 1.0, and 12% β-lactoglobulin in phosphate buffer, pH 7. [24] The amount adsorbed after one hour decreased in the order polysulfone > methylated silica > silica for all concentrations used. Partial desorption of the protein following rinsing with buffer was accomplished only from the silica surface. The β-lactoglobulin adsorbed to polysulfone and methylated silica showed no tendency to desorb. This suggested that multiple states of adsorption of β-lactoglobulin might be occurring at the silica surface.

Recently, Al-Malah *et al.*[25] constructed adsorption isotherms from a temperature study of the apparent equilibrium adsorptive behavior of

β-lactoglobulin for seven different silicon surfaces chemically modified to exhibit varying hydrophobicities. Plateau values of adsorbed mass were observed to increase with increasing degree of solid-surface silanization with dichlorodimethylsilane. The protein adsorption results recorded for the polymer, glass, and stainless steel surfaces could not be explained by solid-surface hydrophobicity alone. Glass was observed to adsorb the greatest mass of β-lactoglobulin, and no. 304 stainless steel the lowest; however, glass and the polymers, being transparent, presented a different optical problem for ellipsometry compared to that posed by silicon and stainless steel. No temperature dependence of β-lactoglobulin adsorption was observed between 310 and 328 K, probably in part because the temperatures studied are below the denaturation temperature of 343 K for β-lactoglobulin indicated by differential scanning calorimetry.[26]

A detailed study of the adsorption and desorption of the plasma proteins bovine serum albumin (BSA), γ-globulin, and fibrinogen was carried out at physiological conditions on a series of polymer membranes, using both neutral polymers and cation exchangers.[27] The results obtained indicated that adsorption takes place simultaneously in two separate and distinct ways. The two types of adsorption appeared to be monolayer, were noninteracting, and occurred on separate membrane sites. One type of adsorption was characterized as being relatively hydrophilic, exothermic, and easily reversible, with a heat of adsorption of approximately 42 kJ mol^{-1}. The other type of adsorption was apparently tightly bound, hydrophobic, and endothermic, with heats of adsorption ranging from about 21 to 84 kJ mol^{-1}.

Elgersma et al.[28,29] studied the competition between the adsorption of BSA and immunoglobulin G (IgG) by sequential and simultaneous addition of the proteins to differently charged polystyrene latices as the adsorbents. Electrostatic interactions were systematically studied by using IgGs of different isoelectric points and positively and negatively charged latices and by experiments with various pH values.

Welin-Klinstroem et al.[30] used a null ellipsometer equipped with an automatic sample scanning device for studies of adsorption and desorption of fibrinogen and IgG at the liquid/solid interface on surface wettability gradients on silicon wafers. To follow the processes along the wettability gradient, "off-null" ellipsometry was used. The kinetics of adsorption and nonionic-surfactant-induced desorption varied considerably between fibrinogen and IgG. In the hydrophilic region, very little protein desorption was seen when a nonionic surfactant was used.

In a recent paper, Arwin *et al.*[31] described the basic theory of off-null ellipsometry in its application to the determination of the surface concentration of adsorbed proteins on silicon substrates. For surface concentrations below 5 ng mm^{-2}, a linear relation was observed between the square root of the intensity and the surface concentration of the proteins with an accuracy of the order of 3% or better.

The adsorption of insulin on metal surfaces from aqueous solutions was monitored by *in situ* ellipsometry by Arnebrant and Nylander.[32] Clean (hydrophilic) chromium and titanium surfaces as well as chromium surfaces treated to be hydrophobic were used. The adsorbed amount was found to be higher on the hydrophilic than on the hydrophobic surfaces.

Razumas *et al.*[12] studied adsorption of insulin on platinum from a physiological saline solution of pH 7.4 at 298 K using both cyclic voltammetry and ellipsometry. They compared the surface adsorption of insulin for different potentials using ellipsometric measurements and found at a potential of 0.4 V (vs. SCE), the surface concentration $\Gamma = 1.27$ mg m^{-2} for insulin. Adsorption was found to increase when zinc ions were added to the protein solutions.

Both ellipsometry and cyclic voltammetry were used by Szucs *et al.*[33] to investigate both the adsorption and the mediated and unmediated electron-transfer processes of cytochrome *c* on gold. The adsorption of cytochrome *c* was studied by ellipsometry on gold electrodes that had not been treated with surface modifiers. The measurements were carried out at 400 mV (vs. NHE), which was approximately equivalent to the open-circuit potential of the system in the absence of protein in the electrolyte. An irreversibly adsorbed layer formed that completely covered the electrode. They found the value of the calculated thickness of the adsorbed layer on bare gold to be about half that expected if cytochrome *c* has been in its native form, which suggested that significant unfolding occurred upon adsorption.

Thus, ellipsometry provides information on surface concentrations and conformational unfolding of proteins for a variety of surfaces of hydrophilic and hydrophobic nature, as well as for surfaces under potential control.

2. Surface-Enhanced Resonance Raman Scattering (SERRS)

The technique of *in situ* Raman spectroscopy at metal electrodes has been recently reviewed by Pettinger.[34] Adsorbed monolayers of proteins have been subjected to interfacial spectroelectrochemical characterization us-

ing techniques such as surface-enhanced resonance Raman spectroscopy.[35–40] In recent years, much effort has been devoted to studies of
electron-transfer reactions of proteins on electrode surfaces because of
their importance in the elucidation of biological reactions at charged
membranes and in developing new biosensors and artificial models of *in
vivo* systems. These studies have contributed a great deal of insight into
the structure and behavior of proteins on reflective metal electrodes.
Hildebrandt and Stockburger[39,40] have examined the structure and the
electron-transfer properties of cytochrome *c* adsorbed on a silver electrode
by SERRS. The sensitivity of surface-enhanced Raman (SER) spectroscopy results from the strong enhancement of Raman scattering observed
for molecules adsorbed on rough silver, gold, or copper surfaces. When
the excitation line is in resonance with an electronic transition of the heme
chromophore, the Raman bands associated with the heme group are
enhanced, while the Raman bands of the protein matrix remain too weak
to be detected. Therefore, the molecular resonance Raman (RR) and the
surface-enhanced Raman (SER) effects combine to give SERR. Thus, the
technique provides a selective probe of the vibrational spectrum of the
heme group. Hildebrandt and Stockburger found the conformational
behavior of the adsorbed cytochrome *c* to be dependent on the electrode
potential. In state I, the native structure was fully preserved, and the redox
potential (+0.02 V vs. SCE) was close to the value for cytochrome *c* in
solution. In state II, structural alterations resulted in an open conformation
of the heme crevice with a corresponding decrease in the redox potentials
to –0.31 and –0.41 V for the five-coordinate high-spin and six-coordinate
low-spin configurations, respectively. They found the molecular binding
sites of the adsorbed cytochrome *c* to the surface to be through the charged
amino acid groups. At potentials below zero charge (~ – 0.6 V), a rapid
denaturation of the adsorbed cytochrome *c* was indicated by drastic and
irreversible changes in the SERR spectrum.

Since little is known about the interfacial behavior of enzymes other
than glucose oxidase (GOD) at an electrode, Taniguchi *et al.*[41] examined
the electron-transfer reactions and adsorption behavior of several flavoenzymes by fluorescence and SERRS measurements. They found that the
intensities were useful in determining the location of flavin adenine
dinucleotide (FAD) in the flavoenzymes, which suggested possible interfacial structures of these flavoenzymes.

3. Infrared Reflection–Absorption Spectroscopy (IRRAS)

The technique of infrared reflection spectroscopy has been used to obtain information about the orientation of enzyme molecules adsorbed on flat metal surfaces.[42] This technique which involves passing the infrared beam through the sample on a reflecting substrate has been reviewed by Hollins and Pritchard[43] and more recently by Nichols.[44] Since the infrared spectra of proteins and other biological molecules are distinctive and well documented, the technique is useful for the study of the protein–surface complex. The principal absorption bands of proteins are the amide A band at 3300 cm^{-1} due to the N–H group and the amide I and amide II bands at 1650 and 1550 cm^{-1}, respectively, due to the -CONH- group. However, these bands are very similar for all proteins and peptides, and this prevents distinguishing among different proteins on the basis of their infrared spectra alone. In addition, IRRAS also provides direct information about the mean orientation of the adsorbed species in relation to the metal surface. Only groups oriented with vibrational transition dipole moments perpendicular to the surface are infrared active due to the surface dipole selection rule.[23]

In a series of papers by Liedberg et al.,[23,45,46] investigations using Fourier transform infrared reflection–absorption spectroscopy were made of films formed by adsorption of β-lactoglobulin and fibrinigen onto a variety of metal and metal oxide surfaces. They found that the amide I band, which is essentially due to the C=O stretching modes, was significantly distorted upon adsorption of the protein air-dried films in relation to the same bands in the spectra of the proteins in the aqueous and solid phases. Blue shifts of about 10 to 40 cm^{-1} were observed with adsorption of the proteins from solutions with pH values in the range of 4.5 to 10.0, whereas only a weak pH dependence was observed in the lineshape and exact frequency of the amide I bands in the spectra of the solutions. These findings strongly suggest that a conformational transition of the protein molecules occurred upon adsorption and that the proteins existed in an unfolded conformation on the gold surface. These results were consistent with ellipsometry findings at pH 6.0 and 10.0 by the same authors.[23] However, a more native protein conformation was found at pH 4.5 from the ellipsometry measurements. This may have resulted from the different environments used with the different techniques. At the higher pH values, loss of the CO_2^- stretching mode was observed in the reflection–absorption spectra. It was suggested that strong coordination of the CO_2^- group occurred through an ester-type bond with the gold surface.

4. Attenuated Total Reflection Infrared Spectrometry (ATR)

The ATR technique (also known as internal reflection spectroscopy) places the sample in contact with an internal reflection element (IRE) such as a cylindrical prism which has been coated on the underside with the reflecting metal or a transparent electrode material such as germanium.[47] The totally reflected infrared beam gives rise to an evanescent wave in the electrolyte that interacts with molecules in the first few hundred nanometers of the surface of a material. A description of this technique, which has been applied to *in situ* studies of solid/liquid interfaces, may be found in a review by Plieth *et al.*[47] and Bockris and Khan.[48] Since modern Fourier transform infrared (FTIR) spectrometers are rapid-scanning devices, the ATR technique lends itself to monitoring the kinetics of adsorption of proteins. This application has been stimulated by the development of medical implant polymers and the problem of adsorption of proteins involved in thrombogenesis on these polymers. The technique allows aqueous protein solutions or even *ex vivo* whole blood to be flowed over the surface of the internal reflection element (IRE), which may be coated with a thin layer of the experimental polymer.[49] The spectra of the adsorbed protein layer, which can be computed from a series of interferograms recorded continuously in time, provide information on the rapid changes in the composition of the adsorbed protein layers in the first several minutes.

Infrared internal reflection spectroscopy was used by Brash and Lyman[50] to study the adsorption of the plasma proteins, albumin, γ-globulin, and fibrinogen, on a variety of hydrophobic polymer surfaces. The results indicated that all the proteins investigated behaved rather similarly on a variety of hydrophobic surfaces. Under static conditions the proteins appeared to be rapidly adsorbed as monomolecular layers from solutions varying in concentration from a few milligram percent to the concentration levels of normal plasma. They deduced these monolayers to be closely packed arrays in which the protein molecules appeared to retain their native globular form.

Infrared studies of the amide I band of transferred films of β-lactoglobulin B spread on 0.5 M KCl were obtained via the multiple internal reflection (MIR) technique by Loeb.[51] The spectra were recorded following transfer of films maintained at different surface pressures. At a high area per molecule, or low surface pressure of 2 dynes cm^{-1}, the spectra were consistent with a denatured protein. On the other hand, at high surface pressure of 16 dynes cm^{-1}, the spectra indicated a species whose secondary structure was similar to that of the native protein. The transition to denatured protein resulted when the film was reexpanded following compression. At an intermediate surface pressure of 6 dynes cm^{-1}, approximately equal amounts of native-type and denatured type spectral components were measured.

Ishida and Griffiths[52] used the ATR technique in investigations of adsorption of proteins (albumin and β-lactoglobulin) and polysaccharides (gum arabic and alginic acid) onto a germanium surface. They found that the proteins exhibited a higher affinity for the surfaces and were more firmly bound than the polysaccharides. Since biofilm formation at aqueous/solid interfaces is known to be preceded by the adsorption of proteins, and bacteria often produce extracellular polysaccharides that adhere to the conditioned substrate, the ATR technique was employed to study polysaccharide adsorption onto protein conditioning films.[53] The proteins albumin, β-lactoglobulin, and myoglobin were used to initiate the fouling process and were found to adsorb tenaciously to the surface of the germanium IRE. The adsorption of alginic acid was enhanced in the presence of the protein conditioning film, independent of the identity of the protein. On the other hand, the neutral polysaccharide dextran was excluded from the conditioned substrate but adsorbed onto the bare germanium surface.

5. Total Internal Reflection Fluorescence (TIRF) Spectroscopy

Asanov and Larina[54] investigated the effect of electrochemical polarization of the SnO$_2$ surface on the behavior of albumin molecules at the solid/liquid interface using a spectroelectrochemical technique based on a combination of total internal reflection fluorescence (TIRF) spectroscopy and electrochemical control of the potential of the surface under study. The principles of TIRF have been described by Hlady *et al.*[55] The amount of adsorbed protein and the rates of adsorption and desorption were found to be strongly affected by the electrochemical polarization of the surface. Desorption of irreversibly adsorbed protein at negative potentials was greatly enhanced by rapid potential cycling between negative and positive values. Slow changes in the same potential range resulted in only very limited desorption. Asanov and Larina proposed a model based on the assumption that under slow potential changes protein molecules have time to reorient and remain adsorbed. However, under fast potential change the large molecules are unable to respond rapidly enough to make the transition into a new adsorbed state.

6. Optically Transparent Thin-Layer Spectroelectrochemistry

Detrich et al.[56] reported initial studies of the electrochemical determination of the relationship between pH and formal potential of hemoglobin and the kinetics of the heterogeneous electron transfer at an unmodified indium oxide electrode. In this work, they used electrochemical cells with a conventional three-electrode design. An optically transparent thin-layer electrochemical (OTTLE) cell was used for spectropotentiometric measurements. Working electrodes were tin-doped indium oxide deposited on glass. Potential-step chronoabsorptometry was used for initial measurements of the heterogeneous electron-transfer rate parameters for the reduction of hemoglobin at the unmodified electrode. The authors found the formal rate constant of hemoglobin varied over a wide range depending on the type of electrode, modification of the electrode surface, and other experimental conditions.

7. Photon Correlation Spectroscopy (PCS)

The technique of photon correlation spectroscopy (PCS) has been the subject of several books[57,58] and reviews.[59,60] Fair and Jamieson[61] obtained adsorption isotherms of BSA and γ-globulin on polystyrene latices. PCS was used to analyze portions of the protein/latex suspensions to determine the hydrodynamic thickness of the adsorbed layer as a function of the amount of adsorbed protein. The hydrodynamic radius was calculated from the diffusion coefficient via the Einstein equation. The results were compared with solution depletion measurements made on the same samples. The authors found no evidence for multilayer adsorption at higher coverages. Conformational "flattening" of the adsorbed protein in the lower plateau regime was established. The authors proposed a mechanism for bimodal adsorption in which a partially organized "glassy" monolayer and a close-packed two-dimensional surface "crystal" of protein are formed at low and high surface concentrations, respectively.

8. Small-Angle X-Ray Scattering (SAXS)

Small-angle X-ray scattering (SAXS) is a technique that is widely used to obtain microstructural information over the size range of a few nanometers to a few hundred nanometers.[62,63] The main features of the technique have

been described by Mackie *et al.*[64] They investigated the behavior of
β-lactoglobulin on polystyrene latex surfaces for a range of pH values.
Comparisons were made with adsorption isotherms obtained with the
same solutions by the solution depletion method. Results of SAXS measurements showed that the bulk of the protein was confined to within 1.8
nm of the surface at all pH values and that the protein content of the
adsorbed layer increased with decreasing pH, which was consistent with
the adsorption densities obtained at this bulk concentration. At lower
concentrations, the results obtained by SAXS[64] were inconsistent with
those obtained using PCS[61] in that the latter provided no evidence of
"conformational flattening" in the lower plateau region. The values obtained by PCS for spherical systems differed from those obtained by
SAXS by up to 1.5 nm. Mackie *et al.* argued that a small number of
nonflattened particles on the polystyrene surface would have an effect on
the diffusion coefficient measured by PCS, and hence on the hydrodynamic radius, particularly since the coated particle would no longer be
strictly spherical, but would probably not affect the SAXS profile.

9. Radioactive Labeling (Radiotracer)

The radiochemical or radiotracer method has been an effective technique
for the examination of interfacial dynamics through measurement of the
rates of adsorption/desorption and the rates of exchange of radioactive
surface species with unlabeled solution molecules.[65] It involves the use of
a chemical species which has been labeled with a radioactive isotope. The
behavior of this chemical species is then monitored with a radiation
detector. The technique is unique in that it measures only the specifically
labeled molecule, and it is based on a straightforward relationship between
surface count rates and surface concentration (Γ). Beta radiation appears
to be the most convenient of the three types of radiation (α, β, and γ), both
in terms of minimal safety hazards and ease of detection. A range of
compounds may be used, provided a radioactive form of the compound is
available, since the radiolabeled compounds do not have to be electroactive. However, radiochemical and some spectroscopic methods suffer
from limitations in the *in situ* experiments due to interference from the
presence of the solution layer. Roughening the electrode, for example, by
platinization, is often necessary to achieve a signal distinguishable from
the bulk solution count rate. However, this is not necessary with tritium,
which provides a superior sensitivity over the other emitting isotopes.
Recent research has focused on the development of radiochemical meth-

ods to measure adsorption at smooth polycrystalline and single-crystal surfaces.[65]

Arnebrant and Nylander[3] used the technique to study the sequential and competitive adsorption behavior of ^{14}C-labeled β-lactoglobulin with κ-casein on hydrophobic and hydrophilic chromium surfaces. The adsorption was also followed by *in situ* ellipsometry measurements, providing a basis for comparison of the two techniques as described in Section II.1.

Van Enckevort *et al.*[66] used radiochemical techniques in their studies of adsorption of BSA at the stainless steel/aqueous solution interface for a wide range of solution concentrations. It was shown that the adsorbed protein was essentially nonexchangeable and the adsorption process was not reversed upon dilution. The adsorption isotherm consisted of a plateau region that extended to the most dilute solutions studied, provided adequate time was allowed. They found adsorption from sufficiently dilute solutions to be diffusion-controlled. In ideal cases, they were able to estimate both the kinetics of diffusion-controlled adsorption and saturation coverages from molecular parameters.

10. Solution Depletion

The solution depletion method for studies of adsorption in solution uses material in the form of powder or pellets with specific dimensions and surface areas. This material may be physically or chemically treated to produce hydrophobic, or positively or negatively charged hydrophilic surfaces which is exposed to the solution. The decrease in concentration of the adsorbate from the bulk solution is then measured. The precision of this method is dependent on the analytical technique used. However, for biological material, the technique does not allow measurement of subtle changes in surface denaturation. The technique has been used by a number of researchers.[61,67,68]

Adsorption isotherms of β-lactoglobulin on polystyrene latex particles were studied by Mackie *et al.*[64] for pH values of 4.65, 7.2, and 9.0. The latex and attached protein were then removed by centrifugation, and the concentration of protein remaining in solution was calculated by assaying the supernatant by the Lowry method.

Kim and Lund[69] used solution depletion techniques to study the adsorption of β-lactoglobulin on nonporous stainless steel particles with an average diameter of 25.5 μm. They found that a monolayer adsorbed very tightly to the stainless steel surface and that a significant fraction of the protein was irreversibly adsorbed.

Adsorption of BSA was examined by Muramatsu and Kondo[70] on positively and negatively charged microcapsules. The overall adsorption profiles were governed by the sign of the net charge of the protein and the adsorbent. The amount of adsorption was smallest on the surface that had an isoelectric point similar to that of BSA. Thus, the charge balance was assumed to play a role in the adsorption and the molecular structure of BSA. The technique lends itself very nicely to a variety of surfaces.

11. Scanning Tunneling Microscopy (STM) and Atomic Force Microscopy (AFM)

Scanning tunneling microscopy (STM) and atomic force microscopy (AFM) are two fairly recently developed techniques, and there has been considerable interest in their utilization for structural studies of bio-molecules.[71–73] However, there are some difficulties with these techniques, including problems associated with probe-induced sample movement and with image validation.[74] The force exerted by the probe tip on the sample can induce movement of the adsorbed molecules, which appears to be an important contributory factor to the lack of reproducibility during STM studies of biomolecules. The problem of image validation became apparent from reported artifacts. These artifactual features are quite often indistinguishable from features due to the sample biological molecule. Leggett et al.[74] have explored two contrasting methods: the application of a physically robust, conducting overcoat and covalent coupling to the substrate. Bovine liver catalase molecules were sprayed onto mica and coated with a platinum/carbon film. This application of a platinum/carbon film was found to produce rigidly immobilized protein molecules and yielded the best resolution. However, covalent coupling techniques provide methods by which assemblies of proteins can be prepared and investigated by STM.

The atomic force microscope was invented in 1986 by Binnig et al.[75] and appeared to have great potential for viewing three-dimensional protein structure and protein–surface interactions. As a result, there have been several studies on the imaging of proteins with AFM.[76–80] Eppell et al.[80] have used AFM to image the globular blood plasma glycoprotein von Willebrand factor, which plays an important role in thrombus formation and cell adhesion on biological and synthetic biomaterial surfaces. They were able to measure the elliptical cross sections of the protein. They suggested, on the basis of their results, that AFM can be used to map changes in charge density of a molecule by looking for alterations in the measured heights of the various parts of the molecule in its charged versus uncharged state. Because the atomic force microscope lacks chemical specificity, any additional physical parameters that can be associated with the tip–surface interaction will be beneficial in the interpretation of AFM images.

III. ELECTROCHEMICAL TECHNIQUES USED FOR STUDIES OF SURFACE ADSORPTION

1. Organic Compounds and Amino Acids

There have been extensive studies made over the years on the surface adsorption behavior of organic compounds at solid electrodes, and this work has formed the basis for many of the electrochemical studies made with amino acids and proteins. This work has been well reviewed by a number of authors.[19,20,81–85] Research in this area continues to be active, as shown by recent publications using a variety of techniques to examine adsorption of organic compounds at electrode surfaces. Bockris and Jeng[86] have published an extensive paper in which they used the *in situ* techniques of radiotracer measurements, ellipsometry, and FTIR measurements to study electrosorption of organic compounds on polycrystalline platinum electrodes. Although the electrosorption process was found to be slow, a bell-shaped curve of coverage, θ, versus potential, V, was generally obtained in the double-layer region with submonolayer adsorption. Measurements were made for a wide variety of organic compounds, and the effects of different functional groups were compared. Arvia and co-workers[87,88] have recently used anodic stripping voltammetry with a conventional flow cell technique at 298 K to study adsorption interactions of different organic systems on electrodispersed and smooth platinum electrodes. The technique of chronocoulometry was used by Iannelli *et al.*[89] to study the quantitative adsorption of pyrazine at the polycrystalline gold/solution interface. From their measurements, they were able to calculate film and surface pressures, relative Gibbs surface excess, free energies of adsorption, and electrosorption valency as a function of both electrode potential and surface charge density. Krauskopf and Wieckowski[65] used infrared and radiotracer techniques to study the interaction of acetic acid with platinum electrodes. The infrared spectra showed both carboxylate oxygens to be oriented toward the metal surface.

Since amino acids form the building blocks of proteins, it is important to compare the surface adsorption behavior and electrochemical oxidation reactions of amino acids at solid electrodes in order to fully understand the behavior of proteins under similar experimental conditions and the role of different functional groups in the adsorption behavior of these compounds.

Brabec and Mornstein[90] investigated a number of L-α-amino acids at a paraffin-wax-impregnated spectroscopic graphite electrode (WISGE) using linear-sweep, cyclic voltammetry, phase-sensitive alternating current, and differential pulse voltammetric techniques. They found that, among the amino acids usually occurring in proteins, only tyrosine, tryptophan, histidine, cystine, cysteine, and methionine were oxidized at the WISGE electrode. The oxidation of tyrosine and tryptophan at the WISGE electrode was determined from coulometric measurements to be a two-electron process. When higher concentrations ($>2 \times 10^{-4}$ M) of tyrosine and tryptophan were used, adsorption of the oxidation products of these amino acids occurred.

Safonova et al.[91] investigated the adsorption and electrooxidation of methionine and of amino acids containing a hydroxyl group—serine, threonine, and tyrosine. Cyclic voltammograms recorded over the potential range 0.05–1.4 V (vs. RHE) on a platinized platinum electrode in acid and alkaline solutions showed that all the amino acids were oxidized at the potentials of the oxygen region. A maximum in the current in the double-layer region at 0.7 V was found for serine, with an enhanced effect in alkaline solution. A similar behavior was found for threonine, but not for methionine or tyrosine.

The electrochemical oxidation of sulfur amino acids has been investigated by Reynaud et al.[92] using AC measurements and cyclic voltammetry at gold and vitreous carbon electrodes. They found that electrochemical oxidation of the amino acids proceeded by both adsorption and diffusion at the metal electrodes and suggested that the process was catalyzed by a labile metal oxide. On the other hand, oxidation of methionine at the vitreous carbon electrode occurred only in the adsorbed state, as shown by the presence of an intermediate species from X-ray photoelectron spectroscopy (XPS) studies. In similar studies of amino acids containing no sulfur atoms, Malfoy and Reynaud[93] found only tryptophan and tyrosine to be specifically oxidizable at gold, platinum, and carbon electrodes. The oxidation phenomena were very similar for all the electrodes and gave a mixture of monomeric species (oxindolalanyl with tryptophan and hexacyclodienone alanyl with tyrosine) and poly-meric species (filming effect: polytryptophan with tryptophan, polyphenylene oxide alanyl with tyrosine). Histidine was found to be oxidizable only at a carbon electrode.

In a number of papers, Horanyi and Rizmayer[94–97] have examined the adsorption of ^{14}C-labeled primary amino acids on platinized platinum electrodes using *in situ* radiotracer studies. This provides direct informa-tion about the adsorption phenomena and the behavior of adsorbed species under various electrochemical experimental conditions. They found that in acid medium there was no significant difference in the potential dependence of the adsorption behavior of glycine and acetic acid, and the adsorption properties of the molecules were entirely determined by the carboxyl group.[94,96,97] Only loosely adsorbed species were observed at low bulk concentrations. A rapid exchange of adsorbed labeled glycine with nonlabeled species suggested that the protonated ^+H_3N-CH_3 group does not influence the adsorption behavior of glycine. Studies of D,L-α-alanine and γ-aminobutyric acid showed that two types of adsorbed species were formed in acid medium, one loosely adsorbed and one strongly chemisorbed. Similarly, the potential dependence of adsorption starting from low potentials resembled that of the simple aliphatic acids at low concentrations. However, a significant hysteresis was observed in the potential dependence, and this was attributed to the formation of strongly chemisorbed species. In alkaline medium, only strong chemi-sorption was found to occur. In order to obtain further evidence concerning the role of the -COOH group in the adsorption of amino acids, the adsorption of labeled CH_3-NH_2 and NH_2-CH_2-CH_2-NH_2 was studied. No adsorption of these compounds was observed in the potential range in which adsorption of glycine and γ-aminobutyric acid occurred. It is interesting to compare the results reported by Horanyi and Rizmayer[94,96] for glycine and acetic acid with those reported by Corrigan *et al.*,[98] who obtained surface concentrations for acetic acid as a function of potential from radiotracer measurements and potential-difference infrared spectros-copy (PDIR) (Fig. 1). The maximum adsorption for acetic acid was found to be at 0.4 V (vs. SCE). The PDIR measurements indicated that both carboxylate oxygens were oriented toward the metal surface.

Studies were also made by Horanyi[99] with methionine, and the potential dependence of the adsorption in alkaline solution was found to be very similar to that observed for other amino acids. The presence of the -S-CH_3 group did not significantly modify the characteristic adsorption

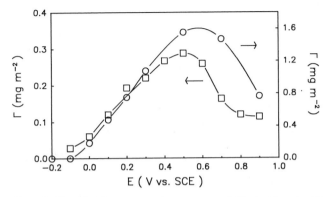

Figure 1. Plots of surface concentration versus potential for 1mM acetic acid from radiochemistry measurements ($\square$) and for 10mM acetic acid from potential-difference infrared spectroscopy ($\circ$) at a platinum electrode in aqueous 0.1M HClO$_4$. (Adapted from Corrigan et al.[98])

behavior for amino acids in alkaline medium. However, a dramatic change was observed in acid medium, and it was assumed that the interaction of the S atom with the platinum surface played a predominant role in the adsorption process. Serine was found to adsorb strongly in acidic medium, and this was attributed to the presence of the OH group.[100]

Horanyi et al.[101] extended these studies to the electrosorption of a ^{14}C-labeled tetrapeptide (Arg-Lys-Asp-Val) at smooth gold and platinized electrodes in acid and alkaline media. A significant difference was found between the adsorption behavior of the peptide in acid and in alkaline media in the case of the gold electrode. In an alkaline medium a strong chemisorption was observed, whereas in acid medium loosely adsorbed species were formed, similar to the behavior observed with the amino acids. In all cases, strong chemisorption was observed over a wide potential range (200–1000 mV vs. RHE).

A series of studies was made by Roscoe and co-workers on the adsorption and electrochemical oxidation mechanisms of the amino acids glycine,[102] α- and β-alanine,[103] and α-, β-, and γ-aminobutyric acid[104] at a platinum electrode in order to determine the role that the position of the amino group plays in the surface adsorption properties and subsequent oxidation of these amino acids. The investigations were made in aqueous solutions at pH 1, 7, and 13 using steady-state current–potential measurements, cyclic voltammetry, and open-circuit potential decay. The capacitance behavior and the high Tafel slopes suggested the production of free radicals at the surface of the electrode, accompanied by a second reaction involving loss of CO_2, which was determined to be the rate-determining step.

$$CH_2(NH_2)COO^- \rightarrow CH_2(NH_2)COO(M) + e^-$$

$$CH_2(NH_2)COO(M) \rightarrow CH_2(NH_2)(M) + CO_2$$

The adsorbed intermediate species was then hydrolyzed anodically to form the next lower aldehyde:

$$CH_2(NH_2)(M) + OH^- \rightarrow NH_3 + CH_2O + e^-$$

Product analysis was carried out using a variety of techniques including polarography, gas chromatography, and Kjeldahl analysis. No dimerized products were detected by gas chromatography. These mechanisms differ from the dimerization process typical of the radical reactions associated with the Kolbe mechanism for such species as acetic acid.[105]

The results from these studies have shown that the adsorption and oxidation reactions of the amino acids are dependent to an extent on the type of electrode surface as well as the pH. In order to compare these results directly with the interfacial behavior of the proteins, it is important to consider the behavior of the functional groups on the side chains of the amino acids and their positions within the three-dimensional structures of the proteins. These features will be discussed in Section IV.

2. Proteins

The emphasis in this section will be on the interfacial behavior of proteins at electrode surfaces. Thus, studies of direct electron transfer from redox proteins will be discussed because of their relevance to surface adsorption properties. Details of mediated electron transfer will not be described here. Thus, this section will review electrochemical measurements made on the interfacial behavior of a variety of proteins at a number of different solid electrode surfaces.

In an attempt to develop an enzyme electrode specifically sensitive to the lactate ion, Durliat and Comtat[106] made a study to determine whether the enzyme L(+)-lactic dehydrogenase from aerobic yeast was adsorbed onto the platinum electrode. The purpose was to define the role played by mass transport in the process and to ascertain whether the enzyme retained its activity in the adsorbed state. The supporting electrolyte was a phosphate buffer solution (pH 7.2, 0.5M ionic strength at 294 K). The enzyme was first adsorbed onto platinum at a potential of 250 mV vs. SCE and then transferred to the cell for measurement. The resulting oxidation current was found to be higher than the residual current in the potential range +0.5 to +1.0 V, with a maximum at a potential of 0.85 V, preceded by a point of inflection at 0.7 V. Also, a decrease was observed in the reduction current for the platinum oxide formed during the anodic phase of the sweep. The anodic potentials decreased with sweep time, and the residual current was again attained after about 10 polarization cycles. The authors suggested that –SH groups play an important role in the adsorption process. Spectroscopic measurements indicated that the enzyme stayed on the surface with retention of its activity. They calculated the maximum concentration of adsorbed enzyme to be $2 \times 10^{-12} - 4 \times 10^{-12}$ mol cm^{-2}, and the surface area of the adsorbed protein to be between 80 and 160 nm^2 molecule^{-1} for two different sets of experiments: the former in an unstirred solution and the latter with the rotating electrode at 250 rpm. The dimension of the protein for a spherical shape would be about 20 nm.[2]

The electrochemical oxidation of the three proteins ribonuclease A, bovine serum albumin (BSA), and concanavalin A was investigated by Reynaud et al.[2] on gold, platinum, vitreous carbon, and carbon paste electrodes using AC voltammetry and cyclic voltammetry. They used a Britton–Robinson buffer (0.5M) with sodium perchlorate as supporting electrolyte. Although they found all three proteins to be adsorbed on these surfaces, oxidation peaks on voltammograms were only well developed in the case of the carbon paste electrode. The peak potential was the same for all three proteins at +0.92 V (vs. SCE) at pH 1.7 and corresponded to the oxidation potentials of the amino acids tyrosine and tryptophan. Thus, the authors concluded that tyrosinyl and tryptophanyl residues were the first to be oxidized over the pH range 1–7. This was supported by the disappearance of tyrosine absorption bands in UV spectra of solutions of ribonuclease that were subjected to repetitive multisweep electrolysis within a potential range of –0.5 to +1.6 V at pH 1.7. The authors reported that at more positive oxidation potentials, other amino acid residues such as histidine and cystine were also oxidized. It was suggested that protein adsorption takes place prior to oxidation and mainly at negative potentials (less than the potential of zero charge) via the positively charged amino functional groups of the proteins at the low pH values used in the study. XPS scanning of S–S bridges on a carbon paste electrode (graphite) following repetitive potential sweeps between –0.5 and +1.05 V suggested that the adsorbed oxidized product was denatured as a few S–S bridges were oxidized. When similar measurements were made following potential sweeping between –0.5 and +1.65 V and the oxidized product was desorbed in the presence of nitrogen, sulfur was not detected by XPS.

Brabec[107] studied the electrochemical oxidation of ribonuclease, albumin, and histone fraction HI at a spectroscopic graphite electrode impregnated with paraffin wax (WISGE) using linear-sweep (LS), cyclic, and differential pulse (DP) voltammetry. He found the proteins to be electrochemically oxidized at the WISGE, yielding a faradaic peak on voltammograms in the vicinity of 0.7–0.8 V vs. SCE in a Britton–Robinson buffer, pH 7.4, at 25°C. This peak did not have the cathodic counterpart expected for nonfaradaic reorientation or adsorption/desorption processes. Also, the oxidation peak of the protein was shifted to more positive potentials with increasing voltage scan rate. The latter two results indicated that the electrode process responsible for the formation of the oxidation protein peak was irreversible. In order to verify these results, Brabec studied the behavior of free amino acids at the WISGE. He found that only tyrosine and tryptophan yielded electrooxidation currents in the vicinity of the protein peak potential. Histidine, methionine, cystine, and cysteine produced smaller electrooxidation currents at the WISGE and at potentials approximately 0.4–0.5 V more positive than the protein peaks. In order to obtain better defined voltammetric peaks of proteins at the WISGE, the more sensitive DP voltammetry was used instead of LS voltammetry. Albumin, ribonuclease, and insulin (proteins containing no or relatively few tryptophan residues as compared with the content of tyrosine residues) yielded only a single oxidation DP voltammetric peak at the WISGE. Lysozyme, which contains twice as many tryptophan as tyrosine residues, produced an additional peak at potentials ~50 mV more positive. The peaks were identified as due to tyrosine and tryptophan, respectively. The dependence of the protein peak heights on the protein concentration approached a limiting value, which suggested a participation of the protein adsorption in the electrode process. Studies were made in order to determine whether the denaturation of the protein with acid or urea would lead to a change in the DP-voltammetric behavior of the protein at the WISGE. The voltammetric peak of ribonuclease in $0.1N$ H_2SO_4 was 30% higher than that obtained at pH 7.4. The addition of $8M$ urea to a neutral solution of ribonuclease led to an increase of the ribonuclease peak by about 40%. It was concluded that the protein undergoes conformational change such that an increased number of the tyrosine residues become accessible for the reaction with the graphite electrode.

The anodic behavior of serum albumin, casein, hemoglobin, peroxidase, trypsin, hexokinase, alcohol dehydrogenase, and gelatin was investigated by Khidirov[108] using a potentiodynamic method with a very slow sweep rate of 2 and 4 mV s^{-1} on lead dioxide and manganese dioxide electrodes. The anodic behavior of the proteins on the MnO_2 electrode was studied in a solution of 0.1M phosphate buffer at pH 8.0 (291 K in an argon atmosphere) with a reversible hydrogen electrode in the same solution. The potential of the working electrode was varied from the steady-state value to the potentials corresponding to those for oxygen evolution. Reverse polarization with the same scanning rate was carried out in the same solution to the initial potential. The curve was first recorded in the supporting electrolyte and then in the presence of protein. The static, and initial, potential of 1.2 V established in the supporting electrolyte solution was virtually unchanged when small quantities of protein (up to 0.2 mg ml^{-1}) were introduced into the solution. Although casein slightly decreased the oxygen deposition current, only small oxidation currents were observed in the presence of the other proteins. The profile of the potentiodynamic curve remained similar to that obtained in the supporting electrolyte solution. The behavior of proteins on a lead dioxide electrode was investigated both in a 0.1M phosphate buffer (pH 8.0) and in 0.1N and 1N H_2SO_4 supporting electrolyte. In all these solutions the rate of oxidation of proteins on a PbO_2 electrode was greater than on an MnO_2 electrode.The rate of oxidation in 1N H_2SO_4 solution was gelatin > peroxidase > trypsin > alcohol dehydrogenase > albumin > hexokinase. For the peroxidase solutions, the oxidation currents increased following repeated polarizations with both electrodes. However, for the other proteins, the rate was suppressed following repeated polarizations at the MnO_2 electrode.

Razumas *et al.*[12] found that insulin adsorbed on platinum from a physiological saline solution of pH 7.4 (298 K) throughout the potential range of water stability. The smallest amount of protein adsorption was found to be around the potential of zero charge of platinum. The authors compared the surface adsorption of insulin for different potentials using ellipsometric measurements and determined a surface concentration of Γ = 1.27 mg m^{-2} for insulin at a potential of 0.4 V (vs. SCE). The Γ_{max} was attained within 20–30 s after mixing the solutions. A small decrease in

Γ_{max} following substitution of 0.15M NaCl solution for the protein solution was observed only at potentials of 0.3 and 0.4 V (the values reached were 0.93 and 1.34 mg m^{-2}, respectively). The lowest values for Γ_{max} were found between –0.1 and –0.3 V. The authors suggested that in solution the insulin monomers form dimeric and hexameric aggregates which are more hydrophilic than the monomer. Adsorption was found to increase when zinc ions were added to the protein solutions. The surface concentrations were found to be 1.58 and 1.52 mg m^{-2} for 2 Zn-insulin and 4 Zn-insulin, respectively. This increased adsorption was attributed to additional aggregation of the hexamers as a result of complex formation with zinc ions. The strong hydrophobicity of the protein in solutions lacking metal ions results in minimum values of Γ_{max}.

Direct electron transfer between a monolayer of a quinoprotein oxidoreductase, fructose dehydrogenase (FDH), and Pt, Au, and GC electrodes was investigated by Khan et al.[109] In order to achieve direct and reversible electron transfer, an FDH monolayer was prepared on these electrodes by a voltage-assisted adsorption method. Fructose dehydrogenase (5 mg ml^{-1}) was dissolved in 0.1M phosphate buffer, pH 6.0, containing 0.1% Triton X-100. A potential of 0.5 V was applied to the electrode for a few minutes, followed by thorough washing with McIlvane buffer of pH 4.5 and storage in this buffer until the experiments were made. The electron transfer between adsorbed FDH and the electrode proceeded directly and reversibly at all the electrodes. A sensitive fluorescence protein assay method was used to measure the surface concentration of FDH.

The reduction of H_2O_2 on horseradish peroxidase (HRP) and horseradish peroxidase–gold sol (HRP-Au) modified electrodes with and without an electron-transfer mediator was studied by Zhao et al.[110] They were able to immobilize HRP on colloidal gold, which was then deposited on a flat electrode surface. They found that the enzyme could be reduced at a convenient rate at 0 V (Ag/AgCl) without an electron-transfer mediator. Since the enzyme retained its activity, the amplification effect owing to the enzyme-catalyzed turnover of substrate facilitated these observations. The immobilization of a redox enzyme on colloidal gold was thought either to help the protein assume a favored orientation or to make possible conducting channels between the prosthetic groups and the electrode surface, or both. Either possibility would reduce the effective electron-transfer distance, thereby facilitating charge transfer. The buffer used was 0.05M phosphate at pH 6.8 with 10 mM KCl. No deaeration was carried out. Cyclic voltammograms were obtained in quiescent conditions, whereas steady-state amperometry experiments were conducted at 0 V (Ag/AgCl) in stirred buffer. The authors suggested that for the simple adsorption of HRP on a glassy carbon electrode, only the first layer of the adsorbed HRP molecules could accept electrons directly from the electrode surface. The observation that addition of mediator gives only a small increase in catalytic current in the case of simple adsorption on a flat glassy carbon surface suggested that no more than a monolayer of HRP was adsorbed on the surface. A determination of the amount of adsorbed protein based on the total enzymatic activity of HRP on the electrode indicated that the surface coverage was less than 5% of a monolayer. On the other hand, the authors pointed out that a colloidal gold surface is very different from flat bulk gold as colloidal gold particles have very high surface-to-volume ratios. Uncontaminated gold sol particle surfaces have high surface energy and therefore are very active. The authors suggested that the small size of the colloidal gold particles (~30 nm) gave the protein molecules more orientational freedom and so increased the possibility that the prosthetic group was closer to the metal particle surface. They determined that about 12 layers of gold sol particles were deposited onto a glassy carbon electrode surface. Spectroscopic data for the enzymatic activity of HRP adsorbed on colloidal gold before deposition on the electrode surface indicated that the active enzyme coverage on the gold sol particle surfaces is only about 40% of a theoretical monolayer. This was consistent with data obtained by others for adsorption of γ-globulin onto latex particles.[61] The likelihood of multilayer adsorption of protein molecules was thought to be negligible.[61]

The surface adsorption behavior of the globular proteins bovine β-lactoglobulin A, hen egg-white lysozyme, and bovine pancreatic ribonuclease A at a platinum electrode as a function of temperature was studied by Roscoe and co-workers[111,112] using cyclic voltammetry. The surface charge density was attributed to a flattening or denaturing of the protein at the electrode surface to allow adsorption of carboxyl groups, accompanied by electron transfer at these anodic potentials. The calculated surface concentrations, Γ, based on the number of carboxyl groups and two electron-transfer processes per carboxyl group gave good agreement with calculated values obtained by others using such techniques as radioactive labeling and ellipsometry. The technique employed in this work, which measures the charge density due to adsorption of proteins at platinum electrode surfaces, appears to be very sensitive to structural changes of the protein both in the bulk solution and at the metal/solution interface.

Although direct electron-transfer reactions of cytochrome c at bare silver electrodes had been reported to proceed by slow, irreversible heterogeneous kinetics,[113] which was attributed to electrode fouling by irreversible adsorption of protein,[114] Reed and Hawkridge[115] were able to obtain direct heterogeneous electron-transfer reactions between cytochrome c and silver electrodes. Cyclic voltammetry, derivative cyclic voltabsorptometry, and single-potential-step chronoabsorptometry were used in the investigations with varying concentrations of chromatographically purified cytochrome c solutions in both $0.05M$ Na_2SO_4 electrolyte and $0.05M$ Tris/cacodylic acid buffer, at pH 7.0. The kinetics of these reactions were found to be quasireversible at polished silver electrodes and at electrochemically roughened silver surfaces. The authors were able to demonstrate that neither electrode surface modification nor the inclusion of mediators was necessary in order to study the electron-transfer reactions of cytochrome c at silver electrodes. Apparently, the commonly used procedure of lyophilization of cytochrome c produces denatured forms that inhibit the direct transfer kinetics of cytochrome c at silver electrodes.

Buchi and Bond[116] studied the electrochemistry of cytochrome c using conventionally sized graphite substrates and carbon microdisk electrodes. They found an unusual concentration dependence of the voltammetric results at both the macro- and micro-sized electrodes, which they attributed to a modification of the electrode resulting from partial blocking of the surface. When the fraction of the electrode blocked was high, the electrochemical behavior was influenced strongly by the nonlinear mass transport to the remaining active sites of the electrode surface. They suggested that the partial blocking of the surface was due to adsorption of cytochrome c itself, in either the native or an electroinactive form, or by surface-active trace impurities present in the lyophilized protein.

Szucs et al.[33] studied both the adsorption and the mediated and unmediated electron transfer from cytochrome c on a gold electrode using ellipsometry and cyclic voltammetry. They found that the protein formed an irreversibly adsorbed layer that completely covered the electrode. The adsorbed cytochrome c was able to mediate the reduction of cytochrome c in solution via electron transfer through the unfolded protein layer. They did not observe oxidation of cytochrome c in the potassium phosphate monobasic buffer solution (0.1M, pH 6.0). The estimation of the adsorbed amount of protein was carried out by measuring the charge transfer between the protein and the electrode. The adsorption of cytochrome c was studied by ellipsometry at 400 mV (vs. NHE) on gold electrodes that had not been treated with surface modifiers. This potential was approximately equivalent to the open-circuit potential of the system in the absence of protein in the electrolyte. Cyclic voltammograms showed that both oxidation and reduction of cytochrome c occurred at a potential close to the standard redox potential for this molecule (i.e., 260 mV vs. NHE). However, with subsequent cycling, oxidation and reduction peaks disappeared completely. Structural change of the protein upon adsorption on bare gold caused a change in the redox potential from 260 to –150 mV. Since the potential at which the bulk proteins were reduced also changed to this lower value, it appeared to the authors that the protein in solution could be reduced only when the adsorbed protein was already reduced. They concluded that the adsorbed layer on bare gold had become an immobilized mediator for electron transfer between the underlying metal and cytochrome c in solution.

Willit and Bowden[117] examined the redox behavior and adsorption thermodynamics of cytochrome c on fluorine-doped tin oxide electrodes using cyclic voltammetry over a temperature range of 281–308 K. The protein was strongly adsorbed from neutral to slightly alkaline low-ionic-strength solutions, with almost a monolayer coverage resulting under optimum conditions. The experimental surface concentration, Γ, was determined to be 13–15 pmol cm^{-2}, except in pH 8, 131 mM ionic strength solutions, where the cytochrome c coverage was 7 pmol cm^{-2}. The variation in electroactive surface coverage with solution pH and ionic strength suggested that the adsorption was primarily controlled by electrostatic interactions. A difference of 2 kJ mol^{-1} in ΔG°_{ads}, the free energy of adsorption, was found between the oxidized and the reduced form of the protein. Also, ferricytochrome c adsorbed to tin oxide more strongly than ferrocytochrome c.

Daido and Akaike[118] investigated the electrochemistry of cytochrome c using indium tin oxide (ITO) electrodes. The electrochemistry was highly dependent not only on the purity of cytochrome c but also on the ionic strength of the electrolyte. Electrodes with adsorbed protein were prepared by allowing the electrodes to stand at constant potential or open circuit in phosphate buffer solutions containing 40 μM cytochrome c at pH 7.4. After a scheduled period, the electrodes were rinsed with the corresponding buffer solution in the absence of protein. The apparent surface excess, Γ, of cytochrome c was estimated from the anodic peak areas of the cyclic voltammograms and reached a maximum value of ~7 pmol cm^{-2}, which compared favorably to the coverage of the close-packed monolayer, estimated to be approximately 9.4 pmol cm^{-2}.

Ohtani and Ikeda[119] studied the direct electrochemistry of cytochrome c at a γ-alumina-coated electrode. They observed a fast redox reaction of cytochrome c on a glassy carbon electrode in a phosphate buffer in the absence of additives when the electrode surface was not properly cleaned after alumina polishing. This was suggested by the authors to be the first example where an insulating layer of an oxide on an electrode surface promoted direct electrochemistry of cytochrome c. As a result of these findings, they prepared a γ-Al$_2$O$_3$-coated glassy carbon electrode and observed a well-defined voltammetric response of cytochrome c with an enhanced current. They estimated a monolayer coverage of the protein from the Brunauer–Emmett–Teller (BET) surface area of γ-Al$_2$O$_3$ (258 m^2 g^{-1}) and an occupied surface area of 20 nm^2 molecule^{-1}.

Haladjian et al.[120] investigated the behavior of cytochrome c_3 at the basal-plane pyrolytic graphite electrode by cyclic voltammetry and differential pulse voltammetry. They observed modifications of the adsorbed film of cytochrome c upon scanning the potentials in both the anodic and cathodic ranges. The redox potentials were found to be pH-dependent, which suggested that irreversibly denatured forms containing a hemin-like group were in close contact with the electrode surface.

Flavin adenine dinucleotide (FAD) is the cofactor in a number of redox enzymes (flavoproteins), including glucose oxidase, which is commonly used in the construction of glucose sensors.[121] An understanding of the behavior of the cofactor near an electrode surface is necessary for correct interpretation of the electrochemistry of flavoproteins.[122] When FAD is bound to a protein, it can exist in three states. The flavoquinone form is fully oxidized, the flavosemiquinone form is one-electron reduced, and the flavohydroquinone form is two-electron reduced. The free cofactor in neutral aqueous solutions exists mainly in the completely oxidized or in the two-electron reduced state, as a result of the very fast disproportionation reaction of the flavosemiquinone. FAD has been studied quite extensively with several electrochemical techniques and different electrode materials,[123–128] and the electrochemistry of FAD has been found to be slow and complicated. Recently, Verhagen and Hagen[122] observed sharp, well-defined voltammograms for FAD with a glassy carbon electrode. The nature of the electrochemical reaction was found to be concentration-dependent, changing from a reversible surface charge transfer to a quasi-reversible diffusion-controlled electron transfer reaction with an increase in FAD concentration from $1\mu M$ to $1mM$ at constant pH. The formation of a surface layer of FAD through adsorption appeared to be essential for fast electron transfer. This suggested a possible mechanism of self-mediation. The authors proposed a mechanism in which adsorption results in formation of a stable monolayer of FAD on the electrode. These adsorbed molecules would then act as mediators through which electron transfer could occur from FAD molecules in solution to the electrode. The surface concentration of adsorbed FAD was determined to be $\Gamma = 3.0 \times 10^{-11}$ mol cm.$^{-2}$ This was estimated to correspond to ~18% electrode surface coverage, assuming the isoalloxazine ring to be adsorbed in a flat position.

IV. BEHAVIOR OF PROTEINS IN SOLUTION AND AT SURFACES

There have been a number of papers and reviews on the subject of the conformational behavior of proteins, both in bulk solution and at interfaces. In order to understand and interpret results obtained from electrochemical studies of the interfacial behavior of proteins, it is important to be able to relate these results to the conformational behavior of proteins. This section will give a general overview of globular protein behavior and the theories associated with it.

1. Conformational Behavior In Solution

An extensive series of review papers and research papers has been published by Tanford on protein denaturation.[129-134] The stability of protein structure has been discussed in several articles by Privalov, with particular attention paid to the small globular proteins.[135-137] Conformational behavior is dependent on the inter- and intramolecular forces experienced by these macromolecules. Weakly polar interactions in proteins play an important role and have been discussed by Burley and Petsko.[138] Gurd and Rothgeb[139] have described the motions observed in these nonquiescent macromolecules. Studies of the calorimetrically determined dynamics of complex unfolding transitions in proteins have been reviewed by Freire et al.[140]

Globular proteins constitute an enormous class of proteins and are involved in many of the chemical processes in cells, including synthesis, transport, and metabolism. X-ray diffraction methods have provided structural details of many of the globular proteins. Many globular proteins contain prosthetic groups that enable them to fulfill special functions. These groups may be covalently or noncovalently bonded to the protein and held within the folds of the tertiary structure of the molecule. Every globular protein has a unique tertiary structure, folded in a specific way such that the conformation is suited to the particular functional role that the protein plays. The information for determining the three-dimensional structure of a protein is carried entirely in its amino acid sequence. In some proteins the folding is dominated by the prosthetic group. All globular proteins are constructed so as to have a defined inside and outside, and, invariably, hydrophobic residues are packed mostly on the inside, whereas hydrophilic residues are on the outside, in contact with the solvent. The secondary structures known as β-sheets are usually wrapped into a barrel configuration. The "native," or natural, three-dimensional structure can be further stabilized by cross-linking via disulfide bonds between cysteine residues. These bonds are very specific, and the cross-linking depends on the already established folding. If the temperature is raised sufficiently, or if the pH is made extremely acid or alkaline, or if an organic solvent such as alcohol or urea is added to the solvent, the protein will unfold, resulting in denaturation. It is remarkable that this total scrambling of protein structures is in most cases a fully reversible, thermodynamically driven process, such that, in the case of enzymes, refolding results in complete recovery of enzymatic activity. Although the folded and unfolded states may be described thermodynamically, it is now known that globular proteins fold via complex kinetic pathways and that even the folded structure is far from static. There is much evidence that various kinds of motions are continually occurring within folded protein molecules, including the opening and closing of clefts of molecules, particularly during the catalytic functions of enzymes.

Thermodynamic studies led to the conclusion that protein denaturation is accompanied by an enormous increase in enthalpy, and this was regarded as an indication that protein denaturation is a highly cooperative process involving the whole macromolecule.[137] The "all-or-none" character of denaturation has been generally accepted. However, with further studies of this problem, accompanied by improvements in the precision of measurements, the validity of the two-state concept and of the entire thermodynamic approach to the problem of protein stability has been questioned. Scanning microcalorimetry has played an important role in the determination of thermodynamic functions related to protein stability.

As mentioned above, protein conformation is dependent on the inter- and intramolecular forces experienced by these macromolecules, with weakly polar interactions playing an important role.[138] Unit electronic charges are found on the carboxy and the amino terminal of proteins as well as on the side chains composed of aspartic acid, glutamic acid, cysteine, tyrosine, lysine, arginine, and histidine. At physiological pH, these ionizable groups are at least partially charged and are typically found on the surface of a protein, where they interact with other charged groups and/or become solvated by water molecules in the protein's hydration shell. Because some of these groups are charged only at one end of their structure, the remainder of the nonpolar amino acid side chain is often at least partially buried. An ion pair or salt bridge is formed when charged groups are positioned in the interior of a protein, where they are inaccessible to solvent. If the ionizable group is not buried, then a hydrogen bond interaction results with either a nearby polar group or an electrically neutral group, forming a charge–dipole interaction. Weaker but longer range electrostatic interactions also occur at quite large charge–charge separations. Nonpolar groups self-associate in water because their dispersal throughout the solvent would be entropically unfavorable. Once they come together and water is largely excluded, enthalpically favorable interactions such as London forces dominate. Hence, nonpolar groups are located in the interior of the protein, where they are not exposed to solvent water molecules, and the polar portions of the polypeptide chain are found on the protein's surface. A positive linear correlation between the hydrophobicity measure of Nozaki and Tanford[141] and the van der Waals surface area buried on folding was established by Rose et al.[142] The latter authors determined that the 20 naturally occurring amino acids could be categorized into three distinct groups on the basis of the mean fractional surface area buried on folding and the average residual unburied surface area after folding. The three amino acid classes are (1) very polar, which includes serine, proline, aspartate, asparagine, glutamate, glutamine, lysine, and arginine; (2) moderately polar, which includes alanine, threonine, histidine, and tyrosine; and (3) hydrophobic, which includes (in increasing order of hydrophobicity) glycine, cysteine, valine, isoleucine, leucine, methionine, phenylalanine, and tryptophan.

Although the precise contribution of any given electrostatic interaction in a protein to the free energy of stabilization is uncertain because it is a function of charge distribution, separation, and interaction geometry and solvation, enthalpy estimates for the various electrostatic interactions have been made.[138] Attractive charge–charge interactions with optimal separation contribute a free energy of about –12 to –17 kJ mol^{-1}; charge–dipole interactions contribute a free energy of about –12 kJ mol^{-1}; the hydrogen bond contributes an average enthalpy of about –8 kJ mol^{-1}, if neither group is charged; and the weakly polar interactions contribute enthalpies of between –4 and –10 kJ mol^{-1}. Each of these individual estimated enthalpies is small compared with the sum of all the enthalpic contributions to the stabilization of a protein's tertiary structure. As an example, ΔH at 298 K is about –272 kJ mol^{-1} for lysozyme going from the unfolded to the folded state.[129] The free energy of protein tertiary structure stabilization is typically about –42 kJ mol^{-1}.[135] The various electrostatic interactions are also involved in stabilizing protein–ligand complexes and, as a result, determine the precise geometry of ligand binding.

Intramolecular movement is seen in each functional form of a protein molecule.[139] Electrostatic potentials will vary relatively slightly with small displacements. However, many charged groups in side chains reorient rapidly with effective correlation times in the 5–50 ps range.[143] Consequently, the resultant electrostatic fields will also vary over time as the distances of separation between the charged side chains, and possibly even their degree of solvent exposure, change. The fluctuations in the resultant force vectors will cover a considerable time range. Backbone components in secondary structural components such as helices, sheets, and turns will be relatively fixed on the time scale of 1–10 ps for unhindered single-bond rotations.[139] The bulkier aromatic groups require relatively large displacements of their neighbors to achieve sizable rotational motions, but they become freer as the temperature rises. While polar constituents are subject to constraining interactions, the nonpolar residues are able to pass by each other relatively freely. Exterior side chains, terminal segments, and certain main-chain loops have been found to be relatively mobile. Because of the length of such residues as lysine, arginine, and glutamic acid, in particular, their charged groups can undergo considerable displacements in short periods of time, of the order of 10–11 s. The result of these various types of mobility of the positively and negatively charged groups is that electrostatic fields vary continuously around their average vectors at various points in and around a protein molecule. Mobility is generally increased in the denatured state and may also be significantly altered in various quaternary forms and in complexes with ligands.[139]

When a protein solution is heated, the solubility of the protein is found to decrease over a narrow temperature range, resulting in intensive aggregation.[135] Pancreatic ribonuclease A, which is a typical small, compact globular protein, has been widely used to study denaturation because of its perfect reversibility after the denaturing condition is eliminated. UV absorbance measurements, circular dichroism, optical rotation, viscosity, stability against proteolysis, and rate of hydrogen exchange have all been used to study the denaturation of ribonuclease. The results of all the measurements indicated that the temperature dependence of denaturation consisted of one sigmoidal function and two very smooth functions below and above the sigmoidal one. The sigmoidal changes were attributed to the denaturational changes, and the minor changes below and above this region were called the pre- and postdenaturational changes in the protein. The influence of temperature was apparent from the very beginning of the heating of the solution. Similar behavior has been observed with lysozyme, which is also a reversibly denaturing globular protein.[135]

2. Behavior at Interfaces

MacRitchie[15,144] and Dickinson and Stainsby[145] have reviewed the behavior of proteins at a variety of interfaces. There have also been a number of recent research papers which have taken a theoretical or computational approach: examples include the computation of the electrostatic interaction energy between a protein and a charged surface[146]; the computation of electrostatic and van der Waals contributions to protein adsorption[147]; Monte Carlo simulation of the conformational behavior of a polypeptide chain near a charged surface[148]; protein structure prediction based on statistical potential[149]; development of a model system for the interaction of proteins with organic surfaces[150]; and investigation of nonequilibrium electric surface phenomena in colloid systems.[151]

Examination of the literature supports the observation that adsorption at solid/liquid interfaces has been traditionally more difficult to study than that at liquid/liquid interfaces.[15] All the features described in the previous section play an important role in the behavior of proteins at interfaces. Proteins tend to concentrate at surfaces as a result of their amphipathic nature arising from the mixture of polar and nonpolar side chains.[15,144]

Polysaccharides and polynucleotides do not share this property to the same extent.[52] Although many proteins are very soluble in water, adsorption leads to stable monolayers that are extremely difficult to desorb. However, protein molecules at solid/liquid interfaces do not appear to undergo the extensive conformational changes that occur at liquid/liquid interfaces. This has been attributed to the fact that, whereas proteins denature at liquid/liquid interfaces as a result of the distribution of polar and apolar residues into the appropriate fluid, at solid/liquid interfaces the adsorbing molecule is unable to penetrate the solid phase, and therefore there is less tendency for protein molecules to unfold and expose their apolar residues.[15] It has also been suggested that adsorption may be confined to particular sites. As an example, a bound fraction of 0.11 for prothrombin and BSA adsorbed on silica was determined from the number of carbonyl surface attachments, as measured by infrared difference spectroscopy.[14] Regardless of the amount of protein adsorbed, the fraction remained constant. This indicated that no significant conformational changes occurred at the surface relative to the conformation of the protein in solution. However, in similar measurements with γ-globulin and β-lactoglobulin, the fraction of adsorbed carbonyl groups decreased as the amount of adsorbed protein increased.[152] This was attributed to changes in the distribution of segments brought about by compression of the adsorbed film.

The concentration dependence of protein adsorption has been found to follow a Langmuir-type isotherm, and therefore the adsorbed film is considered to be monomolecular.[27,50] The surface concentration usually approaches a plateau value as the solution concentration is increased, which is consistent with the saturation of available sites. There has been evidence that globular proteins can undergo extensive unfolding to give these "monolayer" films. An interesting feature of globular protein adsorption is the existence of "steps" in the adsorption isotherms under certain conditions; these steps have been attributed to two modes of adsorption, one at low protein concentrations and one at high protein concentrations.[61] Fair and Jamieson[61] have observed these steps in studies of adsorption of BSA on polystyrene latex particles, which gave an initial plateau ($\Gamma = 1.4$ mg m^{-2}) followed by an increase to a final surface coverage of $\Gamma = 2.6$ mg m^{-2} (Fig. 2). They postulated the existence of three regimes of surface structure as a function of increasing protein concentration. The first was a two-dimensional, disordered structure formed by random, independent adsorption of isolated molecules. The second was a glasslike structure, and the third a surface "crystal." In addition to the

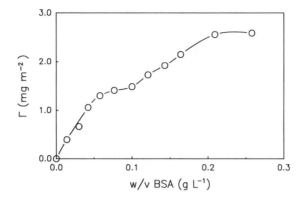

Figure 2. Adsorption isotherm for bovine serum albumin on polystyrene latex particles at 295 K, pH 7.4, obtained by solution depletion methods. (Adapted from Fair and Jamieson.[61])

plateau value concentration, polymer adsorption is often characterized by the thickness of the film and the initial slope of the adsorption isotherm. The thickness data supplement interfacial concentrations to give clues about molecular conformation or orientation.[15] For proteins, the interfacial concentrations can often be rationalized in terms of either adsorbed globular molecules or extended molecules with a fraction of the segments attached to the interface. Because the initial slope reflects the distribution tendency of a polymer between interface and bulk, it can therefore be related to the free energy of adsorption. Desorption is usually very difficult at solid/liquid interfaces, and therefore there has been a tendency to doubt the applicability of Langmuir-type adsorption because of this apparent irreversibility. Adsorption is sensitive to pH, with the maximum adsorption near the isoelectric point of the protein.[27] Greater adsorption occurs at hydrophobic interfaces than at hydrophilic ones, and this has been attributed to a higher free energy of adsorption.[15] As a result, desorption occurs more easily from hydrophilic solids.

V. APPLICATIONS OF ELECTROCHEMICAL TECHNIQUES TO THE STUDY OF THE INTERFACIAL BEHAVIOR OF PROTEINS

The discussions in this section will consider the role of electrochemistry in the examination of the features of proteins important to their behavior at surfaces and will address its application to the medical and industrial fouling problems. The discussion will concentrate on the following globular proteins, whose structures have been well determined: β-lactoglobulin, ribonuclease, lysozyme, κ-casein, bovine serum albumin, and cytochrome c. In the first part of this section, the surface adsorption behavior of these and other proteins at solid electrodes at ambient temperature will be compared with results obtained using other techniques. In the second part, the behavior of these proteins at electrode surfaces as a function of temperature will be described, and the results will be discussed in relation to the conformational behavior of the proteins both in the bulk solution and at the electrode surface and compared with the results obtained using other techniques over similar temperature ranges.

1. Surface Adsorption Behavior of Proteins at Ambient Temperature

(i) Conformational Behavior Studies

The results of electrochemical experiments that have been focused on the elucidation of the conformational behavior of proteins at the electrode surface, particularly in comparison with their conformational behavior in the bulk solution, will be discussed in this section. The purpose is to determine whether electrochemical techniques are able to probe the structure–function relationships of these macromolecules in such a way as to provide a different and complementary perspective which may help unravel the complex behavior of these very important biological compounds. The discussion will be focus on three proteins whose three-dimensional structures have been well characterized by X-ray diffraction and whose interfacial behavior has been extensively investigated by a variety of techniques. Comparisons of the results obtained for these well-characterized proteins from electrochemical experiments under similar experimental conditions provide a basis for evaluating these electrochemical methods in relation to some of the more traditional techniques described in Section II.

The three globular proteins β-lactoglobulin A, hen egg-white lysozyme, and bovine pancreatic ribonuclease A all have similar molecular weights. The whey protein β-lactoglobulin has been the subject of numerous publications owing to the important role it plays in the "fouling" of metal surfaces and in the formation and stabilization of dairy foams and emulsions.[3,4] The structure, determined by X-ray crystallography, consists

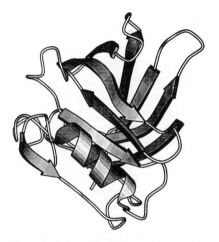

Figure 3. Diagram of β-lactoglobulin. (Mole-
script diagram,[202] kindly provided by L. Sawyer,
University of Edinburgh).

of an antiparallel β-sheet formed by nine strands wrapped around to form
a flattened cone or calyx.[153] Figure 3 shows the calyx entrance at the top
left "behind" the upper layer surmounted by the helix. This globular
protein (18,363 daltons) exists naturally as a dimer of two noncovalently
linked monomeric subunits.[154] There are five cysteine residues/mole, four
of which are involved in disulfide linkages.[154] The single free sulfhydryl
group appears to be equally distributed between cysteine 119 and cysteine
121. The existence of this free sulfhydryl group is of great importance for
changes occurring in milk during heating as it is involved in reactions with
other proteins such as κ-casein. The protein also contains 52 titratable
carboxyl groups at pH values of 6.6 and lower.[155] The formation of dimers
and octamers in this pH region has also been attributed to the presence of
carboxyl groups.[155,156]

Lysozyme and ribonuclease (14,600 and 13,680 daltons, respectively) have similar structures, each with 11 carboxyl groups (including
the α-carboxyl group).[157] They both have four disulfide linkages but
contain no free sulfhydryl groups. Lysozyme is an atypical protein in that
it has an unusually high isoelectric point, close to 11.[15] As a result, the
molecule carries a high positive charge at neutral pH so that a high

electrical potential barrier to adsorption occurs at positively charged surfaces. It has been reported to have 10.5 titratable carboxyl groups.[158] Bovine pancreatic ribonuclease has an isoelectric point of 9.5, and its adsorption behavior has been studied on polystyrene latices.[145]

The surface adsorption behavior of bovine β-lactoglobulin A, hen egg-white lysozyme, and bovine pancreatic ribonuclease A at a platinum electrode were studied by Roscoe and Fuller[111] and Roscoe et al.[112] using cyclic voltammetry. These proteins were chosen for comparison of their interfacial behavior in order to determine the role of the free sulfhydryl group in the surface adsorption mechanism at the platinum electrode. In order to characterize the behavior of β-lactoglobulin A at the platinum electrode, it was necessary first to determine the surface charge density of oxide deposition at the platinum electrode under the same experimental conditions as used with the proteins. Cyclic voltammograms were then recorded after each addition of an aliquot of β-lactoglobulin to the buffer solution in the electrochemical cell. The profile of the cyclic voltammogram changed immediately as shown in Fig. 4, even with the addition of an increment of β-lactoglobulin as small as 0.012 g l^{-1} to the phosphate buffer. The surface charge density resulting from adsorption of the protein, Q_{ADS}, was determined from the integrated current–potential response corresponding to anodic oxidation in the presence of protein, Q_{Oo}^{P}, and to oxide reduction in the presence of protein, Q_{Or}^{P}. The calculated difference between the surface charge density for anodic oxidation and that for oxide reduction was attributed to surface adsorption or oxidation of species other than O or OH present in these aqueous solutions. A small difference in the integrated surface charge density from the cyclic voltammograms of the phosphate buffer, calculated from the difference between the anodic oxide charge Q_{Oo}, and the oxide reduction charge Q_{Or}, was subtracted from the surface charge densities determined for the protein solutions. The remaining charge density was attributed to protein surface adsorption:

$$Q_{ADS} = (Q_{Oo}^{P} - Q_{Or}^{P}) - (Q_{Oo} - Q_{Or})$$

From the surface charge densities, Q_{ADS}, the number of electrons transferred, n, calculated from the number of carboxyl groups present on the protein on the basis of a two-electron-transfer process per carboxyl group, the molar mass of the protein, M, and the Faraday constant, F, an upper limit of the surface concentration, Γ, may be calculated:

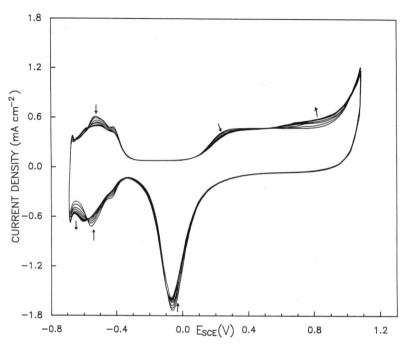

Figure 4. Cyclic voltammogram of 0.1M phosphate buffer (pH 7.0) at a platinum electrode, showing the effect of addition of a 0.012 g l^{-1} aliquot of β-lactoglobulin A in phosphate buffer. (Reprinted from Roscoe *et al.*,[112] with permission.)

$$\Gamma = \frac{Q_{\text{ADS}}\,M}{nF}$$

Similarly for each of the other proteins, cyclic voltammograms were recorded after each addition of an aliquot of protein to the buffer solution in the electrochemical cell.[112] The surface charge density was measured over the region which normally corresponds to a monolayer of OH during oxide formation in aqueous solutions, to the anodic end potential of 0.4 V (vs. SCE). Since this is also the rest potential of lysozyme on platinum,[159] the surface charge densities measured to this anodic end potential may reflect the adsorption of a monolayer of protein and allow correlation with results from other experimental techniques. The calculated values of Γ for the three proteins considered in these studies are shown in Fig. 5. Plateau values in surface charge densities can be seen at low bulk concentrations

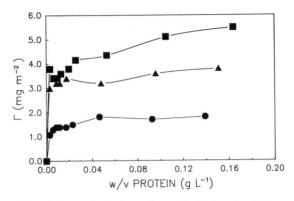

Figure 5. Adsorption isotherms for proteins in phosphate buffer (pH 7.0, 299 K) with an anodic end potential of 0.4 V (vs. SCE): ●, β-lactoglobulin A; ▲, lysozyme; ■, ribonuclease A. (Reprinted from Roscoe and Fuller,[111] with permission.)

for all three proteins (and are particularly well defined in the case of β-lactoglobulin), suggesting that some type of restructuring may be taking place on the surface of the electrode.

From the surface charge density, $Q_{ADS} = 50\,\mu C\,cm^{-2}$, determined with β-lactoglobulin A for an anodic end potential of 0.4 V at 299 K and pH 7, a surface protein concentration of 1.8 mg m^{-2} was calculated; this value is consistent with the results obtained by others using a variety of techniques on hydrophilic surfaces.[3,64,69,160,161] This calculation is based on a transfer of 52 electrons, which is the maximum number on the basis of the total number of carboxyl groups titrated[155] for the dimer of β-lactoglobulin A and a two-electron-transfer process per carboxyl group. It appears that the carboxyl groups are likely the active functional groups over this potential range based on results obtained previously for the anodic reactions of amino acids.[102–104] Because almost all the charged side-chain groups, such as aspartic acid and glutamic acid, are at the outer surface of the globular protein,[145] they can interact directly with the electrode surface. Thus, surface adsorption occurs through the carboxylate anions at pH 7.0 accompanied by electron transfer, and, following this, by decarboxylation, which is the rate-determining step. A transfer of a second electron occurs during the hydrolysis reaction of the adsorbed radical at the electrode surface.

If these reactions are written in an analogous manner in terms of protein (P) adsorption on a metal (M) electrode, then the first step involving protein adsorption is fast, with n representing the number of sites for carboxyl interaction with the metal surface accompanied by the transfer of a total of n electrons. Following this, the decarboxylation reaction is the rate-determining step (rds), with P' representing the modified protein.

$$P + nM \rightarrow P(M)_n + ne^- \qquad (fast) \qquad (1)$$

$$P(M)_n \rightarrow P'(M)_n + nCO_2 \qquad rds \qquad (2)$$

The rate expressions for the forward and reverse reactions, having rate constants k_1 and k_{-1} and free energies of activation $\Delta G_1^{\ddagger}$ and $\Delta G_{-1}^{\ddagger}$, respectively, may be written as

$$i_1 = nzFk_1 (1 - \theta_{G \cdot})^n \, C \exp[-(\Delta G_1^{\ddagger} - \beta VF + \beta r_1 \theta)/RT]$$

$$i_{-1} = nzFk_{-1} (\theta_{G \cdot}) \exp[-(\Delta G_{-1}^{\ddagger} + (1 - \beta)VF - (1 - \beta)r_1 \theta)/RT]$$

where $G \cdot$ refers to the radical species on the electrode surface, C is the concentration of protein in the bulk solution, θ is the fractional surface coverage, n is the number of functional groups on the protein and the number of electrons transferred, β is the symmetry factor, and r_1 allows for variation in the heat of adsorption with coverage.

If step 2 is rate-determining, then step 1 can be considered to be in quasiequilibrium:

$$\frac{\theta}{(1 - \theta)^n} = KC \exp - (\sigma)$$

where

$$K = k_1/k_{-1}$$

$$\sigma = (\Delta G - VF + r_1 \theta)/RT$$

Rearranging and taking natural logarithms gives

$$\ln (C/\theta) = n [-\ln (1 - \theta)] - \ln K + \sigma$$

The value of n may be determined from the slope of a plot of $\ln(C/\theta)$ versus $-\ln(1 - \theta)$, where n represents the number of electrons transferred

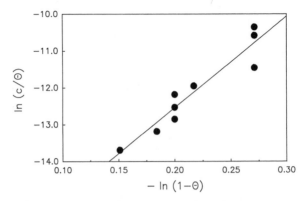

Figure 6. Plot of $\ln(C/\theta)$ versus $-\ln(1 - \theta)$ for β-lactoglobulin in phosphate buffer (pH 7.0, 299 K) with an anodic end potential of 0.4 V (vs. SCE). (Reprinted from Roscoe et al.,[112] with permission.)

and the number of carboxyl groups interacting with the metal electrode surface. The slope of 25 (Fig. 6) obtained from measurements made using an anodic end potential of 0.4 V corresponds well with the number of carboxyl groups (26) on the β-lactoglobulin A molecule. This interpretation has shown consistency for a variety of proteins in ongoing research by the author and co-workers.

Experiments using radioactive isotopes to investigate the adsorption of substances with several functional groups on platinum have shown that, when a carboxyl group was present, the adsorption properties of the molecules were entirely determined by the carboxyl group.[162] Studies by infrared reflection–absorption spectroscopy and ellipsometry on the adsorption of β-lactoglobulin on hydrophilic gold surfaces[23] gave results which were attributed to strong coordination of the carboxyl group to the gold surface through an ester-type bond. The infrared studies showed that the strong symmetric CO stretching mode clearly seen in the aqueous- and solid-phase spectra could not be resolved in the corresponding reflection–absorption spectrum.

Surface concentrations of 1.3 mg m^{-2} for the plateau value at low bulk protein concentrations and 1.8 mg m^{-2} at the higher concentrations, calculated from the surface charge density resulting from adsorption of β-lactoglobulin at 299 K and pH 7.0 on a platinum electrode with an anodic end potential of 0.4 V, agree very well with values in the literature from a variety of techniques. Surface concentrations of β-lactoglobulin, of 0.75 mg m^{-2} and 0.96 mg m^{-2} at hydrophilic and hydrophobic chromium surfaces, respectively, were obtained, following rinsing with buffer, from ellipsometry measurements.[3] Radiotracer methods at each surface yielded values of 0.98 and 1.20 mg m^{-2}, respectively.[3] The bulk protein concentration was 1 g l^{-1} in each of these experiments, a factor of 10 greater than the highest concentration used in the electrochemical measurements, for which plateau values were obtained with bulk concentrations as low as 0.04 g l^{-1}. Values of 1.20 mg m^{-2} (Refs. 160 and 161) and 1.4–1.5 mg m^{-2} (Ref. 161) were reported for hydrophilic and hydrophobic chromium surfaces, respectively, following rinsing and a second addition of protein to give a bulk concentration of 2 g l^{-1}. The method of solution depletion on a stainless steel surface gave a surface adsorption density of 1.5–1.6 mg m^{-2} at pH 6.8 and 300 K.[69] The plateau values of surface concentration on stainless steel versus bulk protein concentration obtained in the latter study were found to agree well with those determined under similar conditions at a platinum electrode for these very dilute protein concentrations (Fig. 7).[112] A similar result of 1.4 mg m^{-2} was obtained for the adsorption of β-lactoglobulin on polystyrene latex as measured by

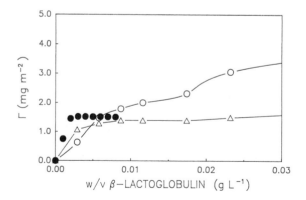

Figure 7. Adsorption isotherms for β-lactoglobulin in phosphate buffer (pH 7.0, 299 K) at a platinum electrode with an anodic end potential of 1.0 V (○) and 0.4 V (△) (vs. SCE) (from Roscoe *et al.*,[112] with permission) and for β-lactoglobulin in phosphate buffer (pH 6.8, 300 K) on stainless steel pellets (●,) adapted from Kim and Lund[69]).

SAXS at pH 7.2.[64] At these temperatures and pHs, the protein should exist in solution as dimers. Calculations based on adsorption of monomers, represented as hard spheres of 3.58-nm diameter, for a hexagonal close-packed array gave the adsorption density maximum as 2.8 mg m^{-2}.[64] A random adsorption value was estimated to be 1.66 mg m^{-2}, taking into account possible lateral movement on the surface, deformation, and unfolding of the protein. For adsorption in the end-on position, an estimated density of 3.32 mg m^{-2} was obtained.[64]

On the other hand, Al-Malah et al.[25] obtained values of surface concentration of β-lactoglobulin of 2 mg m^{-2} and 4 mg m^{-2} for adsorption on hydrophilic and hydrophobic silicon surfaces, respectively, using ellipsometry (Fig. 8). However, although adsorption was from phosphate buffer solutions at pH 7.0, the samples were rinsed and dried in a desiccator overnight prior to measurements by ellipsometry. Wahlgren and Arnebrant[24] obtained lower surface concentrations for β-lactoglobulin adsorption on silica, methylated silica, and polysulfone surfaces using in situ ellipsometry (Fig. 9), although in the case of silica and methylated silica the plateau values for the concentrations used in the bulk solution may not have been reached. The plateau value obtained for adsorption of β-lactoglobulin on polysulfone agrees very well with the calculated maximum of 2.8 mg m^{-2} for a hexagonal close-packed array.[64]

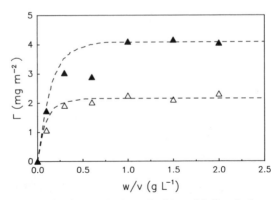

Figure 8. Adsorption isotherms for β-lactoglobulin on hydrophilic ($\triangle$) and hydrophobic ($\blacktriangle$) silicon surfaces measured by ellipsometry after drying of the films. (Adapted from Al-Malah et al.[25])

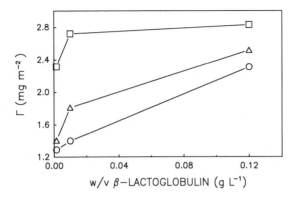

Figure 9. Adsorption isotherms for β-lactoglobulin on silica ($\bigcirc$), methylated silica ($\triangle$), and polysulfone ($\square$) surfaces in phosphate buffer with 0.15M NaCl (pH 7.0, 298 K), measured by *in situ* ellipsometry. (Adapted from Wahlgren and Arnebrant.[24])

The surface charge density, Q_{ADS}, was also obtained from measurements over the full anodic under potential deposition (upd) region to the end potential of 1.0 V (vs. SCE) for each of the three proteins at 299 K and pH 7.0 as a function of protein concentration (Fig. 10).[111] As the

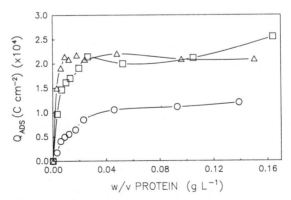

Figure 10. Surface charge density, Q_{ADS}, at a platinum electrode versus the concentration (w/v) of protein in phosphate buffer (pH 7.0, 299 K) with an anodic end potential of 1.0 V (vs. SCE): $\bigcirc$, β-lactoglobulin A; $\triangle$, lysozyme; $\square$, ribonuclease A. (Reprinted from Roscoe and Fuller,[111] with permission.)

solution or bulk concentration of β-lactoglobulin was increased to 0.01 g l^{-1}, the surface charge density increased to a level of about 50 μC cm^{-2}, at which there was a short plateau, similar to the plateau level observed for the 0.4-V end potential, before increasing again as the bulk concentration was increased from 0.02 to 0.03 g l^{-1}. At this point, a plateau value of 100 μC cm^{-2} was reached, and this value remained essentially constant with increasing bulk concentration of protein. Greater surface charge densities were measured for lysozyme and ribonuclease. The surface charge density for lysozyme increased rapidly and reached a plateau value of ~210 μC cm^{-2} at a protein concentration of only 0.02 g l^{-1}. The surface coverage varied from $\theta = 0.5$ for β-lactoglobulin to $\theta \approx 1$ for lysozyme and ribonuclease for an anodic end potential of 1.0 V on the basis of the surface sites available as determined from the upd H peaks in the cyclic voltammogram.[163] From the surface charge density, $Q_{ADS} = 100$ μC cm^{-2}, with this anodic end potential for the β-lactoglobulin solution, calculation of the surface concentration as described previously gave a value of 3.6 mg m^{-2}. Although the surface concentration based on the geometric adsorption of monomeric hexagonally close-packed hard spheres is calculated to be 2.8 mg m^{-2}, and that for end-on adsorption is 3.3 mg m^{-2},[64] which would correspond within experimental error to the observed value for Q_{ADS} at the high anodic end potential, it is more likely that the protein denatures to some extent on the surface of the electrode. Therefore, the electrochemical measurements probably represent several layers of proteins unfolded to different extents and intertwined such that the carboxyl groups approach and adsorb on the electrode surface accompanied by electron transfer. Portions of the protein may extend out toward the bulk solution, particularly those regions containing positively charged residues.

   The plateau regions suggest that further denaturation or surface rearrangement occurs to accommodate the higher concentration of protein. This behavior is similar to that observed in studies employing photon correlation spectroscopy,[61] in which three regimes of surface structure resulted from a stepwise adsorption of a globular protein. Evidence of protein denaturation on surfaces has also been obtained by small-angle X-ray scattering (SAXS) studies on adsorption of β-lactoglobulin on polystyrene latex at pH 7.2, which showed $\Gamma = 1.4$ mg m^{-2} with thicknesses substantially less than the diameter of a monomer.[64] However, the rather high surface concentrations in the work of Roscoe et al.[112] with an anodic end potential of 1.0 V probably result from the potentially-driven compression of the flattened proteins, allowing greater interaction of carboxylate groups at the positively charged electrode surface.

Studies by infrared reflection–absorption spectroscopy and ellipsometry on the adsorption of β-lactoglobulin on hydrophilic gold surfaces suggested that the protein is adsorbed in the native (octamer) conformation at pH 4.5 but that a transition toward a flat unfolded conformation appears to occur at the higher pH values of 6.0 and 10.0.[23] In order to account for the high refractive index and low thickness of a film optically equivalent to the protein layer at pH 6.0, it was suggested[23] that the electrostatically triggered surface denaturation was accompanied by a compression of the adsorbed protein by the protein/metal electric field. It is important to note that the gold surfaces were maintained at open-cell rest potentials with measured potentials of 348 mV, 200 mV, and 49 mV versus a saturated calomel electrode for pH values of 4.5, 6.0, and 10.0, respectively. These potentials on the gold surface are substantially lower than the potentials applied to the platinum electrode surface in the work of Roscoe *et al.*[112]

The effect of pH on the surface adsorption of β-lactoglobulin was also investigated with solutions at pH 2.0, 6.0, 7.0, 8.0, and 11.0 (Fig. 11).[112] Again, the surface charge density appeared to go through a small plateau region at bulk protein concentrations of 0.01–0.02 g l^{-1} in most of the measurements. The maximum plateau level of 170 μC cm^{-2} was reached with a bulk protein concentration of 0.04 g l^{-1} at pH 2.0. In general, a decrease in surface adsorption as a function of pH can be seen in the order pH 2.0 > pH 6.0 > pH 8.0 > pH 11.0 ~ pH 7.0.

β-Lactoglobulin undergoes complex conformational changes as a function of pH. Below pH 3.5, the dimer dissociates reversibly to monomers as a result of the strong electrostatic repulsive forces that develop in this acidic region.[164] This dissociation, however, is not accompanied by gross changes in molecular shape or conformation as there is no change in the optical rotatory dispersion (ORD), circular dichroism (CD), or tryptophyl fluorescence emission maximum in this pH region.[164] A slight swelling of the spherical monomer may result from the high net charge at this pH.

At pH 4.5 and low temperatures, β-lactoglobulin A associates to form an octamer.[165] A stable dimer (~36,700 daltons) forms at pH values near its isoelectric point of pH 5.2. Results of hydrodynamic measurements, small-angle X-ray scattering, and X-ray crystallography support a structure of the dimer as a prolate ellipsoid with a dyad axis of symmetry, having an overall length of 6.95 nm and a width of 3.6 nm.[164]

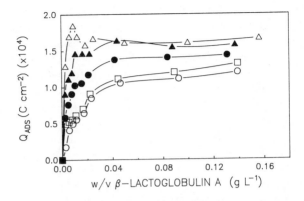

Figure 11. Surface charge density of β-lactoglobulin A at a platinum electrode with an anodic end potential of 1.0 V (vs. SCE) versus the concentration (w/v) of the protein in phosphate buffer (299 K) at pH 2.0 ($\triangle$), 6.0 ($\blacktriangle$), 7.0 ($\bigcirc$), 8.0 ($\bullet$), and 11.0 ($\square$). (Reprinted from Roscoe *et al.*[112] with permission.)

At pH 6.0 the protein is in the N state (native state), and it undergoes a conformational change as the pH is raised.[132] The surface adsorption behavior at pH 6.0 was not significantly different from that observed at pH 2.0 (Fig. 11). As the pH is raised, a transition to the R state (reversible denaturation) occurs.[132] This N–R transition is referred to as the Tanford transition. This reversible conformational change in the alkaline region, which is centered around pH 7.5, exposes and ionizes one abnormal carboxyl per monomer.[164] Increased reactivity of the single sulfhydryl group is also observed, and the transition is also accompanied by a general molecular expansion, as indicated by a decrease in sedimentation coefficient. The β-segments appear to unfold more readily than the α-helical portion of the side chain during changes in alkaline pH. Dissociation of the dimer also occurs; however, the equilibrium constant is essentially unchanged in the pH range 6.9–8.8.[164]

At pH 7.0, Roscoe *et al.*[112] found that surface adsorption as measured by surface charge density was at the minimum value. This may be due partly to an increase in the volume of the β-lactoglobulin structure during this denaturation from the compact globular structures at the lower pHs, resulting in a smaller packing density and hence decreased surface concentration. At pH 8.0, their experimental results showed an increase in the surface charged densities as compared with those obtained at pH 7.0. This may be due to the combined effects of the release of the carboxyl group and the increased denaturation as indicated by increased activity of the free sulfhydryl group.[164] Dissociation of the dimer as well as an increased molecular volume of the protein will also have an effect.

In solutions of pH greater than 9.0, β-lactoglobulin dimers dissociate rapidly, and the monomers suffer an irreversible denaturation to a random coil.[166] The reduced adsorption observed in the work of Roscoe et al.[112] at pH 11.0 may result from the increased volume of the open structure formed during this stage in the denaturation process. The finding that the surface charge densities measured as a function of pH for the protein solutions show the reverse effect to that observed in the studies made on small amino acids[102,103] probably results from the complex nature of the protein structures due to the variation in volume and charge over the pH range investigated. However, these results compare well with those obtained by SAXS measurements, in which protein adsorption was found to increase with a decrease in pH for the values 9.0, 7.2, and 4.65.[64] Studies by infrared reflection–absorption spectroscopy and ellipsometry on the adsorption of β-lactoglobulin on hydrophilic gold surfaces suggested that the protein is adsorbed in the native (octamer) conformation at pH 4.5 but that a transition toward a flat, unfolded conformation appears to occur at the higher pH values of 6.0 and 10.0.[23] As mentioned previously, it was suggested in the latter work that the electrostatically triggered surface denaturation was accompanied by a compression of the adsorbed protein by the protein/metal electric field. The open-cell rest potentials on the gold surface were substantially lower than the potentials applied to the platinum electrode surface in the work of Roscoe et al.[112]

The complexity of the conformational behavior of β-lactoglobulin as a function of pH makes the analyses more difficult under conditions of varying pH. However, the observed increase in adsorption with decrease in pH is consistent with the results of studies of the interaction of β-lactoglobulin with other interfaces[64] and suggests that electrochemical measurements may provide information sensitive to the conformational behavior of the protein under these conditions.

Calculation of the surface concentration for lysozyme from the surface charge densities measured at 299 K with an anodic end potential of 0.4 V gave $\Gamma = 3.2$ mg m^{-2}.[111] This result agrees very well with those obtained using other techniques. A comparison of the reactivity of lysozyme on three metals, platinum, zirconium oxide, and titanium oxide, showed a "plateau phase" in the surface concentration, Γ, which developed faster on platinum than on the other metals.[159] The ratio between the number of molecules per unit area from ellipsometry, N, and the number of molecules from a calculation based on a compact "end-on" configuration was reported to be 1.5 for platinum,[159] consistent with the adsorption of more than one layer.[159] Based on the small measured thickness of the adsorbed protein layer (1.1 nm), and the markedly high refractive index, it was suggested that the lysozyme molecules adsorb in a flat, compact, and probably unfolded configuration with a second, less dense and less strongly attached lysozyme layer that can be desorbed upon rinsing. Because of the large refractive index, $nF \approx 2$, for protein layers adsorbed onto platinum, it was suggested that the structure of the adsorbed molecules is different from that of molecules adsorbed on other metal surfaces. The surface concentration of 4.1 mg m^{-2} combined with the very small thickness (1.1 nm) gave an unreasonably high "measured" density of 3.7 g cm^{-3}. A more reasonable surface concentration of 3.2 mg m^{-2} was obtained when a more common refractive index for the protein layer of $n_F = 1.6$ and a layer thickness of 2.3 nm were used, which were consistent with a compact monolayer of partly unfolded molecules. It was concluded that electrostatic interactions between the protein and the metal seemed to be of great importance in determining the amount of protein adsorbed and that the adsorbed lysozyme molecules form a compact and very dense layer on platinum. Lysozyme has also been shown to form multilayers at liquid interfaces.[167]

Calculation of the surface concentration for ribonuclease A from the surface charge densities measured by Roscoe and Fuller[111] gave $\Gamma = $ 3.4–4.2 mg m^{-2} for low concentrations of proteins with an anodic end potential of 0.4 V. There is a similarity between the results obtained for ribonuclease A and for lysozyme at 299 K and pH 7.0 with an anodic end potential of 0.4 V. This may be a reflection of the similarities in their structures. Like lysozyme, ribonuclease has a hydrophobic core on one side of the active site crevice and a less substantial second wing, which gives it a V shape.[157] All the charged polar groups are on the surface, each protein having 11 carboxyl groups and no free sulfhydryl groups. A thorough study of the adsorption of bovine pancreatic ribonuclease on polystyrene latices was made by Norde.[168] The change in protein conformation upon adsorption was found to be related to the conformational stability in bulk solution. Adsorption was found to increase with increasing temperature. Hydrogen-ion titration showed that the dissociation of

Table 1
Surface Concentrations of a Variety of Proteins

Protein	Γ (mg m^{-2})	Surface	Technique	Reference(s)
β-Lactoglobulin	1.8^a	Platinum	Electrochemistry	112
	1.4	Polystyrene	Small-angle X-ray scattering	64
	1.2	Hydrophilic chromium	Ellipsometry	161
	0.98	Hydrophilic chromium	Radiotracer	3
	1.20	Hydrophobic chromium	Radiotracer	3
	1.20	Hydrophilic chromium	Ellipsometry	4,160
	1.4–1.5	Hydrophobic chromium	Ellipsometry	4
	1.5–1.6	Stainless steel	Solution depletion	69
Ribonuclease	3.4–4.2	Platinum	Electrochemistry	111
Lysozyme	3.2	Platinum	Electrochemistry	111
	3.2	Platinum	Ellipsometry	159
κ-Casein	3.5	Platinum	Electrochemistry	200
	3.4	Hydrophilic chromium	Ellipsometry	3

^{a}Plateau value = 1.4 mg m^{-2}.

carboxyl groups was suppressed by adsorption, and therefore it was deduced that structural rearrangements accompanying adsorption lead to a larger proportion of carboxyl groups being located close to the polystyrene surface. This agrees with the present results, which indicated that the carboxyl groups appear to be the surface interactive species.

The results of the studies discussed above as well as of other studies are summarized in Table 1.

(ii) Effect of Interfacial Behavior on Direct Redox Electron Transfer

Although direct electron transfer has captured the interest of electrochemists for some time, recently it has been observed at a greater variety of surfaces. Extensive studies have been made on mediated and unmediated electron transfer, and these have been discussed in reviews.[169,170] The present section will focus on those studies which have contributed to the understanding of the interfacial and conformational behavior of cytochrome c at electrode surfaces and on how this behavior contributes to the redox electron-transfer process.

Cytochrome c has a molar mass of 12,384 and contains a single polypeptide chain and a heme group. It functions as an electron carrier in the respiratory chain in the final step of electron transport during oxidative phosphorylation.[171] The prosthetic group is combined with the protein through thioether bonds and coordination of the iron atom with a histidine group on the protein chain. Hence, the heme is embedded in the hydrophobic crevicelike fold of the globular polypeptide, with only an edge exposed to solvent. It is believed that electron transfer occurs at the edge of the heme.[172]

The area of direct (i.e., unmediated) electrochemistry has received active interest from the viewpoint of clarification of the electron transport mechanism in biological systems and of application to biochemical sensors and biocatalyst electrodes. However, the electrochemistry of cytochrome c is characterized by the existence of marked adsorption phenomena. Electron transfer from cytochrome c to metal electrodes such as platinum, gold, nickel, silver, mercury, and p-type silicon has been reported to be very slow, and the irreversible electrochemistry has been considered to be due to adsorption-induced denaturation.[118,169,170] However, direct and quasireversible electrochemistry has been reported with surface-modified metal electrodes, including gold, platinum, and silver, and with an edge-graphite electrode. The electron transfer is considered to occur by the attractive interaction between positively charged lysine residues and the functional groups of the modifiers and graphite. Similar electrochemistry has been reported with semiconducting metal oxides.[114,117,173,174]

A number of authors have pointed out the need for highly purified cytochrome c in order to obtain reproducible results with redox electron transfer. Daido and Akaike[118] found that denatured impurities of cytochrome c deposited on the indium tin oxide electrode, and these impurities were assumed to block the electron transfer. Willit and Bowden[174,175] have pointed out that the electrochemistry depends on the purity of the protein and the pretreatment of the electrodes. They have also found cytochrome c to be strongly adsorbed by these tin oxide electrodes while retaining its native redox potential. In contrast to the adsorption by metals accompanied by denaturation, they suggested that this adsorption was interesting in its similarity to the complex formation with the physiological redox partners. An examination of the redox and adsorption thermodynamics of cytochrome c on fluorine-doped tin oxide electrodes was made by Willit and Bowden[117] using cyclic voltammetry over a temperature range of 281 to 308 K. Cytochrome c was found to adsorb strongly in neutral to slightly alkaline low-ionic-strength solutions, with near monolayer coverage resulting under optimum conditions. Variation in electroactive surface coverage with solution pH and ionic strength indicated that the adsorption was primarily controlled by electrostatic interactions. Because ferricytochrome c undergoes a pH-dependent conformational change with a K_a of *ca.* 9.3, solutions more alkaline than pH 8 were not used with cytochrome c so that the protein would retain its native conformation. Also, only solutions yielding substantial and stable protein coverage on the electrode were used. This requirement prohibited the use of ionic strengths greater than 100mM at pH 7 and of all pH 6 solutions because the cytochrome c coverage was unacceptably small and nonpersistent at these ionic strengths, presumably due to decreased electrostatic attraction. Electroactive surface coverage was found to be of the order of a monolayer. Assuming a surface roughness factor of 2 and a single molecule adsorption area of *ca.* 15 nm^2, Willit and Bowden calculated the density of a close-packed monolayer of cytochrome c as *ca.* 20 pmol cm^{-2} (or 2.5 mg m^{-2}).* The experimental coverage, Γ, was found to be 1.6–1.9 mg m^{-2}, except in pH 8 and 131mM ionic strength solutions, for which the cytochrome c coverage was 0.87 mg m^{-2}. However, Koller and Hawkridge[176] obtained a value of 1.24 mg m^{-2} with a tin-doped indium oxide electrode at pH 7.0 and 8.0, consistent with the theoretical value, but a lower value of 0.74 mg m^{-2} at pH 5.3. They found some interesting temperature effects in fast-scan cyclic-voltammetric measurements at pH 8.0. In the quasireversible cyclic voltammograms, a maximum current was obtained at 313 K, which was attributed to a conformational change in the cytochrome c accompanied by the onset of thermal denaturation with increasing temperature.

In variable temperature studies, Willit and Bowden[117] found that the electroactive coverage remained essentially constant for several hours as the temperature was varied between 281 and 301 K. However, upon increasing the temperature to 308 K, irreversible decreases in both the rate of electron transfer and the electroactive surface coverage were observed. They found evidence of a 2 kJ mol^{-1} difference in ΔG°_{ads}, the free energy

---

*In order to compare the surface concentrations for cytochrome c with those for other proteins, the results obtained by these authors have been converted to units of milligrams per square meter using 12,384 as the molar mass of cytochrome c.

of adsorption, between the oxidized and the reduced forms of the protein, and a stronger adsorption of ferricytochrome c than ferrocytochrome c on tin oxide. In all of these studies, negative potential shifts ranging from 20 to 50 mV indicated that the oxidized form of cytochrome c bound more strongly than the reduced form. The authors pointed out that the oxidized form of cytochrome c is known to be less stable and more susceptible to structural alteration than the reduced form; that is, ferricytochrome c is more susceptible to thermal denaturation, pH-induced denaturation, and proteolytic digestion than the ferrous form. These differences have been attributed to the differences in the dynamic behavior of the two redox forms. Thus, although cytochrome c is considered to be a quite rigid protein in both redox forms, the oxidized form does exhibit a greater degree of dynamic freedom and structural flexibility. Therefore, Willit and Bowden suggested that the oxidized form would be expected to be more susceptible to structural alteration as a result of interfacial forces at solid/aqueous interfaces. Their finding that ΔS°_{ads} for ferricytochrome c was larger by $ca.$ 25 J K^{-1} mol^{-1} than that for ferrocytochrome c supported this proposal. Their conclusion was also consistent with general views that proteins with greater structural flexibility undergo more significant structural alterations upon adsorption and therefore bind more strongly than more rigid proteins.

When Daido and Akaike[118] investigated the electrochemistry of cytochrome c using an indium tin oxide (ITO) electrode, they found that the electrochemistry was highly dependent not only on the purity of the cytochrome c, as discussed earlier, but also on the ionic strength of the electrolyte. ITO electrodes with adsorbed cytochrome c [type VI (ferric form) from the Sigma Chemical Company] were prepared by allowing the electrodes to stand at constant potential or open circuit in phosphate buffer solutions containing $40\mu M$ cytochrome c at pH 7.4. The ionic strength was adjusted by varying the concentrations of NaH_2PO_4 and Na_2HPO_4. After a scheduled period, the electrodes were rinsed with the corresponding buffer solution in the absence of protein. The apparent surface excess Γ of cytochrome c was estimated from the anodic peak areas of the cyclic voltammograms. It reached a maximum value of ~0.9 mg m^{-2} at an ionic strength of $0.2M$ (Fig. 12), which corresponds reasonably well to the coverage of a close-packed monolayer, estimated to be approximately 1.2 mg m^{-2}. The surface coverage of a single molecule was taken as 17.6 nm^2 based on an effective diameter of 4.2 nm for the solvated cytochrome c. The experimental values compared favorably with those reported by Willit

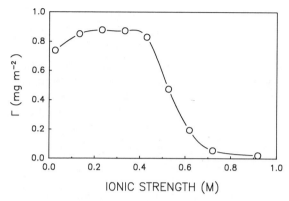

Figure 12. Surface excess, Γ, of purified cytochrome c on an indium tin oxide (ITO) electrode, determined from anodic peak areas, as a function of ionic strength of phosphate buffer solutions (pH 7.4). (Adapted from Daido and Akaike.[118])

and Bowden[117] for an In_2O_3 electrode at similar pH. At ionic strengths greater than $0.7M$, Γ was almost negligible,[118] again consistent with the results of Willit and Bowden.[117] Daido and Akaike[118] suggested that at ionic strengths below $0.3M$ purified cytochrome c was adsorbed strongly and maintained its native conformation. Unpurified cytochrome c containing polymeric and deaminated forms was preferentially adsorbed in the denatured state. These denatured forms blocked electron transfer between cytochrome c and the electrode. Guanidinated cytochrome c was found to be strongly adsorbed and behaved like denatured cytochrome c.

Szucs *et al.*[33] found that cytochrome c formed an irreversibly adsorbed layer on the gold electrode that completely covered the electrode and which could mediate the reduction of cytochrome c in solution via electron transfer through the unfolded protein layer. Oxidation of cytochrome c in solution was not observed. They used potassium phosphate monobasic buffer ($0.1M$) adjusted to pH 6.0 with potassium hydroxide. Although they investigated the behavior of cytochrome c both in the presence and in the absence of the promoter 4,4'-dipyridyl disulfide, the present discussion will focus mainly on the results obtained with bare gold electrodes. The adsorption of cytochrome c was studied by ellipsometry on bare gold electrodes at 400 mV (vs. NHE), which was approximately equivalent to the open-circuit potential of the system in the absence of protein in the electrolyte. The value of the calculated thickness of the adsorbed layer on bare gold was about half that expected if cytochrome c were in its native form. This suggested that a significant unfolding occurred upon adsorption. Cyclic voltammograms at sweep rates of 100 mVs^{-1} showed both oxidation and reduction of cytochrome c at a potential close to the standard redox potential for this molecule (i.e., 260 mV vs. NHE). However, the oxidation and reduction peaks disappeared completely with subsequent cycling. The adsorbed amount on a bare surface was calculated to be 1.72 mg m^{-2}, with a thickness of 1.8–2.2 nm and a refractive index of 1.48–1.5 from ellipsometry measurements. A similar amount was found on the modified surface. From geometric considerations, the maximum value of the adsorbed amount was estimated for a sphere 3.5–4.0 nm in diameter, and assuming the most compact hexagonal packing on the surface, to be 1.50–1.93 mg m^{-2}. This estimate suggested that the surface was covered with a complete monolayer of the protein on bare and modified gold. A summary of surface concentrations of cytochrome c is given in Table 2.

Szucs et al.[33] found that the structural change upon adsorption on bare gold caused a change in the redox potential of cytochrome c from 260 to –150 mV. Since the reduction of bulk proteins was also changed to this lower potential, the authors concluded that the protein in solution could be reduced only when the adsorbed protein was already reduced. Thus, the adsorbed layer on bare gold acted as an immobilized mediator for electron transfer between the underlying metal and cytochrome c in solution, but only for the reduction process. Because this adsorbed mediator had a redox potential that is lower than the redox potential of the native

Table 2
Surface Concentrations of Cytochrome c

pH	Γ (mg m^{-2})	Surface	Technique	Reference
6.0	1.72	Bare gold	Ellipsometry	33
5.3	0.74	Indium tin oxide	Electrochemistry	176
7.0	1.24	Indium tin oxide	Electrochemistry	176
8.0	1.24	Indium tin oxide	Electrochemistry	176
7.4	0.87	Indium tin oxide	Electrochemistry	118
7.0	1.61–1.86	Fluorine-doped tin oxide	Electrochemistry	117
8.0	0.87	Fluorine-doped tin oxide	Electrochemistry	117

molecules in solution, it could not mediate the oxidation process. The authors suggested that this may explain why the oxidation peak was completely missing on bare gold after a long adsorption time, when the surface was completely covered with unfolded molecules. Although the exact nature of the adsorbed cytochrome c on bare gold was not known, the authors suggested that changes in the redox potential may have resulted from exposure of the heme group to solvent[177] or substitution of the methionine axial group in position 80,[178] both of which had been shown to cause negative shifts in the reduction potential. Ellipsometry on the promoter-modified surface showed that the adsorbed cytochrome c was not unfolded, and the electrochemical response was reversible and stable. However, the authors concluded that electron transfer through the adsorbed cytochrome c layer was feasible regardless of the conformational state of the adsorbed protein.

Cytochrome c_3 differs from cytochrome c in that it has four heme groups embedded in a single polypeptide chain of a comparable molar mass of about 12,000. The hemes in cytochrome c_3 are exposed to solvents to a greater degree than in the other cytochromes. Haladjian $et\ al.$[120] investigated the behavior of cytochrome c_3 at the basal-plane pyrolytic graphite electrode in 10mM Tris chloride buffer at pH 7.6 using cyclic voltammetry and differential pulse voltammetry. Modification of the adsorbed film of cytochrome c_3 was observed upon scanning the potentials in both the anodic and the cathodic range. The pH dependence of the redox potentials suggested that irreversibly denatured forms containing a hemin-like group were in close contact with the electrode surface. These results suggested that cytochrome c_3 underwent severe modifications when it adsorbed. The authors suggested that a hemelike group was formed which came in close contact with the pyrolytic graphite (PG) surface. The redox properties of this hemelike group were very similar to those of free hemin. From the study of repetitive alternating current voltammetric scans, it was suggested that only one heme (i.e., that closest to the electrode surface) would be responsible for the redox behavior. Although cytochrome c_3 can be considered a very stable protein in solution, it appeared to denature when adsorbed on the PG surface. This tendency toward instability was found to increase at acid or alkaline pH. The authors suggested that electrical communication between the electrode and the redox centers of diffusing cytochrome c_3 molecules could be established through the heme groups of adsorbed molecules, behaving as electron-relaying centers. Consequently, electron transfer was able to take place over relatively large distances through the protein interior. This hypothesis agreed with the existence of an intramolecular electron exchange for cytochrome c_3 proposed by Dolla and Bruschi.[179]

These results indicate that direct electron transfer from a redox center is possible in certain situations either from protein directly adsorbed on the surface or with the adsorbed protein acting as mediator for the proteins in solution. Thus, electrochemistry has been able to provide insight into the mechanisms and conformational behavior of redox proteins at electrode surfaces.

(iii) Applications to Medical Fouling Problems

Lack of biocompatibility and the surface fouling of implant devices and biosensors are important problems in medical applications. Thus, the question arises whether electrochemical techniques are successful in providing information that may be useful in solving these complex problems. The proteins relevant to these systems are human and bovine serum albumin, γ-globulin, fibrinogen, von Willebrand factor, immunoglobulin G (IgG), and other blood proteins.

Very little of the research that has been done on these proteins has involved the use of electrochemical techniques. Instead, ellipsometry, FTIR/ATR spectroscopy, radioactive labeling, and photon correlation spectroscopy have been used. Many of the studies have been directed toward the development of biocompatible polymer surfaces. The first event that takes place after contact of blood or plasma with an artificial surface is the rapid adsorption of proteins from the blood onto the material surface. It is generally assumed that all subsequent events, such as platelet adhesion and surface activation of blood coagulation, are determined by the composition and structure of the initially adsorbed protein layer. It is known from *in vitro* experiments that the adhesion of platelets is promoted when fibrinogen has been adsorbed on a material surface[16] and that platelet adhesion is reduced when preadsorbed albumin is present on the surface.[180] In a study of the adsorption behavior of three of the more abundant plasma proteins, fibrinogen, γ-globulin, and albumin, on a polyethylene surface, the ratios of their infrared internal reflection bands showed that adsorption was greatest in the case of fibrinogen (Fig. 13).[50] When protein solutions are put in contact with solid surfaces, adsorption is rapid, and surfaces are substantially covered within a few seconds. Denaturation of adsorbed proteins may occur. Brash[181] found that the thickness of an adsorbed layer of fibrinogen on glass increased with increase in pH and

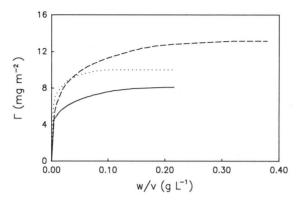

Figure 13. Adsorption of plasma proteins on polyethylene at 310 K measured by infrared internal reflection spectroscopy: —, fibrinogen; ···, γ-globulin, - - -, albumin. (Adapted from Brash and Lyman.[47])

retracted as the pH was lowered. Fibrinogen adsorption on glass was found to increase with plasma concentration in isotonic Tris buffer, pH 7.35 until a peak was observed around 1% plasma; adsorption decreased with further increases in protein concentration. Similar behavior was observed with whole blood. Recently, there has been interest in the use of poly(ethylene glycol) (PEG) coatings for the reduction of protein and cell adsorption and thrombosis on surfaces. Osterberg et al.[182] found that certain polysaccharide coatings are as effective as, if not slightly more effective than, PEG in preventing fibrinogen adsorption. However, each protein–surface system must be considered individually. Results from a number of studies made by Brash and co-workers have been reviewed.[181]

The adsorption behavior of bovine serum albumin (BSA) has been examined on a number of different surfaces. Van Enckevort et al.[66] obtained a plateau value of 2.1 mg m^{-2} for the adsorption of BSA from 0.02 g l^{-1} solutions onto stainless steel surfaces, measured by radiochemical techniques (Fig. 14). They found that the adsorbed protein was essentially nonexchangeable. Using reflectometry measurements, Elgersma et al.[29] have investigated the effect of pH in the range 4–8 on the adsorption behavior of BSA on polystyrene-covered silicon wafers, since pH can affect the electrical charge on the protein, which in turn affects its structural stability. The maximum plateau value for adsorption over the

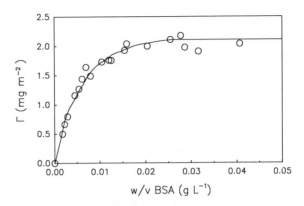

Figure 14. Adsorption isotherm for bovine serum albumin (BSA) on stainless steel from cacodylate buffer (pH 6.8, ionic strength 0.06M, 277 K) measured by radiochemical techniques. (Adapted from van Enckevort et al.[66])

pH range studied was found at pH 5, which corresponds to the isoelectric point of BSA (Fig. 15). A very similar behavior was observed, using solution depletion methods, for adsorption of BSA onto polyamide microcapsules.[70] This was attributed to a reduction in the intermolecular interactions near the pI value, allowing a more compact configuration on the

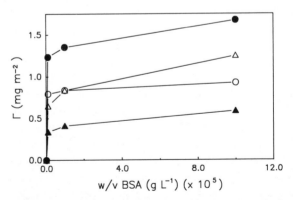

Figure 15. Adsorption isotherms for bovine serum albumin (BSA) in phosphate buffer on polystyrene-covered silicon wafers measured by reflectometry at pH values of 4.0 (○), 5.0 (●), 6.0 (△), and 8.0 (▲). (Adapted from Elgersma et al.[29])

surface with fewer lateral electrical interactions between the adsorbed molecules.

A comparison of the adsorption behavior of some serum proteins on a variety of different surfaces is given in Table 3.

One of the few electrochemical investigations on proteins directly related to medical applications is that reported by Razumas et al.,[12] who studied adsorption of insulin on platinum from a physiological saline solution of pH 7.4 (298 K). The smallest amount of protein adsorption was found to be around the potential of zero charge of platinum. The authors compared the surface adsorption of insulin at different potentials using ellipsometric measurements and determined a surface charge density of Γ = 1.27 mg m^{-2} for insulin at a potential of 0.4 V (vs. SCE). The Γ_{max} was attained within 20–30 s after the solutions were mixed. A small decrease

Table 3
Surface Concentrations of Some Serum Proteins

Protein	Γ (mg m^{-2})	Surface	Technique	Reference
Bovine serum albumin	0.88	Silica, pH 7.4	Solution depletion	14
	1.0	Polyethylene/polystyrene sulfonic acid, pH 7.4	Solution depletion	27
	2.6	Polystyrene, pH 7.4	Solution depletion	61
	1.5	Polystyrene, pH 5.0	Reflectometry	29
	2.1	Stainless steel, pH 6.8	Radiotracer	66
Bovine prothrombin	4.36	Silica, pH 7.4	Solution depletion	14
Human albumin	0.5	Polystyrene	IR internal reflection	47
	0.8	Polyethylene	IR internal reflection	47
	1.6	Silastic	IR internal reflection	47
	0.8	Teflon FEP^a	IR internal reflection	47
Fibrinogen	1.7	Polystyrene	IR internal reflection	47
	1.3	Polyethylene	IR internal reflection	47
	1.6	Silastic	IR internal reflection	47
	1.4	Teflon FEP	IR internal reflection	47
γ-Globulin	0.7	Polystyrene	IR internal reflection	47
	1.0	Polyethylene	IR internal reflection	47
	1.8	Silastic	IR internal reflection	47
	0	Teflon FEP	IR internal reflection	47

^{a}FEP, fluorinated ethylene–propylene copolymer (E.I. duPont Co.)

in Γ_{max} following substitution of 0.15M NaCl solution for the protein solution was observed only at potentials of 0.3 and 0.4 V (the values reached were 0.93 and 1.34 mg m^{-2}, respectively). The lowest values for Γ_{max} were found between –0.1 and –0.3 V. The authors suggested that in solution the insulin monomers form dimeric and hexameric aggregates that are more hydrophilic than the monomer. Adsorption was found to increase when zinc ions were added to the protein solutions. The surface concentrations were found to be 1.58 and 1.52 mg m^{-2} for 2 Zn-insulin and 4 Zn-insulin, respectively. They suggested that complex formation with zinc ions leads to additional aggregation of the hexamers. The strong hydrophobicity of the protein in insulin solutions lacking metal ions results in minimum values of Γ_{max}.

This is an area in which electrochemical techniques have been under-utilized and could be very useful in providing fundamental information regarding the adsorption phenomena of proteins, which may then be applied to other surfaces.

2. Surface Adsorption of Proteins as a Function of Temperature

(i) Conformational Behavior Studies

Because proteins undergo conformational changes as a function of temperature, it was important to determine whether electrochemical techniques could be used to detect these changes at the electrode surface and correspondingly relate these changes to the behavior exhibited by the protein in the bulk solution with changes in temperature. Again, electrochemical measurements have been carried out on well-characterized proteins, and the results are compared here with those obtained by other researchers using a variety of other techniques. The discussion will begin with the three globular proteins β-lactoglobulin, lysozyme, and ribonuclease, whose behavior at ambient temperatures was described previously.

Heat affects the conformation of proteins, and most proteins undergo fairly rapid heat denaturation somewhere between 333 and 363 K.[154] A series of experiments was made by Roscoe and Fuller[111,112] in which the surface adsorption behavior of bovine β-lactoglobulin A, hen egg-white lysozyme and bovine pancreatic ribonuclease A was studied at a platinum electrode as a function of temperature using cyclic voltammetry. The protein solution and the electrochemical cell were separately immersed in the thermostated bath and allowed to equilibrate for 30 mins. The protein was then added to the electrochemical cell in measured aliquots, and the steady-state cyclic voltammograms were recorded.

At pH 7.0 and room temperature, β-lactoglobulin exists as dimers. Dissociation to monomers occurs at 323 K and has been referred to as the predenatured state.[154] Upon heating at temperatures above 333 K, sulfhydryl groups are exposed by unfolding of the protein. They become available for reaction with various reagents and can be used as an index of the denaturation process.[154] Lysozyme and ribonuclease both have four disulfide linkages but contain no free sulfhydryl groups. Thus, comparison of the interfacial behavior of these proteins with that of β-lactoglobulin[112] allows one to determine the role of the free sulfhydryl group in the surface adsorption mechanism at the platinum electrode. In order to compare the electrochemical results for β-lactoglobulin with those obtained by other techniques, experiments were made in which the free sulfhydryl group was monitored as a function of temperature, over the range 273–358 K, using Ellman's reagent and spectroscopic measurements.[183,184] The results indicated that the sulfhydryl group is not the dominant factor in the surface adsorption process. However, its presence could be used to monitor structural changes of β-lactoglobulin in the bulk solution which affect the exposure of the free sulfhydryl group. Thus, a measure of the free sulfhydryl group was used to monitor the unfolding, denaturation, and agglomeration of the protein in the bulk solution in order to investigate the effects of these structural changes on the behavior of the protein at the electrode surface.

Plateau values of surface charge density were obtained for increasing concentrations of β-lactoglobulin A in the bulk solution.[111] At temperatures ranging from 273 to 343 K, the surface charge density for protein adsorption increased due to the unfolding of the β-lactoglobulin molecule (Fig. 16). At temperatures in the range 348–358 K, where denaturation of the protein occurs, the surface charge density was found to diminish (Fig. 17), probably as a result of agglomeration of the protein. The surface adsorption behavior of β-lactoglobulin as a function of temperature suggested gradual conformational change, with a slightly more rapid change at temperatures above 323 K, as shown in Fig. 18. This is the temperature at which β-lactoglobulin dissociates into monomers, referred to as a prenatured state.[154] Spectroscopic measurements using Ellman's reagent showed a corresponding increase in the absorbance at 412 nm due to exposure of the sulfhydryl group with increasing temperature (Fig. 18).[111] A comparison of the surface charge density with the percent free sulfhy-

Sharon G. Roscoe

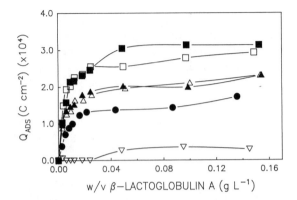

Figure 16. Surface charge density, Q_{ADS}, of β-lactoglobulin A at a platinum electrode versus the concentration (w/v) of protein in phosphate buffer (pH 7.0) with an anodic end potential of 1.0 V at various temperatures. ∇, 273 K; $\bullet$, 303 K; $\triangle$, 313 K; $\blacktriangle$, 323 K; $\square$, 333 K; $\blacksquare$, 343 K. (Reprinted from Roscoe and Fuller,[111] with permission.)

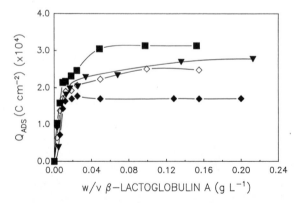

Figure 17. Surface charge density, Q_{ADS}, of β-lactoglobulin A at a platinum electrode versus the concentration (w/v) of the protein in phosphate buffer (pH 7.0) with an anodic end potential of 1.0 V at various temperatures. $\blacksquare$, 343 K; $\blacktriangledown$, 348 K; $\diamond$, 353 K; $\blacklozenge$ 358 K. (Reprinted from Roscoe et al.[112] with permission.)

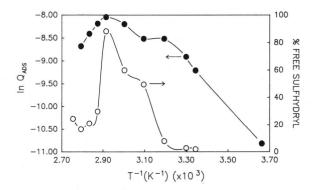

Figure 18. Plots of the logarithm of surface charge density of β-lactoglobulin A (0.05 g l^{-1} in phosphate buffer, pH 7.0) at a platinum electrode versus T^{-1} (●) and of % free sulfhydryl versus T^{-1} for β-lactoglobulin heated at the specified temperature for 30 min (○). (Reprinted from Roscoe *et al.*[112] with permission.)

dryl as measured spectroscopically over the temperature range 273–343 K suggested that protein unfolding, as monitored by increased exposure of the free sulfhydryl group, allowed greater access of the protein's active sites to the electrode surface. Casal *et al.*[185] reported from infrared measurements that, although the secondary structure of β-lactoglobulin does not change in the temperature range 293–323 K at pH 7, there is a distinct change at 331 K. The temperature dependence of the amide I band in the infrared spectrum, showing discontinuities at 331 K and 343 K, compares favorably with the surface charge density measurements (Fig. 19).[111]

Using ellipsometry, Arnebrant *et al.*[4] found a different adsorption behavior for β-lactoglobulin at the hydrophilic chromium surface; adsorption increased only slightly from about 2.0 to 2.2 mg m^{-2} over the temperature range 298–346 K. However, when the temperature reached 348 K, the surface concentration increased abruptly to 8 mg m^{-2}. The authors concluded that at these temperatures some conformational changes of the protein must occur before aggregation involving sulfhydryl groups begins at the surface.

When the protein solution was held at 343 K for 30 min in a separate container and was then allowed to cool to the temperature of the electro-

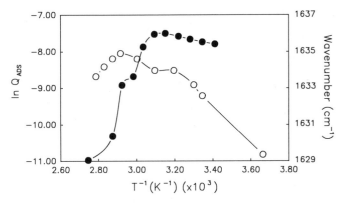

Figure 19. Plots of the wavenumber of the amide I band contour in the infrared spectra of β-lactoglobulin versus T^{-1} (●, adapted from Casal *et al.*[185]) and of the logarithm of the surface charge density of β-lactoglobulin A (0.05 g l^{-1} in phosphate buffer, pH 7.0) at a platinum electrode versus T^{-1} (○, adapted from Roscoe *et al.*[112])

chemical cell, thermostated in a water bath at 299 K, the plateau values of surface charge density with an anodic end potential of 0.4 V (vs. SCE) were very similar to those obtained for a protein solution maintained at 299 K (Fig. 20).[112] Spectroscopic measurements of the concentration of the free sulfhydryl group in these experiments indicated that unfolding of this protein at pH 7.0 in this temperature range was reversible.

At each temperature, an equilibrium constant was calculated from the ratio of the concentration of exposed sulfhydryl to the concentration of β-lactoglobulin remaining with no exposed sulfhydryl:

$$K = [SH]/([\beta LG]_o - [SH])$$

From a plot of ln K versus T^{-1} and using the van't Hoff equation,

$$d(\ln K)/d(T^{-1}) = -\Delta H^o/R$$

values of $\Delta H^o = 118$ kJ mol^{-1} and $\Delta S^o = 360$ J K^{-1} mol^{-1} were determined for the unfolding process exposing this free sulfhydryl group over the temperature range 299–343 K. From these values, ΔG^o for the protein unfolding was calculated to be 10.6 kJ mol^{-1}. This compares favorably with the value of 12.5 kJ mol^{-1} for the free energy of the hydrophobic effect of transferring proline from an organic solvent into water.[186] Since

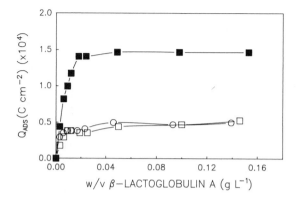

Figure 20. Surface charge density, Q_{ADS}, of β-lactoglobulin A at a platinum electrode as a function of the concentration (w/v) of the protein in phosphate buffer (pH 7.0) with an anodic end potential of 0.4 V at various temperatures. ○, 299 K; ■, 343 K; □, denatured at 343 K, then cooled to 299 K. (Reprinted from Roscoe *et al.*[112] with permission.)

the unfolding of β-lactoglobulin would involve proline, which is in fact located close to this free sulfhydryl group, it seems likely that the experimental value for unfolding of β-lactoglobulin reflects a similar hydrophobic effect.

Thermal denaturation of β-lactoglobulin is considered to be a two-state transition.[135,187] Results from a differential scanning calorimetric study of the thermal denaturation of β-lactoglobulin in phosphate buffer at pH 6.8 at temperatures up to 373 K showed that 343 K was a critical temperature for the denaturation of this protein.[26] The value of 230 kJ mol^{-1} obtained for the enthalpy of denaturation of β-lactoglobulin[26,188] is twice the value obtained in the present investigation for the enthalpy of protein unfolding prior to denaturation, over the temperature range 299–343 K. This lower enthalpy value would be expected for these small conformational changes that occur at temperatures lower than the critical temperature required for complete protein denaturation. The infrared spectra at 293 and 331 K were found to be almost the same (Fig. 19), which indicated that the secondary structure of β-lactoglobulin does not change in this temperature range.[185] However, at 333 K the infrared spectrum changed to that observed at pH 11, which strongly suggested that the thermal denaturation proceeds through the same intermediate as that formed during alkaline denaturation. The surface charge density plateau values resulting from adsorption of β-lactoglobulin at temperatures above 343 K decreased with increasing temperature in the range 348–358 K. Spectroscopic measurements over this same temperature range also showed a decrease in the exposure of the free sulfhydryl group (Fig. 18). This probably resulted from further conformational changes accompanied by agglomeration at these temperatures. When β-lactoglobulin at pH 7.0 is heated above 343 K in the absence if other proteins, polymerization and aggregation occur.[189] Aggregation results from increased sulfhydryl reactivity due to an increased intra- and intermolecular randomization of disulfide cross-links during heat treatments to 370 K.[190–192] The infrared spectra at 363 K showed that β-lactoglobulin does not unfold to a random coil before aggregating and polymerizing in the final stage of the thermal denaturation process.[185]

The surface adsorption behavior of lysozyme with an anodic end potential of 1.0 V over the temperature range 273–343 K (Fig. 21) showed a steady increase of surface charge density with temperature. Similar results were obtained with ribonuclease A, as shown in Fig. 22. Negligible

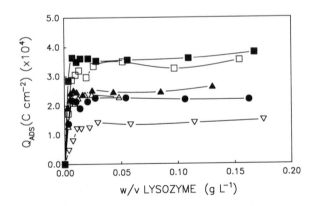

Figure 21. Surface charge density, Q_{ADS}, of lysozyme at a platinum electrode as a function of the concentration (w/v) of the protein in phosphate buffer (pH 7.0) with an anodic end potential of 1.0 V at various temperatures. $\triangledown$, 273 K; $\bullet$ 303 K; $\triangle$, 313 K; $\blacktriangle$, 323 K; $\square$, 333 K; $\blacksquare$, 343 K. (Reprinted from Roscoe and Fuller[111] with permission.)

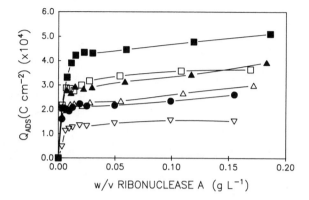

Figure 22. Surface charge density, Q_{ADS}, of ribonuclease A at a platinum electrode as a function of the concentration (w/v) of the protein in phosphate buffer (pH 7.0) with an anodic end potential of 1.0 V at various temperatures. ▽, 273 K; ●, 303 K; △, 313 K; ▲, 323 K; □, 333 K; ■, 343 K. (Reprinted from Roscoe and Fuller[111] with permission.)

absorbance was observed at 412 nm with the use of Ellman's reagent with these proteins over this temperature range, as expected since neither protein contains free sulfhydryl groups. A change in slope in a plot of ln Q_{ADS} versus T^{-1} occurred at 333 K for both lysozyme and ribonuclease, which also suggested that structural changes probably occurred with these molecules at these temperatures. As discussed in Section V.1(i), ribonuclease is very similar in its general structure to lysozyme,[157] both proteins having a hydrophobic core on one side of the active site crevice and a less substantial second wing, giving them a V shape. All the charged polar groups are on the surface, each protein having 11 carboxyl groups and no free sulfhydryl groups. These features may have played an important role in the similar behavior of the two proteins. Surface adsorption was not inhibited by lower temperatures (273 K) in the case of lysozyme and ribonuclease as it appeared to be in the case of β-lactoglobulin. This is an important factor to be considered in terms of the effects of surface fouling.

(ii) Industrial Surface Fouling

It is of interest then to determine whether electrochemical techniques can be used to obtain information applicable to the major industrial problem of surface fouling. In this section, experiments that have been undertaken specifically for this purpose will be examined. The focus will be on investigations related to the fouling problems encountered in the dairy industry. The proteins β-lactoglobulin and κ-casein have been chosen for these investigations as they have been previously identified as major contributors to surface fouling. The electrochemical measurements have also been extended to investigate the behavior of unpasteurized whole milk under similar experimental conditions and heat treatments.

The problems in the dairy industry related to the fouling of metal surfaces during processing have resulted in an interest in the competitive aspects of the surface adsorption of the milk proteins κ-casein and β-lactoglobulin.[3,193] Thermally induced interactions of these two proteins have been extensively studied, and it is generally accepted that a complex of β-lactoglobulin, the most abundant whey protein, and the surface-active κ-casein is formed during heat treatment of milk.[194,195] Interactions which stabilize the complex involve hydrophobic and ionic groups. However, sulfhydryl interactions are considered to be the most important.[196,197] The deposition of casein micelles and β-lactoglobulin together on stainless steel has also been considered to result from an interaction between κ-casein and β-lactoglobulin.[198] Experiments by Tissier et al.[5] have shown that β-lactoglobulin is the dominant protein in the mechanism of deposit formation at the inner surface of a plate sterilizer. A correlation was observed between the maximum denaturation rate of β-lactoglobulin and the maximum formation of deposits. An electrochemical approach to the investigation of both the adsorption behavior and the conditions for optimization of removal of β-lactoglobulin from this model system was used by Fuller and Roscoe.[199] Good agreement was observed between the surface concentrations obtained for adsorption of β-lactoglobulin at the platinum electrode surface (the model system)[111] and those obtained for adsorption of the protein on stainless steel pellets[69] under similar experimental conditions but using solution depletion methods. The surface adsorption behavior of κ-casein A and κ-casein A/β-lactoglobulin A mixtures was also studied by Roscoe and Fuller[200] using cyclic voltammetry, and the study was extended to the investigation of surface adsorption of unpasteurized whole milk.[201] The surface adsorption of the proteins increased with increasing concentration of the bulk solution until a plateau level in surface charge density was reached. κ-Casein gave a higher surface charge density than β-lactoglobulin, which was consistent with the surface concentration measurements obtained by others using ellipsometry.[3] Protein adsorption was also found to occur readily at the open-circuit electrode, which was then transferred to the electrochemical cell for subsequent characterization of the surface charge density by potential sweeping. Plateau-level surface charge densities were attained even with dip times as short as 2 min for measurements made over a temperature range of 299–358 K. The plateau values observed when the electrode was dipped sequentially in solutions of κ-casein and β-lacto-globulin, and vice versa, were similar to those obtained with κ-casein alone. This suggested that κ-casein was adsorbing competitively with β-lactoglobulin under these experimental conditions.

Complete oxidation and removal of protein resulted from potential cycling over the range –0.75 to 1.0 V (vs. SCE) for protein adsorbed over the temperature range 299–348 K by dipping of the electrode in a solution of κ-casein and by sequential dipping in solutions of κ-casein and β-lactoglobulin, and vice versa. However, incomplete oxidation and removal of protein resulted when the protein was adsorbed at temperatures of 353–363 K. Instead, a lower plateau value was attained which remained constant even with extensive potential cycling (Fig. 23).

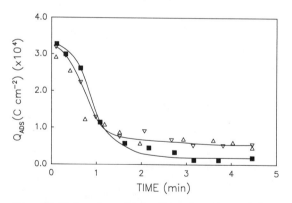

Figure 23. Surface charge density, Q_{ADS}, of protein at a platinum electrode versus the time of potential sweeping after the electrode was dipped in protein solution(s) in phosphate buffer (pH 7.0, 358 K) and measured at the same temperature. ■, Dipped in a κ-casein solution (0.13 g l^{-1}); △, dipped first in the κ-casein solution and then in a β-lactoglobulin solution (0.16 g l^{-1}), 2 min dip each; ▽, dipped first in the β-lactoglobulin solution followed by the κ-casein solution, 2 min dip each. (Reprinted from Roscoe and Fuller[200] with permission.)

Interesting results were obtained when protein removal was measured at a lower temperature than protein adsorption. This was particularly evident when the temperature was reduced to 273 K, as removal of protein adsorbed at 368 K was accomplished by potential sweeping at this lower temperature (Fig. 24). The results suggested that protein adsorption on the open-circuit electrode at the higher temperature, at which denaturation and agglomeration occurs in the bulk solution, probably included many layers, some of which were probably loosely bound. When the electrode was transferred to the electrochemical cell containing the buffer solution at the lower temperatures of 299 and 273 K, very fast thermal equilibration must have occurred. A rapid response to the change in temperature appeared to be reflected by rapid conformational changes in the adsorbed protein resulting in a much reduced surface interaction at the cold electrode surface. The surface charge density measurements reflected this structural reorientation. These findings are consistent with previous results which

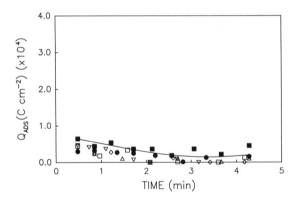

Figure 24. Surface charge density, Q_{ADS}, of protein at a platinum electrode versus the time of potential sweeping after the electrode was dipped in protein solution(s) in phosphate buffer (pH 7.0, 368 K) and measured at 273 K. ■, dipped in a κ-casein solution (0.13 g l^{-1}); △, dipped first in the κ-casein solution and then in a β-lactoglobulin solution (0.16 g l^{-1}), 2 min dip each; ▽, dipped first in the κ-casein solution followed by the β-lactoglobulin solution, 8 min dip each; □, dipped first in the β-lactoglobulin solution followed by the κ-casein solution, 2 min dip each; ◇, dipped first in the β-lactoglobulin solution followed by the κ-casein solution, 8 min dip each. (Reprinted from Roscoe and Fuller[200] with permission.)

showed very little adsorption of β-lactoglobulin at 273 K at the anodically polarized electrode.[111] Thus, it appeared that, with this model, the adsorption and the removal of protein were temperature-dependent owing to the conformational behavior of the proteins both in the bulk solution and at the metal surface. By utilizing this conformational behavior, accompanied by potential cycling, protein removal following adsorption was accomplished in a phosphate buffer at pH 7.0 over the temperature range 299–363 K. The surface charge density resulting from electron transfer has been used to quantify adsorption and removal of protein from the electrode surface. Since the results for β-lactoglobulin adsorption at the platinum electrode surface[111] corresponded well with those obtained using stainless steel surfaces,[69] these results may be extrapolated to types of surfaces more commonly used in industrial applications. This technique provides information which may be used for the optimization of conditions for removal of adsorbed proteins.

Similar results were obtained from experiments on unpasteurized whole milk.[201] Complete oxidation and removal of milk proteins adsorbed at temperatures of 299–363 K was accomplished by potential cycling between –0.75 and 1.0 V (vs. SCE) at the lower temperatures of 299 and

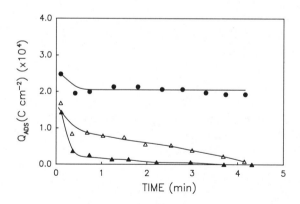

Figure 25. Surface charge density, Q_{ADS}, of adsorbed milk at a platinum electrode versus the time of potential sweeping (–0.75 to 1.0 V vs. SCE) in phosphate buffer (pH 7.0) after the electrode was dipped in unpasteurized milk for 16 min at 363 K and measured at 363 K, ●; 299 K, △; and 273 K, ▲. (Reprinted from Roscoe and Fuller[201] with permission.)

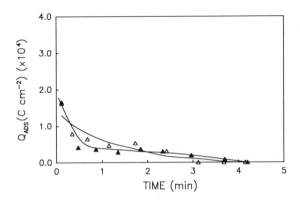

Figure 26. Surface charge density, Q_{ADS}, of adsorbed milk at a platinum electrode versus the time of potential sweeping (–0.75 to 1.0 V vs. SCE) in phosphate buffer (pH 7.0) or sodium sulfate solution (pH 7.0) at 273 K, after the electrode was dipped for 60 min in unpasteurized milk maintained at 363 K. △, Phosphate buffer solution; ▲, sodium sulfate solution. (Reprinted from Roscoe and Fuller[201] with permission.)

273 K (Fig. 25). Since phosphates are not desirable for industrial use, the experiments at 363 K were repeated using a solution of sodium sulfate at pH 7.0, which would be inexpensive, nontoxic, and "environmentally friendly." After the electrode was held in the milk at 363 K for 1 h, complete removal of the milk adsorbates was accomplished by potential cycling for less than 5 min (Fig. 26). These results suggested that surface fouling due to the milk proteins may be prevented or removed by intermittent treatment of surfaces such as stainless steel by potential cycling, particularly at the reduced temperatures of 299 or 273 K, eliminating the need for caustic cleaning solutions.

VI. CONCLUSIONS

Although electrochemistry has already made contributions to the understanding of the interfacial behavior of proteins at solid surfaces, there are many areas and opportunities for further research in order to fully understand the behavior of these complex molecules. Such an understanding will only be achieved by comparing results from many techniques and pooling the knowledge acquired throughout the range of disciplines that this area of research encompasses. The problems of surface fouling encountered in both medical applications and industrial settings can only be fully resolved when a thorough understanding of the heterogeneous processes involved has been achieved. Electrochemistry has a major role to play in the elucidation of these fundamental aspects of protein surface chemistry and in solving the many problems encountered with developing technologies dealing with biological materials.

REFERENCES

[1] K. Kandori, S. Sawai, Y. Yamamoto, H. Saito, and T. Ishikawa, *Colloids Surf.* **68** (1992) 283.

[2] J. A. Reynaud, B. Malfoy, and A. Bere, *Bioelectrochem. Bioenerg.* **7** (1980) 595.

[3] T. Arnebrant and T. Nylander, *J. Colloid Interface Sci.* **111** (1986) 529.

[4] T. Arnebrant, K. Barton, and T. Nylander, *J. Colloid Interface Sci.* **119** (1987) 383.

[5] J. P. Tissier, M. Lalande, and G. Corrieu, in *Engineering and Food*, Vol. 1, Ed. by B. M. McKenna, Elsevier Applied Science, London, 1984, pp. 49–58.

[6] S. A. Kirtley and J McGuire, *J. Dairy Sci.* **72** (1989) 1748.

[7] M. Fletcher and K. C. Marshall, *Appl. Environ. Microbiol.* **44** (1982) 184.

[8] S. McEldowney and M. Fletcher, *J. Gen. Microbiol.* **132** (1986) 513.

[9] M. Fletcher, *Methods in Microbiol.* **22** (1990) 251.

[10] E. A. Zottola, *Biofouling* **5** (1991) 37.

[11] R. E. Baier, A. E. Meyer, J. R. Natiella, R. R. Natiella, and J. M. Carter, *J. Biomed. Mater. Res.* **18** (1984) 337.

[12] V. Razumas, J. Kulys, T. Arnebrant, T. Nylander, and K. Larsson, *Elektrokhimiya* **24** (1988) 1518.

[13] E. J. Castillo, J. L. Koenig, and J. M. Anderson, *Biomaterials* **7** (1986) 89.

[14] B. Morrissey and R. R. Stromberg, *J. Colloid Interface Sci.* **46** (1974) 152.

[15] F. MacRitchie, in *Advances in Protein Chemistry*, Vol. 32, Ed. by C. B. Anfinsen, J. T. Edsall, and F. M. Richards, Academic Press, New York, 1978, pp. 283–325.

[16] M. A. Packham, G. Evans, M. F. Glynn, and J. F. Mustard, *J. Lab. Clin. Med.* **73** (1969) 686.

[17] H. Petscheck, D. Ademis, and A. R. Kantrowitz, *Trans. Am. Soc. Artif. Intern. Organs* **14** (1968) 256.

[18] D. J. Lyman, L. C. Metcalf, D. Albo, K. F. Richards, and J. Lamb, *Trans. Am. Soc. Artif. Intern. Organs* **20** (1974) 474.

[19] J. O'M. Bockris and S. U. M. Khan, *Surface Electrochemistry*, Plenum Press, New York, 1993.

[20] J. Lipkowski and P. N. Ross, eds., *Adsorption of Molecules at Metal Electrodes*, Vol. 1, VCH Publishers, New York, 1992.

[21] A. K. N. Reddy, M. A. V. Devanathan, and J. O'M. Bockris, *J. Electroanal. Chem.* **6** (1963) 61.

[22] W. K. Paik and J. O'M. Bockris, *Surf. Sci.* **28** (1971) 61.

[23] B. Liedberg, B. Ivarsson, P. O. Hegg, and I. Lundstrom, *J. Colloid Interface Sci.* **114** (1986) 386.

[24] M. Wahlgren and T. Arnebrant, *J. Colloid Interface Sci.* **136** (136) 259.

[25] K. Al-Malah, J. McGuire, V. Kridhasima, P. Suttiprasit, and R. Sproull, *Biotechnol. Prog.* **8** (1992) 58.

[26] J. N. de Wit and G. A. M. Swinkels, *Biochim. Biophys. Acta* **624** (1980) 40.

[27] W. J. Dillman, Jr. and I. F. Miller, *J. Colloid Interface Sci.* **44** (1973) 221.

[28] A. V. Elgersma, R. L. J. Zsom, J. Lyklema, and W. Norde, *J. Colloid Interface Sci.* **152** (1992) 410.

[29] A. V. Elgersma, R. L. J. Zsom, J. Lyklema, and W. Norde, *Colloids Surf.* **65** (1992) 17.

[30] S. Welin-Klinstroem, R. Jansson, and H. Elwing, *J. Colloid Interface Sci.* **157** (1993) 498.

[31] H. Arwin, S. Welin-Klinstroem, and R. Jansson, *J. Colloid Interface Sci.* **156** (1993) 377.

[32] T. Arnebrant and T. Nylander, *J. Colloid Interface Sci.* **122** (1988) 557.

[33] A. Szucs, G. D. Hitchens, and J. O'M. Bockris, *Electrochim. Acta* **37** (1991) 403.

[34] B. Pettinger, in *Adsorption of Molecules at Metal Electrodes*, Vol. 1, Ed. by J. Lipkowski and P. N. Ross, VCH Publishers, New York, 1992, pp. 285–345.

[35] D. Hobara, K. Niki, and T. M. Cotton, *Denki Kagaku* **61** (1993) 776.

[36] T. M. Cotton, in *Spectroscopy of Surfaces*, Ed. by R. J. H. Clark and R. E. Hester, John Wiley & Sons, New York, 1988, pp. 91–153.

[37] K. Niki, Y. Kawasaki, Y. Kimura, Y. Higuchi, and N. Yasuoka, *Langmuir* **3** (1987) 982.

[38] G. Smulevich and T. G. Spiro, *J. Phys. Chem.* **89** (1985) 5168.

[39] P. Hildebrandt and M. Stockburger, *Biochemistry* **28** (1989) 6710.

[40] P. Hildebrandt and M. Stockburger, *Biochemistry* **28** (1989) 6722.

[41] I. Taniguchi, H. Eguchi, and S. Tomimura, *Chem. Sens. Tech.* **3** (1991) 233.

[42] R. G. Greenler, *J. Chem Phys.* **50** (1969) 1963.

[43] P. Hollins and J. Pritchard, in *Vibrational Spectroscopy of Adsorbates*, Ed. by R. F. Willis, Springer-Verlag, New York, 1980, p. 125.

[44] R. J. Nichols, in *Adsorption of Molecules at Metal Electrodes*, Vol. 1, Ed. by J. Lipkowski and P. N. Ross, VCH Publishers, New York, 1992, pp. 347–389.

[45] B. Liedberg, B. Ivarsson, and I. Lundstrom, *Biochem. Biophys. Methods* **9** (1984) 233.

[46] B. Liedberg, B. Ivarsson, I. Lundstrom, and W. R. Salaneck, *Prog. Colloid Polym. Sci.* **70** (1985) 67.

[47] W. Plieth, W. Kozlowski, and T. Twomey, in *Adsorption of Molecules at Metal Electrodes*, Vol. 1, Ed. by J. Lipkowski and P. N. Ross, VCH Publishers, New York, 1992, pp. 239–283.

[48] J. O'M. Bockris and S. U. M. Khan, *Surface Electrochemistry*, Plenum Press, New York, 1993, pp. 43–45.

[49] D. R. Scheuing, ed., *Fourier Transform Infrared Spectroscopy in Colloid and Interface Science*, ACS Symposium Series 447, American Chemical Society, Washington, D.C., 1991, p. 13.

[50] J. L. Brash and D. J. Lyman, *J. Biomed. Mater. Res.* **3** (1969) 175.

[51] G.I. Loeb, *J. Polym. Sci., Part C* **34** (1971) 63.

[52] K. P. Ishida and P. R. Griffiths, in *Fourier Transform Infrared Spectroscopy in Colloid and Interface Science,* ACS Symposium Series 447, Ed. by D. R. Scheuing, American Chemical Society, Washington, D.C., 1991, pp. 208–224.

[53] K. P. Ishida and P. R. Griffiths, *J. Colloid Interface Sci.* **160** (1993) 190.

[54] A. N. Asanov and L. L. Larina, in *Charge and Field Effects in Biosystems—3, (3rd International Symposium)*, Ed. by M. J. Allen, Birkhauser, Boston, 1992, pp. 13–27.

[55] V. Hlady, R. A. Van Wagenen, and J. D. Andrade, in *Surface and Interfacial Aspects of Biomedical Polymers*, Vol. 2, Ed. by J. Andrade, Plenum Press, New York, 1985, p. 81.

[56] J. L. Detrich, G. A. Erb, D. A. Beres, and L. H. Rickard, in *Charge and Field Effects in Biosystems—3*, Ed. by M. J. Allen, Birkhauser, Boston, 1992, pp. 41–52.

[57] B. J. Berne and R. Pecora, *Dynamic Light Scattering*, John Wiley & Sons, New York, 1976.

[58] B. Chu, *Laser Light Scattering*, Academic Press, New York, 1974.

[59] A. M. Jamieson and M. E. McDonnell, *Adv. Chem. Ser.* **174** (1979) 163.

[60] J. M. Schurr, *CRC Crit. Rev. Biochem.* **4** (1977) 371.

[61] B. D. Fair and A. M. Jamieson, *J. Colloid Interface Sci.* **77** (1980) 525.

[62] O. Glatter and O. Kratky, eds., *Small-Angle X-Ray Scattering*, Academic Press, New York, 1984.

[63] L. A. Feigin and D. I. Svergun, *Structure Analysis by Small-Angle X-Ray and Neutron Scattering*, Plenum Press, New York, 1987.

[64] A. R. Mackie, J. Mingins, and R. Dann, in *Food Polymers, Gels and Colloids*, Ed. by E. Dickinson, The Royal Society of Chemistry, Cambridge, 1991, p. 96.

[65] E. K. Krauskopf and A. Wieckowski, in *Adsorption of Molecules at Metal Electrodes*, Vol. 1, Ed. by J. Lipkowski and P. N. Ross, VCH Publishers, New York, 1992, pp. 119–169.

[66] H. J. van Enckevort, D. V. Dass, and A. G. Langdon, *J. Colloid Interface Sci.* **98** (1984) 138.

[67] E. Kiss, *Colloids Surf. A* **79** (1993) 135.

[68] J. G. E. M. Fraaije, W. Norde, and J. Lyklema, *Biophys. Chem.* **41** (1991) 263.

[69] J. C. Kim and D. B. Lund, in *Fouling and Cleansing in Food Processing*, Ed. by H. G. Kessler and D. B. Lund, Druckerei Walch, Augsburg, Germany, 1989, pp. 187–199.

[70] N. Muramatsu and T. Kondo, *J. Colloid Interface Sci.* **153** (1992) 23.

[71] A. Engel, *Annu. Rev. Biophys. Biophys. Chem.* **20** (1991) 79.

[72] C. J. Roberts, M. Sekowski, M. C. Davies, D. E. Jackson, M. R. Price, and S. J. B. Tendler, *Biochem. J.* **283** (1992) 181.

[73] Z. Nawaz, T. R. I. Cataldi, J. Knall, R. Somekh, and J. B. Pethica, *Surf. Sci.* **265** (1992) 129.

[74] G. J. Leggett, C. J. Roberts, P. M. Williams, M. C. Davies, D. E. Jackson, and S. J. B. Tendler, *Langmuir* **9** (1993) 2356.

[75] G. Binnig, C. F. Quate, and C. Gerber, *Phys. Rev. Lett.* **56** (1986) 930.

[76] W. Wiegrabe, M. Nonemacher, R. Guckenberger, and O. J. Wolter, *J. Microsc.* **163** (1991) 79.

[77] H. Butt, K. H. Downing, and P. K. Hansma, *Biophys. J.* **58** (1990) 1473.

[78] O. M. Egger, F. Ohnesorge, A. L. Weisenhorn, S. P. Heyn, B. Drake, C. B. Prater, S. A. C. Gould, P. K. Hansma, and H. E. Gaub, *J. Struct. B* **103** (1990) 89.

[79] R. E. Marchant, S. A. Lea, J. D. Andrade, and P. Bockenstedt, *J. Colloid Interface Sci.* **148** (1992) 261.

[80] S. J. Eppell, F. R. Zypman, and R. E. Marchant, *Langmuir* **9** (1993) 2281.

[81] M. W. Breiter, in *Modern Aspects of Electrochemistry*, No. 10, Ed. by J. O'M. Bockris and B. E. Conway, Plenum Press, New York, 1975, pp. 161–210.

[82] B. B. Damaskin, O. A. Petrii, and V. V. Batrakov, *Adsorption of Organic Compounds on Electrodes*, Plenum Press, New York, 1971.

[83] B. E. Conway, in *Electrodes of Conductive Metallic Oxides. Part B,* Ed. by S. Trasatti, *Studies in Physical and Theoretical Chemistry,* Vol. 11, Elsevier, Amsterdam, 1981, pp. 433–519.

[84] E. J. Rudd and B. E. Conway, in *Comprehensive Treatise of Electrochemistry*, Vol. 7, Ed. by B. E. Conway, J. O'M. Bockris, E. Yeager, S. U. M. Khan, and R. E. White, Plenum Press, New York, 1983, pp. 643–680.

[85] S. Sarangapani and E. Yeager, in *Comprehensive Treatise of Electrochemistry*, Vol. 9, Ed. by E. Yeager, J. O'M. Bockris, B. E. Conway, and S. Sarangapani, Plenum Press, New York, 1984, pp. 1–12.

[86] J. O'M. Bockris and K. T. Jeng, *J. Electroanal. Chem.* **330** (1992) 541.

[87] E. Pastor, M. C. Arevalo, S. Gonzalez, and A. J. Arvia, *Electrochim. Acta* **36** (1991) 2003.

[88] M. C. Arevalo, A. M. Castro Luna, A. Arevalo, and A. J. Arvia, *J. Electroanal. Chem.* **330** (1992) 595.

[89] A. Iannelli, J. Richer, and J. Lipkowski, *Langmuir* **5** (1989) 466.

[90] V. Brabec and V. Mornstein, *Biophys. Chem.* **12** (1980) 159.

[91] T. Y. Safonova, S. S. Kidirov, and O. A. Petrii, *Elektrokhimiya* **20** (1984) 1666.

[92] J. A. Reynaud, B. Malfoy, and P. Canesson, *J. Electroanal. Chem.* **114** (1980) 195.

[93] B. Malfoy and J. A. Reynaud, *J. Electroanal. Chem.* **114** (1980) 213.

[94] G. Horanyi and E. M. Rizmayer, *J. Electroanal. Chem.* **64** (1975) 15.

[95] G. Horanyi and E. M. Rizmayer, *J. Electroanal. Chem.* **80** (1977) 401.

[96] G. Horanyi and E. M. Rizmayer, *J. Electroanal. Chem.* **198** (1986) 393.

[97] G. Horanyi, *Electrochim. Acta* **35** (1990) 919.

[98] D. S. Corrigan, E. K. Krauskopf, L. M. Rice, A. Wieckowski, and M. J. Weaver, *J. Phys. Chem.* **92** (1988) 1596.

[99] G. Horanyi, *J. Electroanal. Chem.* **280** (1990) 425.

[100] G. Horanyi, *J. Electroanal. Chem.* **304** (1991) 211.

[101] G. Horanyi, E. M. Rizmayer, E. P. Simon, and J. Szammer, *J. Electroanal. Chem.* **323** (1992) 329.

[102] D. G. Marangoni, R. S. Smith, and S. G. Roscoe, *Can. J. Chem.* **67** (1989) 921.

[103] D. G. Marangoni, I. G. N. Wylie, and S. G. Roscoe, *Bioelectrochem. Bioenerg.* **25** (1991) 269.

[104] I. G. N. Wylie and S. G. Roscoe, *Bioelectrochem. Bioenerg.* **28** (1992) 367.

[105] A. K. Vijh and B. E. Conway, *Chem. Rev.* **67** (1967) 623.

[106] H. Durliat and M. Comtat, *J. Electroanal. Chem.* **89** (1978) 221.

[107] V. Brabec, *Bioelectrochem. Bioenerg.* **7** (1980) 69.

[108] S. S. Khidirov, *Elektrokhimiya* **20** (1984) 1540.

[109] G. F. Khan, H. Shinohara, Y. Ikariyama, and M. Aizawa, *J. Electroanal. Chem.* **315** (1991) 263.

[110] J. Zhao, R. W. Henkens, J. Stonehuerner, J. P. O'Daly, and A. L. Crumbliss, *J. Electroanal. Chem.* **327** (1992) 109.

[111] S. G. Roscoe and K. L. Fuller, *J. Colloid Interface Sci.* **152** (1992) 429.

[112] S. G. Roscoe, K. L. Fuller, and G. Robitaille, *J. Colloid Interface Sci.* **160** (1993) 243.

[113] T. M. Cotton, D. Kaddi, and D. Iorga, *J. Am. Chem. Soc.* **105** (1983) 7462.

[114] E. F. Bowden, F. M. Hawkridge, and H. N. Blount, *J. Electroanal. Chem.* **161** (1984) 355.

[115] D. E. Reed and F. M. Hawkridge, *Anal. Chem.* **59** (1987) 2334.

[116] F. N. Buchi and A. M. Bond, *J. Electroanal. Chem.* **314** (1991) 191.

[117] J. L. Willit and E. F. Bowden, *J. Phys. Chem.* **94** (1990) 8241.

[118] T. Daido and T. Akaike, *J. Electroanal. Chem.* **344** (1993) 91.

[119] M. Ohtani and O. Ikeda, *J. Electroanal. Chem.* **354** (1993) 311.

[120] J. Haladjian, K. Draoui, and P. Bianco, *Electrochim. Acta* **36** (1991) 1423.

[121] A. E. G. Cass, G. Davis, G. D. Francis, H. A. O. Hill, W. J. Aston, I. J. Higgins, E. V. Plotkin, L. D. L. Scott, and A. P. F. Turner, *Anal. Chem.* **56** (1984) 667.

[122] M. F. J. M. Verhagen and W. R. Hagen, *J. Electroanal. Chem.* **334** (1992) 339.

[123] L. Gorton and G. Johansson, *J. Electroanal. Chem.* **113** (1980) 151.

[124] O. Miyawaki and L. B. Wingard, Jr., *Biotechnol. Bioeng.* **26** (1984) 1364.

[125] O. Miyawaki and L. B. Wingard, Jr., *Biochim. Biophys. Acta* **838** (1985) 60.

[126] S. Ueyama, S. Isoda, and M. Maeda, *J. Electroanal. Chem.* **264** (1989) 149.

[127] H. Shinohara, M. Gratzel, N. Vlachopoulos, and M. Aizawa, *Bioelectrochem. Bioenerg.* **26** (1991) 307.

[128] M. M. Kamal, H. Elzanowska, M. Gaur, D. Kim, and V. I. Birss, *J. Electroanal. Chem.* **318** (1991) 349.

[129] C. Tanford, in *Advances in Protein Chemistry*, Vol. 23, Ed. by C. B. Anfinsen, J. T. Edsall, F.M. Richards, and D. S. Eisenberg, Academic Press, New York, 1968, pp. 121–282.

[130] C. Tanford, in *Advances in Protein Chemistry*, Vol. 24, Ed. by C. B. Anfinsen, J. T. Edsall, F.M. Richards, and D. S. Eisenberg, Academic Press, New York, 1970, pp. 1–95.

[131] C. Tanford, *J. Amer. Chem. Soc.* **83** (1961) 1628.

[132]C. Tanford and V. G. Taggart, *J. Am. Chem. Soc.* **83** (1961) 1634.

[133]C. Tanford, in *Symposium on Polymerization Reactions in Biological Systems*, Vol. 8, Associated Scientific Publishers, Amsterdam, New York, 1972, p. 314.

[134]A. Ikai, W. W. Fish, and C. Tanford, *J. Mol. Biol.* **73** (1973) 165.

[135]P. L. Privalov, in *Advances in Protein Chemistry*, Vol. 33, Ed. by C. B. Anfinsen, J. T. Edsall, and F. M. Richards, Academic Press, New York, 1979, pp. 167–241.

[136]P. L. Privalov, in *Advances in Protein Chemistry*, Vol. 35, Ed. by C. B. Anfinsen, J. T. Edsall, and F. M. Richards, Academic Press, New York, 1982, pp. 1–104.

[137]P. L. Privalov and S. J. Gill, in *Advances in Protein Chemistry*, Vol. 39, Ed. by C. B. Anfinsen, J. T. Edsall, F. M. Richards, and D. S. Eisenberg, Academic Press, New York, 1988, pp. 191–234.

[138]S. K. Burley and G. A. Petsko, in *Advances in Protein Chemistry*, Vol. 39, Ed. by C. B. Anfinsen, J. T. Edsall, F. M. Richards, and D. S. Eisenberg, Academic Press, Inc, New York, 1988, pp. 125–189.

[139]F. R. N. Gurd and T. M. Rothgeb, in *Advances in Protein Chemistry*, Vol. 33, Ed. by C. B. Anfinsen, J. T. Edsall, and F. M. Richards, Academic Press, New York, 1979, pp. 73–165.

[140]E. Freire, W. W. van Osdol, O. L. Mayorga, and J. M. Sanchez-Ruiz, *Annu. Rev. Biophys. Biophys. Chem.* **19** (1990) 159.

[141]Y. Nozaki and C. Tanford, *J. Biol. Chem.* **246** (1971) 2211.

[142]G. D. Rose, A. R. Geselowitz, G. J. Lesser, R. H. Lee, and M. H. Zehfus, *Science* **229** (1985) 834.

[143]A. Allerhand, D. Doddrell, V. Glushko, D. W. Cochran, E. Wenkert, P. J. Lawson, and F. R. N. Gurd, *J. Am. Chem. Soc.* **93** (1971) 544.

[144]F. MacRitchie, *Colloids Surf.* **76** (1993) 159.

[145]E. Dickinson and G. Stainsby, in *Colloids in Food*, Applied Science Publisher, London, 1982, pp. 285–330.

[146]B. J. Yoon and A. M. Lenhoff, *J. Phys. Chem.* **96** (1992) 3130.

[147]C. M. Roth and A. M. Lenhoff, *Langmuir* **9** (1993) 962.

[148]F. Elhebil and S. Premilat, *Biophys. Chem.* **42** (1992) 195.

[149]S. Sun, N. Luo, R. L. Ornstein, and R. Rein, *Biophys. J.* **62** (1992) 104.

[150]K. L. Prime and G. M. Whitesides, *Science* **252** (1991) 1164.

[151]S. S. Dukhin, *Adv. Colloid Interface Sci.* **44** (1993) 1.

[152]C. Fenstermaker and B. W. Morrissey, *Trans. Am. Soc. Artif. Intern. Organs* **22** (1976) 278.

[153]M. Z. Papiz, L. Sawyer, E. E. Eliopoulos, A. C. T. North, J. B. C. Findlay, R. Sivaprasadarao, T. A. Jones, M. E. Newcomer, and P. J. Kraulis, *Nature* (London) **324** (1986) 383.

[154]P. Walstra and R. Jenness, *Dairy Chemistry and Physics*, John Wiley & Sons, New York, 1984.

[155]H. A. McKenzie, in *Milk Proteins, Chemistry and Molecular Biology*, Vol. 2, Ed. by H. A. McKenzie, Academic Press, New York, 1971, p. 255.

[156]S. N. Timasheff and R. Townsend, *J. Am. Chem. Soc.* **83** (1961) 470.

[157]R. E. Dickerson and I. Geis, *The Structure and Action of Proteins*, Harper and Row, New York, 1969, p. 69.

[158]C. Tanford and M. L. Wagner, *J. Am. Chem. Soc.* **76** (1954) 3331.

[159]B. A. Ivarsson, P. Hegg, K. I. Lundstrom, and U. Jonsson, *Colloids Surf.* **13** (1985) 169.

[160]T. Nylander, Ph.D. Dissertation, University of Lund, Sweden, 1987.

[161]T. Arnebrant, B. A. Ivarsson, K. Larsson, K. I. Lundstrom, and T. Nylander, *Prog. Colloid Polym. Sci.* **70** (1985) 62.

[162]G. Horanyi, G. Vertes, and E. M. Rizmayer, *J. Electroanal. Chem.* **48** (1973) 207.

[163]H. Angerstein-Kozlowska, in *Comprehensive Treatise of Electrochemistry*, Vol. 9, Ed. by E. Yeager, J. O'M. Bockris, B. E. Conway, and S. Sarangapani, Plenum Press, New York, 1984, p. 15.

[164]H. E. Swaisgood, in *Developments in Dairy Chemistry*, Vol. 1, Ed. by P. F. Fox, Applied Science, London, 1982, p. 1.

[165]S. N. Timasheff and R. Townsend, *Nature* (London) **203** (1964) 517.

[166]H. Roels, G. Preaux, and R. Lonti, *Biochimie* **53** (1971) 1085.

[167]D. E. Graham and M. C. Phillips, *J. Colloid Interface Sci.* **70** (1979) 427.

[168]W. Norde, Ph.D. Dissertation, Agricultural University, Wageningen, 1976.

[169]M. R. Tarasevich, in *Comprehensive Treatise of Electrochemistry*, Vol. 10, Ed. by S. Srinivasan, Y. A. Chizmadzhev, J. O'M. Bockris, B. E. Conway, and E. Yeager, Plenum Press, New York, 1985, pp. 231–295.

[170]E. F. Bowden, F. M. Hawkridge, and H. N. Blount, in *Comprehensive Treatise of Electrochemistry*, Vol. 10, Ed. by S. Srinivasan, Y. A. Chizmadzhev, J. O'M. Bockris, B. E. Conway, and E. Yeager, Plenum Press, New York, 1985, pp. 297–346.

[171]T. E. Meyer and M. D. Kaman, *Adv. Protein Chem.* **35** (1982) 105.

[172]G. R. Moore, Z. Huang, C. G. S. Eley, H. A. Barker, G. Williams, M. N. Robinson, and R. J. P. Williams, *Faraday Discuss. Chem. Soc.* **74** (1982) 311.

[173]P. Yeh and T. Kuwana, *Chem. Lett.* **1977** 1145.

[174]J. L. Willit and E. F. Bowden, in *Redox Chemistry and Interfacial Behavior of Biological Molecules*, Ed. by G. Dryhurst and K. Niki, New York, 1989, pp. 69–79. Plenum Press.

[175]J. L. Willit and E. F. Bowden, *J. Electroanal. Chem.* **221** (1987) 265.

[176]K. B. Koller and F. M. Hawkridge, *J. Electroanal. Chem.* **239** (1988) 291.

[177]E. Stellwagen, *Nature (London)* **275** (1978) 73.

[178]A. L. Raphael and H. B. Gray, *J. Am. Chem. Soc.* **113** (1991) 1038.

[179]A. Dolla and M. Bruschi, *Biochim. Biophys. Acta* **932** (1988) 26.

[180]R. G. Lee and S. W. Kim, *Trans. Am. Soc. Artif. Intern. Organs* **25** (1979) 124.

[181]J. L. Brash, in *Modern Aspects of Protein Adsorption on Biomaterials*, Ed. by Y. F. Missirlis and W. Lemm, Kluwer Academic, Dordrecht, 1991, pp. 39–47.

[182]E. Osterberg, K. Bergstrom, K. Holmberg, J. A. Riggs, J. M. Van Alstine, T. P. Schuman, N. L. Burns, and J. M. Harris, *Colloids Surf. A* **77** (1993) 159.

[183]A. N. Glazer, R. J. De Lange, and D. S. Sigman, in *Laboratory Techniques in Biochemistry and Molecular Biology. Part I. Chemical Modification of Proteins, Selected Methods and Analytical Procedures*, Vol. 4, Ed. by T. S. Work and E. Work, North-Holland, Amsterdam, 1975, p. 113.

[184]J. Janatova, J. K. Fuller, and M. J. Hunter, *J. Biol. Chem.* **243** (1968) 3612.

[185]H. L. Casal, U. Kohler, and H. H. Mantsch, *Biochim. Biophys. Acta* **957** (1988) 11.

[186]C. Tanford, *The Hydrophobic Effect*, John Wiley & Sons, New York, 1973.

[187]Y. V. Griko and P. L. Privalov, *Biochemistry* **31** (1992) 8810.

[188]M. Paulsson and P. Dejmek, *J. Dairy Sci.* **73** (1990) 590.

[189]D. W. S. Wong, *Mechanism and Theory in Food Chemistry*, Van Nostrand Reinhold, New York, 1989, p. 87.

[190]W. H. Sawyer, *J. Dairy Sci.* **51** (1968) 323.

[191]K. Watanabe and H. Klostermeyer, *J. Dairy Res.* **43** (1976) 411.

[192]A. Laligant, E. Dumay, C. C. Valencia, J. L. Cuq, and J. C. Cheftel, *J. Agric. Food Chem.* **39** (1991) 2147.

[193]E. Dickinson, A. Mauffret, S. E. Rolfe, and C. M. Woskett, *J. Soc. Dairy Technol.* **42** (1989) 18.

[194]B. S. Noh and T. Richardson, *J. Dairy Sci. (Suppl. 1)* **71** (1988) 78.

[195]D. G. Dalgleish, *J. Agric. Food Chem.* **38** (1990) 1995.

[196]W. H. Sawyer, *J. Dairy Sci.* **52** (1969) 1347.

[197]P. Smits and J. H. van Brouwershaven, *J. Dairy Res.* **47** (1980) 313

[198]T. J. M. Jeurnink, *Neth. Milk Dairy J.* **45** (1991) 23.

[199]K. L. Fuller and S. G. Roscoe, in *Protein Structure-Function Relationships in Foods*, Ed. by R. Y. Yada, R. L. Jackman, and J. L. Smith, Blackie Academic and Professional Publishers, Glasgow, 1994, pp. 143–162.

[200]S. G. Roscoe and K. L. Fuller, *Food Res. Int.* **26** (1993) 343.

[201]S. G. Roscoe and K. L. Fuller, *Food Res. Int.*, **27** (1994) 363.

[202]P. Kraulis, J., *J. Appl. Cryst.* **24** (1991) 946.

5

Chemisorption of Thiols on Metals and Metal Sulfides

Ronald Woods

CSIRO Division of Minerals, Port Melbourne, Victoria 3207, Australia

I. INTRODUCTION

This chapter reviews the adsorption of thiols at metal and metal sulfide surfaces. The motivation for studying these systems derives from the interaction of this interesting group of organic compounds with native metal and sulfide mineral surfaces being the key chemical step in the separation and concentration of such minerals from their ores by froth flotation. A number of different species can be formed by the interaction of thiols with metals and sulfides, the identity of the adsorbate depending on the particular thiol and mineral, the adsorption conditions, and the extent of reaction. This review focuses on the chemisorbed species formed in these systems. The concept of chemisorption is one of the formation of chemical bonds between the thiol and atoms in the mineral surface without removal of component atoms from their positions in the metal or metal sulfide lattice.

II. FROTH FLOTATION

The froth flotation process involves crushing the ore to liberate separate grains of the various valuable minerals and gangue components, pulping the ore particles with water, and then selectively rendering hydrophobic the surface of the mineral of interest through the addition of an organic

Modern Aspects of Electrochemistry, Number 29, edited by John O'M. Bockris *et al.*
Plenum Press, New York, 1996.

species which is termed the *collector*. Following this procedure, a stream of air bubbles is passed through the pulp; the bubbles attach to and levitate the hydrophobic particles, which collect in a froth layer that flows over the weir of the flotation cell. A second and third valuable component can then be floated from the sink fraction by adjusting the chemical composition of the pulp.

The flotation process was developed in Australia at the turn of the century to treat the primary sulfidic silver/lead/zinc ore at Broken Hill, New South Wales, following the mining of the overlying secondary ore.[1] Oxidation to form the secondary zone had resulted in a concentration of silver and lead, and this section of the ore body could be smelted directly. The underlying sulfide zone, which comprised the bulk of the resource, was less amenable to treatment in the smelters of that era, and a method was required to recover the zinc values if the potentiality of the primary ore was to be realized. Many approaches were pursued to solve the "sulfide problem" before selective flotation was developed. The flotation process was then rapidly adopted by mining companies throughout the world.

A major advance in flotation practice was made in 1925 with the introduction[2] of short-alkyl-chain xanthates (dithiocarbonates) as collectors. These compounds proved to be much more selective than those used previously; they are most efficient for the flotation of sulfides but quite inert for common gangue minerals. The flotation process has been developed throughout the twentieth century to increase the efficiency of separation and to treat ever more complex sulfide ores. It is a key unit process in the recovery of most of the world's copper, lead, molybdenum, nickel, platinum group elements, silver, and zinc and in the treatment of certain gold and tin ores. Fleming and Kitchener[3] pointed out that, in light of the very large tonnages processed each year ($>10^9$ tons), flotation must be considered one of the major practical applications of surface chemistry.

III. DEVELOPMENT OF THEORIES OF MINERAL–COLLECTOR INTERACTION

The 1930s saw the development of theories to elucidate collector–mineral interaction and to explain the selectivity for sulfide mineral flotation of thiol collectors such as the xanthates. Three propositions were put forward, and each was promoted and defended by its proponents with considerable vigor and intensity.

Gaudin[4] and Wark and Cox[5] considered that the attachment of a collector to a sulfide surface was best considered as adsorption. Taggart *et al.*[6] took a different view; they maintained that the interaction of collectors with sulfides occurred "by reason of chemical reactions of well-recognized types." Taggart and co-workers[7] dismissed the adsorption theory and claimed that the term adsorption "is commonly used as a haven of refuge for writers lost in a morass of shaky experiment and muddy thinking." They also asserted that "to attribute to the term adsorption . . . any element of explanatory character is to adopt the mental attitude of the ostrich seeking concealment." In 1946, Wark[8] responded to these views with the opinion:

> For an explanation of adsorption we must await a fuller development of the theory of intermolecular attractions, but one simple generalization is useful in this field, namely that adsorption can usually be expected if a compound which is relatively insoluble is formed between the metal of the mineral and the anion of the collector: this is the Paneth–Horowitz principle, enunciated in 1915. Taggart, who re-discovered the principle a decade later, proceeded to ascribe all adsorption to double decomposition, but this explanation is far too simple.

In the concept promoted by Taggart, the formation of a hydrophobic surface is governed by the solubility of the metal thiol compounds. In order to overcome the difficulty that flotation occurs at collector concentrations that are orders of magnitude less than those at which one would expect a metal thiol compound to deposit from solubility considerations based on bulk species, Taggart and Hassialis[9] considered that the concentration of the metal ion be taken as that in the lattice of the mineral. Sutherland and Wark[10] pointed out the fallacy of this reasoning and reaffirmed the conviction that, to explain the floatability at low collector concentrations, an adsorption process must be operative.

The third approach was advanced by Cook and Nixon, who asserted[11] that the entity undergoing adsorption must be the acid form of the thiol. The justification for this model is that the adsorption of thiol ions would result in a mineral surface that is too highly charged to result in flotation. Furthermore, the thiol ions would became separated from their counter-ions when a gas bubble attached to the surface, and, as Cook and Nixon colorfully claimed, the resulting flotation concentrate would have "so much charge (if it could be formed in bulk by some mysterious non-terrestrial device) that it would explode with greater violence than an equal weight of nitroglycerine." However, there is a problem inherent in this theory; the proposed adsorbate, the thiol acid, is unstable.

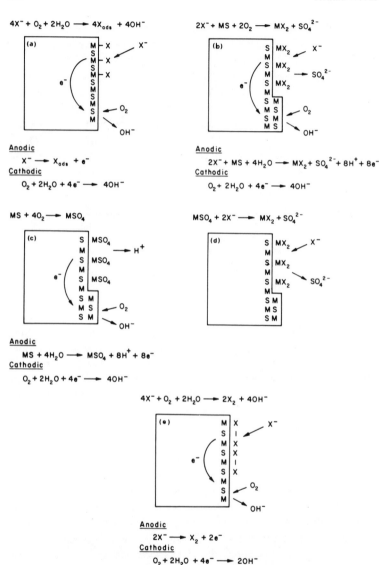

Figure 1. Schematic representation of the mixed-potential mechanism for the interaction of thiol collectors with sulfide minerals in which the anodic process is chemisorption (a), a single-step reaction to form a metal collector compound (b), the latter reaction occurring in two stages, comprising oxidation of the mineral (c), ion exchange with the collector (d), and formation of the dithiolate (e). (From Woods.[14])

Thus, three theories had been developed and each had difficulties; formation of metal thiols did not fit thermodynamics, adsorption of ions did not fit charge requirements, and adsorption of thiol acids did not fit stability criteria.

A solution to this dilemma was provided in 1957 by Nixon,[12] who saw another way in which a neutral surface could be generated, based on corrosion science concepts. It was proposed that "the troublesome electrons" identified by the Cook school "are removed by a reaction with oxygen." It was proposed that the interaction of thiol collectors with mineral surfaces comprised an electrochemical mechanism in which an anodic oxidation involving the collector transferred charge to the mineral, and the simultaneous cathodic reduction of oxygen returned this charge to the solution phase. Most sulfide minerals are semiconductors with high conductivity and hence can readily sustain such processes. Not only does this mechanism result in a neutral surface species, but it also accounts for the finding[13] that the presence of dissolved oxygen is a prerequisite for the flotation of sulfide minerals with thiol ions.

Nixon[12] also suggested that "prominent theories of flotation could be reconciled" by the electrochemical approach. This is illustrated in Fig. 1, in which the requirement of a neutral surface species in Cook and Nixon's model can be met in different ways. The anodic reaction can be adsorption as proposed by Wark and Cox (Fig. 1a). Also, it can be the formation of a metal thiol compound as proposed by Taggart and co-workers. The latter process can occur as a single step as in Fig. 1b or through separate surface oxidation and ion-exchange processes as shown in Figs. 1c and 1d. In addition, the anodic process can be the formation of the dithiolate as suggested by Nixon[12] and illustrated in Fig. 1e.

The mixed-potential mechanism has two important implications. First, the potential across the mineral/solution interface will be an important parameter in determining flotation recovery. Second, the reaction imparting floatability, the anodic process involving the collector, is amenable to investigation using electrochemical techniques.

IV. VOLTAMMETRIC STUDIES OF THE OXIDATION OF THIOLS

The first voltammetric study of flotation-related systems[15] examined the oxidation of ethyl xanthate on platinum, gold, copper, and galena (PbS) electrodes. For platinum and gold, voltammograms displayed an anodic peak arising from the oxidation of xanthate to dixanthogen, which is deposited on the electrode surface, and a cathodic peak from the reverse process.

$$2C_2H_5OCS_2^- \rightarrow (C_2H_5OCS_2)_2 + 2e^- \tag{1}$$

The Tafel slope for the anodic oxidation of xanthate on platinum and gold indicated that reaction (1) proceeded via an initial chemisorption step:

$$C_2H_5OCS_2^- \rightarrow (C_2H_5OCS_2)_{ads} + e^- \tag{2}$$

Voltammograms for copper and for galena (Fig. 2) displayed prewaves on the positive-going scans at potentials below the reversible value for the formation of the metal xanthate and dixanthogen. These prewaves are indicative of the chemisorption process, reaction (2).

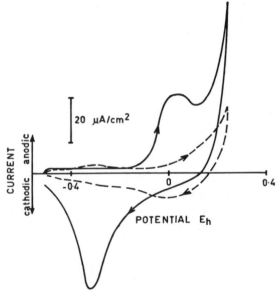

Figure 2. Voltammograms for a galena electrode in 0.1 mol dm^{-3} sodium tetraborate (pH 9.2) (---) and in the same solution containing 9.5×10^{-3} mol dm^{-3} ethyl xanthate (—). Triangular potential scans at 10 mV s^{-1}. (From Woods.[15])

Studies on galena[16-18] in the absence of collectors have shown that, at all pH values, anodic oxidation commences at the potential expected for sulfur formation and hence the formation of lead xanthate by the interaction of xanthate with galena would be expected to occur at the potential for the reaction

$$PbS + 2C_2H_5OCS_2^- \rightarrow Pb(C_2H_5OCS_2)_2 + S + 2e^- \qquad (3)$$

The solubility product[19] of lead ethyl xanthate is 4×10^{-17}, and hence the reversible potential of reaction (3) for an ethyl xanthate concentration of 10^{-2} mol dm^{-3} (the conditions of Fig. 2) is 0.01 V. It can be seen from Fig. 2 that the anodic oxidation reaction commences at ~0.2 V below the value at which reaction (3) is possible. The solubility product increase equivalent to a decrease in potential of 0.2 V is nearly seven orders of magnitude. Thus, the observed underpotential chemisorption of xanthate conforms to the model proposed by Wark and Cox, in which flotation is induced by an adsorbed analog of the bulk metal–collector compound with a much lower solubility.

Later studies[20,21] on the surface oxidation of galena, applying X-ray photoelectron spectroscopy, indicated that the initial product was not elemental sulfur as presented in reaction (3), but rather a metal-deficient sulfide. That is, the development of a lead xanthate phase is represented by

$$PbS + 2xC_2H_5OCS_2^- \rightarrow xPb(C_2H_5OCS_2)_2 + Pb_{1-x}S + 2xe^- \qquad (4)$$

This does not alter the conclusion that chemisorption occurs at underpotentials. The free energy of $Pb_{1-x}S$, and therefore the standard potential of reaction (4), is not known. However, $Pb_{1-x}S$ is metastable and must have a positive free energy compared with the stable species, PbS plus S^0, and hence the standard potential of reaction (4) must be more positive than that of reaction (3) and hence the underpotential of chemisorption must be greater than that presented above.

Surface photovoltage and interfacial capacitance measurements on *in situ* cleaved natural and synthetic lead sulfide crystals were combined with voltammetry in order to study the influence of the semiconductor type of galena on xanthate chemisorption.[22] The characteristics of the voltammogram for highly *n*-type and highly *p*-type galena were found to be the same. It was concluded that xanthate chemisorption induces a donorlike surface state in the forbidden gap of the galena, the occupancy of which is determined by the Fermi energy at the surface. Band diagrams showing flat-band potentials of highly p- and highly n-type lead sulfide electrodes relative to xanthate chemisorption/desorption are shown in Fig. 3. This figure indicates that xanthate can spontaneously adsorb on freshly fractured, highly p-type galena but that the surface of freshly fractured n-type galena must oxidize to increase the free-hole concentration to the point where xanthate can chemisorb. It was also shown[22] that the chemisorption/desorption process on galena is more reversible than reactions leading to the formation of lead xanthate (reaction 4). The authors considered that this was to be expected since reaction (4) requires structural rearrangement at the mineral surface.

Voltammetric studies[23] have also been carried out on the anodic oxidation of methyl and n-butyl xanthates at galena electrodes. Prewaves were observed, analogous to those for the ethyl homolog.

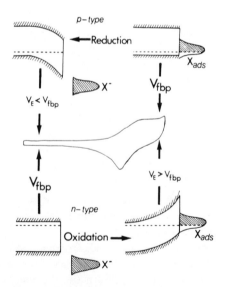

Figure 3. Band diagrams showing flat-band potentials (V_{fbp}) of highly p- and n-type lead sulfide electrodes relative to the xanthate chemisorption/desorption reaction. (From Richardson and O'Dell.[22])

A prewave is also evident on voltammograms for ethyl xanthate oxidation at a lead metal electrode,[24] and its characteristics are indicative of a reversible chemisorption process. The prewave commences at –0.7 V under the same conditions as those for Fig. 2, that is, at an ethyl xanthate concentration of 10^{-2} mol dm^{-3}. This corresponds to an underpotential to the formation of lead xanthate on lead of ~0.2 V, which is the same as that observed for galena.

An analogous correspondence between the underpotential for xanthate oxidation on the metal and the metal sulfide is observed for copper.[25] There have been a number of voltammetric investigations of the ethyl xanthate/copper[25–27] and ethyl xanthate/chalcocite (Cu_2S)[25,27–31] systems, and similar findings have been reported. A single prewave peak is observed for copper, and this occurs at a potential ~0.28 V more negative than that for chalcocite. As with galena, this shift corresponds to the difference in the activity of the metal in the element and in its sulfide.

Voltammograms for ethyl xanthate on chalcocite are presented in Fig. 4. The cathodic current close to the lower potential limit is due to the

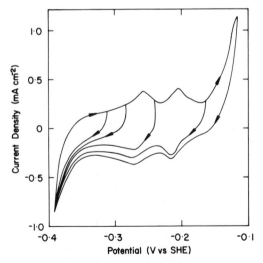

Figure 4. Cyclic voltammograms for chalcocite in 0.05 mol dm^{-3} sodium tetraborate (pH 9.2) containing 10^{-2} mol dm^{-3} ethyl xanthate. Scans at 20 mV s^{-1} to different upper potential limits. (From Woods et al.[27])

reduction of chalcocite to copper. The anodic current close to the upper potential limit arises from the formation of bulk copper xanthate according to the reaction

$$Cu_2S + xC_2H_5OCS_2^- \rightarrow xCu(C_2H_5OCS_2) + Cu_{2-x}S + xe^- \quad (5)$$

The reversible potential of reaction (5), under the conditions of Fig. 4 and taking $Cu_{2-x}S$ to be $Cu_{1.93}S$ (the first metastable nonstoichiometric copper sulfide oxidation product of chalcocite[32]), is –0.15 V. The two anodic peaks in Fig. 4, and the corresponding cathodic peaks, occur at potentials below that at which reaction (5) is possible and have been assigned to the chemisorption and desorption of xanthate. It can be seen that the chemisorption process is quite reversible.

O'Dell et al.[29,30] extended voltammetry on the ethyl xanthate/chalcocite system from conventional mineral electrodes to include studies on particle beds. A bed composed of 1.4 g of mineral particles in the size

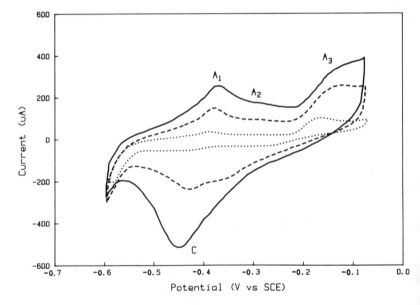

Figure 5. Cyclic voltammograms for chalcocite particulate particle-bed electrode in 0.05 mol dm^{-3} sodium tetraborate (pH 9.2) containing 1.9×10^{-5} mol dm^{-3} ethyl xanthate with the solution flowed through the bed. Potential sweep between –0.6 and –0.05 V vs. SCE. (From O'Dell et al.[30]) Solid line, 10 mV s^{-1}; dashed line, 5 mV s^{-1}; dotted line, 1 mV s^{-1}.

range 590–840 μm was held between two fritted glass disks to ensure physical and electrical contact between particles while allowing solution to be flowed through the bed. Voltammograms obtained (Fig. 5) are analogous to those for massive chalcocite (Fig. 4); peaks A_1 and A_2 correspond to chemisorption and peak A_3 to copper xanthate formation by reaction (5). It can be seen that the influence of scan rate is consistent with this assignment.

The interaction of chalcocite with diethyl dithiophosphate parallels that with ethyl xanthate.[33–35] The voltammograms presented in Fig. 6 exhibit a prewave due to dithiophosphate chemisorption followed by copper dithiophosphate formation.

Voltammograms for the oxidation of diethyl dithiophosphate on gold electrodes also indicate[36] that a chemisorption process precedes gold dithiophosphate and dithiophosphate disulfide formation. The latter reaction is the dithiophosphate analog of reaction (1).

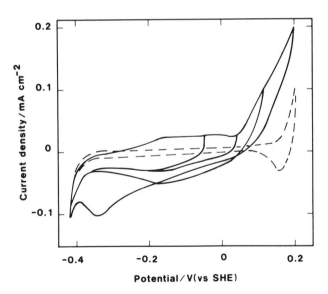

Figure 6. Cyclic voltammograms for a chalcocite electrode in 0.05 mol dm^{-3} sodium tetraborate (pH 9.2) (---) and in the same solution containing 10^{-3} mol dm^{-3} diethyl dithiophosphate (—). Scans at 50 mV s^{-1} taken to different upper potential limits. (From Buckley and Woods.[35])

A chemisorption prewave is also observed for diethyl dithiophosphate oxidation on lead electrodes,[36] as can be seen from Fig. 7. It is, therefore, surprising that no evidence of chemisorption of this thiol was observed for galena.[36]

Voltammograms for silver in the presence of ethyl xanthate display a prewave[37–39] as shown in Fig. 8. The reversible potential of the silver/silver ethyl xanthate couple at a xanthate concentration of 10^{-4} mol dm^{-3} is 0 V, and the major anodic and cathodic peaks arise from the formation and reduction of silver xanthate, respectively. Reversible chemisorption of xanthate is responsible for the currents observed at more negative potentials. Xanthate has also been found[38,40] to chemisorb on silver sites on silver–gold alloys.

Voltammograms reported for chalcopyrite $(CuFeS_2)$[41] and pyrite (FeS_2)[23,41] in the presence of ethyl xanthate are similar to those for platinum and gold in that the only discernible current for xanthate oxidation is that due to dixanthogen formation, reaction (1). However, as with gold, the Tafel slope for the reaction indicated that it proceeds via a chemisorbed xanthate intermediate with a significant coverage. Voltam-

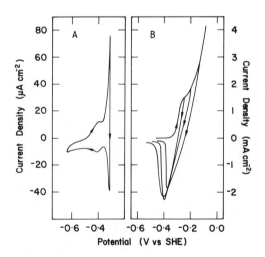

Figure 7. Voltammograms for a lead electrode in an acetate buffer of pH 4.6 containing 10^{-2} mol dm^{-3} diethyl dithiophosphate. Scans at 10 mV s^{-1} taken to different upper potential limits. (From Buckley and Woods.[36])

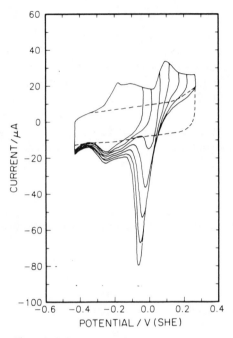

Figure 8. Voltammograms for a silver electrode in 0.05 mol dm^{-3} sodium tetraborate (pH 9.2) (---) and in the same solution containing 10^{-4} mol dm^{-3} ethyl xanthate (—). Scans at 20 mV s^{-1} taken to different upper potential limits. (From Woods et al.[39])

mograms for ethyl xanthate on bornite (Cu$_5$FeS$_4$) were found[42,43] to display anodic currents due to copper xanthate and dixanthogen formation. There was an indication that xanthate was chemisorbed at lower potentials, but it was difficult to distinguish the current due to this process from that for background reactions. The latter arise from oxidation of bornite itself, which results in removal of iron, leaving Cu$_5$S$_4$.[44] In order to better distinguish xanthate reactions, Richardson and co-workers[42,43] combined voltammetry with spectroscopic monitoring of the solution phase and confirmed that xanthate was adsorbed in the low-potential region. This approach, and other nonelectrochemical methods for examining surface layers, will be discussed in Section VII.

V. CHEMISORPTION ISOTHERMS

Voltammetry reveals that the chemisorption of thiols on metal and metal sulfide electrodes is reversible, and this indicates that it should be possible to determine isotherms delineating this process. Chemisorption coverages have been determined as a function of potential for ethyl xanthate on copper,[27] chalcocite,[27] and silver[38,39] and for diethyl dithiophosphate on chalcocite.[35] In these investigations, coverages were determined by holding the electrode at a selected potential until equilibrium conditions had been established, applying a potential scan in the negative-going direction to reduce the adsorbed thiol layer to the thiol ion, and integrating the cathodic charge passed between the set potential and the lower limit. Stripping curves for ethyl xanthate on copper are presented in Fig. 9. For each of the systems studied, the relationship between coverage and potential has been found to fit a Frumkin isotherm of the form presented by Schultze[45]:

$$[\theta/(1 - \theta)] \exp g\theta = Ka_A \exp (\gamma FE/RT) \qquad (6)$$

where θ is the fractional surface coverage, g and K are constants, a_A is the activity of the adsorbate in solution, γ is the electrosorption valency, and

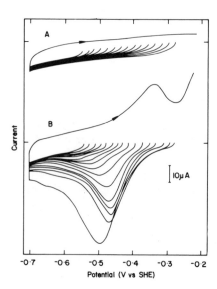

Figure 9. Stripping voltammograms for a copper electrode in 0.05 mol dm^{-3} sodium tetraborate (pH 9.2) (A) and in the same solution containing 10^{-4} mol dm^{-3} ethyl xanthate (B). Potential held for 4 min at different set potentials followed by potential scan applied in the negative-going direction at 20 mV s^{-1}. (From Woods *et al.*[27])

the other terms have their usual meaning. The term $g\theta$ arises from variation of the free energy of adsorption with coverage, through either heterogeneity of the surface or interactions between adsorbed molecules. The electrosorption valency, γ, reflects changes in the double layer associated with the adsorption process, in addition to the number of electrons involved in the charge-transfer reaction. The latter is unity for the adsorption of xanthate or dithiophosphate.

Figure 10 presents the coverages obtained for ethyl xanthate on chalcocite together with the derived isotherm. It can be seen that the coverage at each potential is the same at the two pH values considered; this is as expected for reaction (2), since no hydrogen or hydroxide ions are involved. It is also apparent that there is a good correspondence between the experimental data and the derived isotherm.

The values of the various constants in the Frumkin isotherm for the systems studied are presented in Table 1. The finite g values reflect the multiple peaks observed in the corresponding voltammograms and suggest that there is a range of different adsorption sites on the relevant surfaces. It is interesting to note that γ has a high value for chalcocite for both xanthate and dithiophosphate chemisorption and this is probably related to the semiconducting nature of the mineral.

The potential dependence of ethyl xanthate coverage has also been determined[38,40] for silver–gold alloys. In this system, xanthate chemisorbs onto silver sites in the alloy surface, and, as with silver itself, adsorption occurs at potentials below that at which silver xanthate is formed. Figure 11 presents the coverages determined for this system. The solid lines represent the Frumkin isotherm previously derived for silver. The isotherms for the alloys were obtained by introducing a silver activity term, taking values of activity derived from the thermodynamic data collated by

Table 1
Values of Constants in the Frumkin Isotherms for Selected Systems

Thiol	Surface	K	g	γ
Ethyl xanthate	Copper	5.3×10^{11}	0	1.2
Ethyl xanthate	Chalcocite	8.4×10^9	4	1.6
Ethyl xanthate	Silver	4.25×10^7	4	1.1
Diethyl dithiophosphate	Chalcocite	1.04×10^9	4	1.6

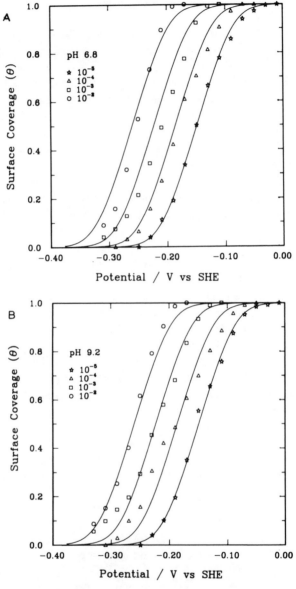

Figure 10. Potential dependence of coverage of chemisorbed xanthate on chalcocite in solutions of pH 6.8 (A) and 9.2 (B) containing various concentrations of ethyl xanthate (mol dm^{-3}). The solid lines are the derived isotherms (Eq. 6). (From Woods et al.[27])

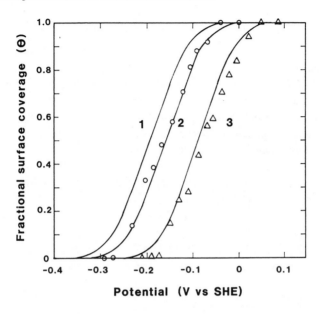

Figure 11. Potential dependence of the fractional coverage of chemisorbed xanthate on silver and silver–gold alloys in solutions of pH 9.2 containing 10^{-3} mol dm^{-3} ethyl xanthate. Solid lines are isotherms for silver (1), 50:50 wt. % Ag–Au alloy (2), and 20:80 wt. % Ag–Au alloy. Data points are experimental fractional coverages for the 50:50 (O) and 20:80 alloys (Δ). (From Woods et al.[40])

White et al.[46] It can be seen that the coverage data fit the isotherms within experimental error.

Nowak[47] analyzed the data reported by Leppinen[48] for the uptake of ethyl xanthate on precipitated lead sulfide. He correctly pointed out that the coverages reported by Leppinen were too high a by factor of two since they were based on a xanthate site occupancy that is equivalent to one xanthate per every second lead atom in the surface. He analyzed the recalculated coverages in terms of a Frumkin isotherm of the form

$$[\theta/(1-\theta)]\exp g\theta = K[\text{X}^-(\text{aq})]^2/[\text{S}^{2-}] \qquad (7)$$

The adsorption in this case was a precursor to lead xanthate formation by the reaction

$$PbS + H^+ + 2(C_2H_5OCS_2)^- \rightarrow Pb(C_2H_5OCS_2)_2 + HS^- \qquad (8)$$

Nowak[47] considered that this demonstrated that the adsorbed species formed by the interaction of xanthate with lead sulfide was molecular $Pb(C_2H_5OCS_2)_2$. However, Leppinen[48] saw no disagreement with previous findings since adsorption in the absence of oxygen only occurred at pH values of ≤ 6 and xanthate concentrations of $\geq 10^{-3}$ mol dm^{-3}. Leppinen found that the measured potential in his system was related to the sulfide ion concentration according to the Nernst equation for the S^0/S^{2-} redox couple. He was therefore able to apply his coverage/sulfide ion relationship to determine coverage as a function of potential for conditions outside his experimental range. In Fig. 12, his derived coverage values are compared with those determined[49] by stripping voltammetry for the same conditions. It can be seen that there is reasonable agreement, considering the assumptions involved. Recently, the data have been fitted[50] with a Frumkin isotherm of the form of Eq. (6), and the results are included in

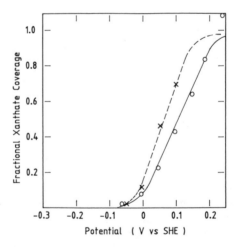

Figure 12. Potential dependence of coverage of chemisorbed xanthate on galena determined in solutions of pH 9.2 containing 2.3×10^{-5} mol dm^{-3} ethyl xanthate (○, Ref. 49) and coverages derived for 2×10^{-5} mol dm^{-3} ethyl xanthate (x, Ref. 48). The lines are the derived Frumkin isotherms (Eq. 6). (From Buckley and Woods.[50])

Fig. 12. The fit is acceptable for both sets of data. The constants found for the experimental coverages of Buckley and Woods[49] are $K = 5.8 \times 10^3$, g = 8, and $\gamma = 1.4$. These values are less precise that those listed in Table 1 since they were derived from coverage/potential data for only one xanthate concentration.

Consideration of the available electrochemical and spectroscopic information led to the conclusion[50] that there is no valid basis for the hypothesis put forward by Nowak[47] that the xanthate entity chemisorbed on galena is lead xanthate in molecular form, but there is strong evidence that the chemisorbed xanthate is attached to metal atoms in the sulfide surface.

VI. SELF-ASSEMBLED MONOLAYERS

Adsorbed thiol collectors on metal and metal sulfide surfaces are examples of *self-assembled monolayers* (SAMs). An important field of surface chemistry has been established in recent years regarding SAMs, because of the significance of these layers in a wide range of applications (adhesion, lubrication, microelectronics, chemical sensors, etc.) as well as because their investigation provides a means of understanding ordered organic layers that are analogs of biological systems such as lipid bilayers. A major group of compounds studied in this field is the alkanethiols. Other organic sulfides and disulfides have also been studied, including, in a few instances, compounds that are applied as flotation collectors. Metal sulfides have not been selected as substrates in SAM research, but metals that are relevant to the flotation field, in particular gold but also silver and copper, are commonly employed.

The focus of attention in investigations of SAMs has been on the structure of the adsorbate, and a wide range of spectroscopic and other methods have been applied to elucidate the nature of adsorbed species. In addition, the skills of the organic chemist have been utilized to construct special molecular architectures and incorporate a variety of elements and functional groups for the spectroscopists to target. The kinetics and mechanisms of formation have received much less attention than structural aspects, SAMs usually being considered to form "spontaneously."

The literature on SAMs will be considered in this chapter only when there is information that elucidates the adsorption of thiols relevant to flotation, and the results of these publications will be cited in the following sections. For detailed information on SAMs in general, the reader is referred to recent reviews by Ulman[51] and Dubois and Nuzzo.[52]

VII. CHARACTERIZATION OF THIOL ADSORPTION BY NONELECTROCHEMICAL TECHNIQUES

1. UV-Vis Spectroscopy

Thiol compounds such as xanthates have distinctive spectra at wavelengths in the UV-vis range and relatively high extinction coefficients. These properties have been exploited in many nonelectrochemical studies, UV-vis spectroscopy having been applied to the determination of the uptake of such compounds on sulfide minerals. Richardson and co-workers[29,30,42,43,53] extended this approach by studying the interaction of ethyl xanthate with mineral beds under potential control. They were able to show[29,30,53] that xanthate was abstracted from solution when the potential was taken into the region in which currents are observed on voltammograms. For potential steps, they found a good correlation between the charge passed and the decrease in absorbance at the characteristic wavelength for xanthate. Furthermore, they demonstrated that xanthate was released back to the solution phase when the potential was returned to values below the adsorption region. These findings substantiated the conclusion reached from voltammetry, namely, that the interaction of minerals with thiols involves a charge-transfer chemisorption as exemplified by reaction (2).

The identification of the prewave observed on voltammograms for ethyl xanthate on chalcocite and copper with chemisorbed xanthate was disputed by Mielczarski *et al.*[54] on the basis of Fourier transform infrared (FTIR) and XPS studies. These authors concluded that the surface species in the underpotential region could not be xanthate itself and suggested that it was probably a decomposition product of this compound. A similar conclusion with respect to silver was reached in later work.[55]

Further UV-vis spectroscopic studies were carried out on the ethyl xanthate/silver[38,39] and ethyl xanthate/copper systems[56] in order to resolve this controversy. Figure 13 shows the results obtained for the silver system. The intensity of the absorbance at 301 nm, the characteristic value for xanthate, is presented for a 1 mV s^{-1} potential cycle between –0.43 and 0.14 V, applied after a 2-min period at –0.43 V. It can be seen that xanthate began to be abstracted from solution at ~–0.35 V, the potential at which

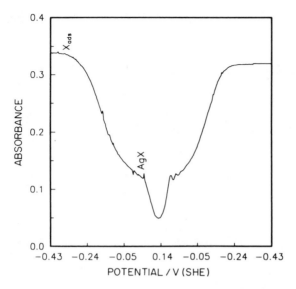

Figure 13. Absorbance at 301 nm of 0.05 mol dm^{-3} sodium tetraborate (pH 9.2) containing 2×10^{-5} mol dm^{-3} ethyl xanthate being circulated through a silver wool electrode. The potential was held at −0.43 V for 2 min, and then a triangular potential cycle at 1 mV s^{-1} was applied between −0.43 and 0.14 V. (From Woods et al.[39])

the voltammetric prewave commences (see Fig. 8). When the potential was close to the upper limit of the cycle, the rate of abstraction increased because the formation of silver ethyl xanthate was then also taking place. On reversing the scan, reduction of the silver xanthate took place followed by desorption of the chemisorbed layer. Figure 14 shows the full spectra recorded before and after the potential scan. It can be seen from the height of the 301-nm peak that 95% of the xanthate reappeared in solution following the electrochemical procedure.

It was found[39,53,56] that, if oxygen was not rigorously removed from solution, a small peak developed in the spectrum at ~350 nm. The peak was assigned to perxanthate; this species is known to be formed by the reaction of xanthate and peroxide[57]:

$$C_2H_5OCS_2^- + H_2O_2 \rightarrow C_2H_5OCS_2O^- + H_2O \tag{9}$$

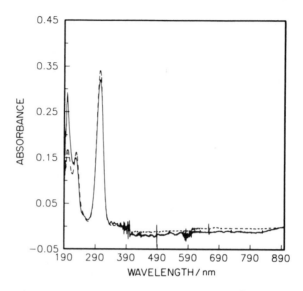

Figure 14. UV-vis spectra recorded for 0.05 mol dm^{-3} sodium
tetraborate (pH 9.2) containing 2×10^{-5} mol dm^{-3} ethyl xanthate
being circulated through a silver wool electrode prior to (---) and
following (—) the potential cycle applied to record Fig. 13. (From
Woods et al.[39])

It was considered that hydrogen peroxide is formed at silver and chalcocite
surfaces when oxygen is present as this compound is an intermediate in
the cathodic reduction of oxygen. Clearly, then, there is interaction between the two reactions that make up the mixed potential system, namely,
xanthate oxidation and oxygen reduction. Thus, care must be taken in
considering the individual processes in isolation.

Combining UV-vis spectroscopy with electrochemical investigations
is particularly useful in situations in which it is difficult to separate thiol
oxidation from background currents arising from oxidation of the substrate. As mentioned in Section IV, this approach allowed the adsorption
of xanthate by bornite to be measured as a function of potential. It was
found[42,43] that, on a 1 mV s^{-1} potential cycle with an ethyl xanthate
concentration of 3.2×10^{-4} mol dm^{-3}, the thiol began to be abstracted from
solution at –0.15 V, and the rate of abstraction increased at 0.05 V and
again at 0.15 V. It was concluded that the latter arose from dixanthogen

production, reaction (1), since the reversible potential of this process at the relevant xanthate concentration is 0.15 V. The first two potential regions were considered to correspond to chemisorption and copper xanthate formation, respectively. This assignment is consistent with the surface being Cu_5S_4 in this potential region as a result of surface oxidation of bornite.[44] The formation of copper xanthate from Cu_5S_4 by the analog of reaction (5) is expected to occur at a potential ~0.15 V more positive than that for chalcocite. Thus, for the relevant xanthate concentration, the reversible potential for copper xanthate formation is ~0.05 V. Assuming that chemisorption of ethyl xanthate commences at the same underpotential as on chalcocite, this process should commence at ~–0.15 V, which is the experimental value.

As pointed out in Section IV, voltammetric studies of the oxidation of xanthate on chalcopyrite revealed only the formation of dixanthogen by reaction (1). However, the situation is analogous to that for bornite in that the mineral itself is oxidized in the potential region of interest, and hence only large xanthate oxidation currents can be unequivocally identified on voltammograms. The oxidation process occurring with chalcopyrite is the removal of iron to leave a CuS_2 surface.[58,59] Richardson and Walker[43] carried out studies on a particulate chalcopyrite electrode similar to those made with bornite. They found that ethyl xanthate began to be abstracted from a 5×10^{-5} mol dm^{-3} solution at –0.05 V, with a significant increase in rate at 0.25 V. The latter was identified with dixanthogen formation, the reversible potential of this process being 0.22 V at the experimental xanthate concentration. It is not possible to calculate the potential at which the analog of reaction (5) should occur for CuS_2 since the free energy of formation of this metastable surface copper sulfide is unavailable. However, the potential must be greater than that for CuS. The standard potential for the reaction

$$CuS + C_2H_5OCS_2^- \rightarrow Cu(C_2H_5OCS_2) + S + e^- \qquad (10)$$

is –0.07 V, taking the free energy of CuS to be that of the metastable species formed in the oxidation of chalcocite.[60] Note that a more positive value is obtained if the free energy of formation of the stable mineral, covellite, is taken. The reversible potential of reaction (10) under the conditions employed by Richardson and Walker[43] is 0.21 V, and hence formation of copper xanthate should not precede dixanthogen formation. However, chemisorption could commence at ~0 V, assuming the same underpotential is manifest as with chalcocite. As with bornite, the calculated potential for chemisorption on chalcopyrite corresponds closely with the value observed experimentally for abstraction of xanthate from solution.

2. Fourier Transform Infrared (FTIR) Spectroscopy

Infrared spectroscopy provides a valuable means of detecting the presence of thiol species at surfaces, identifying the chemical nature of surface compounds, and determining the structure and orientation of the adsorbed entity. This technique has been extensively applied to mineral–thiol collector interactions since the pioneering work of Greenler[61] and Leja et al.[62] on the lead sulfide/ethyl xanthate system. In xanthate/mineral systems, metal thiol compounds and dixanthogen have been identified depending on the nature of the mineral and the experimental conditions, with the initial layer often having distinctive spectra. For example, the first monolayer of ethyl xanthate on galena was considered[61,62] to have a 1:1 xanthate/lead stoichiometry as distinct from the 2:1 stoichiometry that occurs in bulk lead xanthate.

Leppinen et al.[63] developed an electrochemical cell in which FTIR spectroscopy could be carried out in situ on mineral particles under potential control. This approach provided a much better means of controlling the surface condition and allowed the results to be related to findings derived from complementary electrochemical techniques. In addition, ex situ FTIR spectroscopy has been applied to the study of the interaction of thiols with sulfide and metal surfaces as a function of potential.

Before discussing the results of FTIR investigations of the adsorption of thiol collectors under potential control, it is pertinent to consider the study carried out by Ihs et al.[64] on the nature of ethyl and octyl xanthate adsorbed on freshly evaporated gold surfaces. This work provides valuable information on the influence of orientation on the relative FTIR peak heights for adsorbed xanthate. The findings are important in appraising the conclusions made by previous authors regarding the identity of the adsorbate derived from interaction of xanthate with other substrates. The spectra observed by Ihs et al.[64] for ethyl xanthate, as well as the spectra calculated from bulk optical constants for gold(I) ethyl xanthate, are shown in Fig. 15. The simulation of the spectra was based on a three-layer model in which the adsorbed layer is assumed to be isotropic (randomly oriented molecules) with a thickness of 1 nm. The difference between the relative peak heights in the spectra in Figs. 15a and b and those in the simulated spectrum (Fig. 15d) was considered to arise from orientation

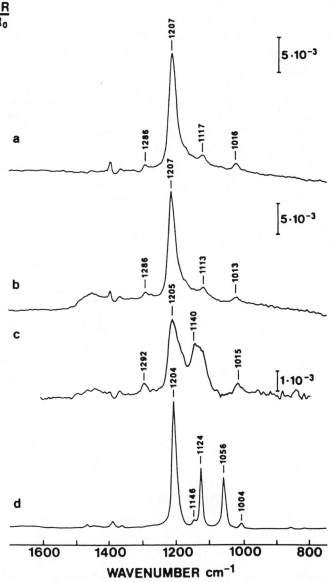

Figure 15. FTIR reflection–absorption spectra of ethyl xanthate adsorbed on gold. The concentrations and adsorption times are 10^{-3} mol dm^{-3} and 10 min (a), 10^{-5} mol dm^{-3} and 10 min (b), and 10^{-5} mol dm^{-3} and 1 min (c). The simulated spectrum for ethyl xanthate is shown in (d). (From Ihs *et al.*[64])

effects. A coordination in which both sulfur atoms in the xanthate radical are involved in binding to the gold surface will give a transition dipole moment of the v_{as}(S–C–S) vibration parallel to the surface. According to the surface selection rule, the corresponding bands will then disappear in the reflection–absorption spectrum. Similar intensity behavior was observed[65] for the adsorption of p-methylbenzyl and p-(trifluoromethyl)benzyl xanthates on gold. The appearance of a broad peak near 1140 cm^{-1} (Fig. 15c) at lower ethyl xanthate coverages suggested that the xanthate molecules were more randomly ordered at the lower surface coverage.

Ihs *et al.*[64] also investigated the adsorption of octyl xanthate on gold and found that the ratios of absorbance peaks in the 1000- to 1300-cm^{-1} region of the FTIR spectrum, arising from C–O and C–S vibrations, were

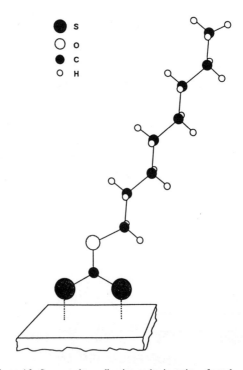

Figure 16. Suggested coordination and orientation of octyl xanthate on gold at full monolayer coverage. The alkyl chain is tilted 29° away from the surface normal and rotated 40° out of the plane of the paper around the axis of the chain. (From Ihs *et al.*[64])

similar to those for ethyl xanthate. However, with the long-chain xanthate, they were able to obtain additional information from the high-frequency bands just below 3000 cm^{-1} that correspond to CH_2 and CH_3 vibrational modes. The characteristics of these spectra substantiated the assignments for adsorbed xanthate and allowed the orientation and conformation of the adsorbed molecule to be determined. The data fitted the model illustrated in Fig. 16. The alkyl xanthate is coordinated to the gold surface through both sulfur atoms; the alkyl chain at full coverage has a fully extended all *trans*, zigzag conformation and is oriented at an angle of 29° to the surface normal. This orientation and conformation are consistent with previous studies of alkanethiols at metal surfaces.[51]

Leppinen *et al.*[55] correlated *ex situ* FTIR spectra with voltammograms for the interaction of ethyl xanthate with silver, gold, and silver–gold alloys. Spectra characteristic of silver xanthate and of dixanthogen developed at the potentials of the major anodic waves on silver and gold, respectively. Both these species were observed on the alloys. At lower potentials, spectra were recorded that had a dominant peak near 1200 cm^{-1}. These spectra were assigned to "unknown" species because they did not display the other major characteristic adsorption bands and they were apparent on surfaces that had been held at low potentials. The appearance of such peaks on silver at potentials below the prewave region was shown[38,39] to arise from changes in potential during transfer of the electrode from the cell to the spectrometer. When the silver surface was washed *in situ* with deoxygenated water immediately after opening the circuit, the spectrum only appeared at potentials above the beginning of the prewave (Fig. 17).

Talonen *et al.*[37] studied the interaction of ethyl xanthate with gold, silver, and copper as a function of potential using an *in situ* FTIR specular external reflection spectroscopic technique. With gold, they confirmed the conclusion reached from electrochemical investigations that dixanthogen was the major oxidation product. In contrast, Ihs *et al.*[64] did not observe any peaks that could be assigned to this compound in their *ex situ* recorded spectra. The absence of dixanthogen could be due to volatilization of this species in the spectrometer prior to recording of the spectra. Dixanthogen is known[55,66] to evaporate rapidly from metal surfaces in the gas stream used to maintain the spectrometer free from water vapor.

Talonen *et al.*,[37] using an *in situ* FTIR technique, found that the spectra for silver in the prewave potential region displayed a major peak near 1200 cm^{-1} and minor peaks at 1034 and 1018 cm^{-1}. They attributed

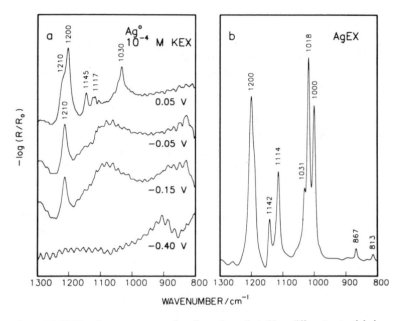

Figure 17. FTIR reflectance spectra of a silver electrode held at different potentials in sodium tetraborate (pH 9.2) containing 10^{-4} mol dm^{-3} ethyl xanthate (a) and silver ethyl xanthate (b). (From Woods et al.[40])

these peaks to the chemisorption of xanthate on the silver surface. In later studies based on an *ex situ* FTIR technique, Leppinen et al.[55] described the spectrum in this potential region as showing "extraneous absorption bands near 1200 cm^{-1} due to an unknown surface species." This conclusion was based on the dominance of the 1200-cm^{-1} peak and a lack of correlation between the spectrum and voltammetric currents. Further work on this system[38,39] indicated that, as with gold, the lack of correlation arose from a drift in potential following opening of the circuit and transfer of the electrode to the spectrometer. When the electrode was washed with deoxygenated solution immediately after opening the circuit, a good correlation was obtained. This can be seen by comparing Fig. 17 with Fig. 8. It is reasonable to assume that the departure of the relative intensities of the absorbance peaks for chemisorbed xanthate on silver from the expected ratio is due to orientation effects similar to those reported for gold by Ihs et al.[64]

The significant difference between the FTIR spectra of adsorbed xanthate and dixanthogen was exploited[66] to determine the coverage of chemisorbed ethyl xanthate as a function of potential even when significant quantities of dixanthogen were also present. Chemisorption was found to commence about 0.15 V below the potential at which dixanthogen deposits and to reach a fractional coverage of ~0.2 at the reversible value of the xanthate/dixanthogen couple (Fig. 18). It can be seen from the figure that the coverage increased in parallel with the anodic current arising from dixanthogen deposition to reach a monolayer at ~0.4 V above the reversible xanthate/dixanthogen potential. The solid line in Fig. 18 is the "best fit" of the coverage data with a Frumkin isotherm. The poor correspondence could arise from the nonequilibrium conditions obtaining in this system.[66]

Mielczarski *et al.*[54] reported that, after holding the potential of a copper electrode in the prewave potential region for the oxidation of ethyl

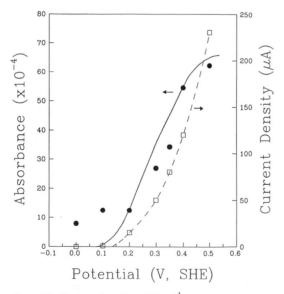

Figure 18. The intensity of the 1200-cm^{-1} FTIR reflectance peak (●) and current density recorded after 2 min (□) for a gold electrode held at different potentials in 0.05 mol dm^{-3} sodium tetraborate (pH 9.2) containing 5×10^{-4} mol dm^{-3} ethyl xanthate. (From Woods *et al.*[66])

xanthate, *ex situ* FTIR spectra showed no evidence of xanthate on the metal surface. On the other hand, Talonen *et al.*,[37] applying their *in situ* method, identified the prewave peak with the chemisorption of xanthate. Recent studies[56] have shown that *ex situ* FTIR spectra characteristic of xanthate are observed at underpotentials to copper xanthate formation under conditions of high fractional coverage and that the chemisorbed layer is easily oxidized so that the xanthate spectrum is rapidly degraded.

Leppinen and co-workers[48,63,67–69] carried out FTIR studies of the adsorption of ethyl xanthate on lead sulfide in which the potential of the system was controlled with sulfide ions. The coverage of ethyl xanthate was determined as a function of potential from measurements of the intensity of the major absorption band.[68] Adsorption of xanthate from a 10^{-3} mol dm^{-3} solution at pH 6 was found to commence at –0.12 V and reach monolayer coverage at 0.06 V (adjusting the coverage value as suggested by Nowak[47]). The latter potential is close to the reversible value of 0.07 V for lead xanthate formation by reaction (3). Thus, these results are in agreement with voltammetric results that showed underpotential deposition of a monolayer of chemisorbed xanthate to take place prior to the formation of bulk lead xanthate. It was concluded[69] that, in the chemisorbed layer, ethyl xanthate is coordinated to the surface lead atoms remaining in the lattice of lead sulfide and that the xanthate is different from that in the crystal lattice of bulk lead xanthate. However, it was not possible to determine if the surface complex had a 1:1 or 2:1 stoichiometry. In this regard, it should be noted that, if the lead atom remains in the galena lattice, there is only just sufficient space for a 1:1 stoichiometry. It would be possible to have a 2:1 stoichiometry on every second surface lead atom, but this would seem unlikely.

Leppinen *et al.*[63] carried out an *in situ* FTIR study of the interaction of ethyl xanthate with galena, chalcocite, chalcopyrite, and pyrite particles contained in a carbon paste electrode and compared changes in IR signal intensity with voltammetric behavior. With galena, increase in IR intensity occurred in the same potential range as the voltammetric prewave, thus confirming previous findings. It was also found that a significant amount of xanthate adsorption persisted at potentials below the prewave region. This was attributed to reaction of xanthate with oxidized lead species formed during grinding of the mineral prior to electrode preparation. It was shown[19] that oxidation of galena could result in loss of sulfur as thiosulfate, and subsequent electrochemical reduction would then give rise to lead metal, which interacts with xanthate at a potential ~0.5 V below that for galena.

Leppinen et al.[63] also found a good correlation between FTIR spectra and voltammograms for the ethyl xanthate/chalcocite system, with the spectrum of chemisorbed xanthate being similar to that for bulk copper xanthate.

Figure 19 shows the reflectance spectra reported by Leppinen et al.[63] for the interaction of ethyl xanthate with chalcopyrite. The peaks at 1240 and 1260 cm^{-1} are characteristic of dixanthogen, and the spectra indicate that this species was the initial surface product. Copper xanthate has characteristic peaks near 1190 cm^{-1}, and it can be seen from Fig. 19 that this compound is deposited additionally at the higher potentials applied. This order of product formation is as expected from studies of chalcopyrite oxidation (see Section VII.1). Leppinen et al.[63] found that the development of FTIR intensity correlated with the growth of voltammetric currents. No evidence was reported for adsorption of xanthate at lower potentials than dixanthogen deposition as would be expected from the UV-vis results of Richardson and Walker.[43] The absence of a clearly discernible xanthate spectrum in the potential range expected for chemi-

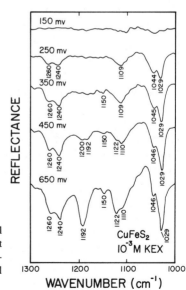

Figure 19. FTIR reflectance spectra of ethyl xanthate adsorbed on chalcopyrite at different potentials in 0.05 mol dm^{-3} sodium tetraborate (pH 9.2) containing 10^{-3} mol dm^{-3} ethyl xanthate. (From Leppinen et al.[63])

sorption could be due to the much lower surface density of xanthate in the chemisorbed layer on chalcopyrite than on other minerals. As pointed out in Section VII.1, the surface composition of chalcopyrite at the relevant potential is expected to have a stoichiometry of ~CuS_2, and hence the number of surface copper sites for xanthate chemisorption is only one-fourth of that in the case of chalcocite.

Dixanthogen was the only surface species identified[63] for ethyl xanthate oxidation on pyrite, with the appearance of FTIR reflectance intensity correlating with the growth of voltammetric currents. Again, no evidence for a chemisorbed xanthate species was obtained. This finding contrasts with that of Mielczarski[70] for the interaction of ethyl xanthate with marcasite, which is polymorphic with pyrite. His *in situ* IR spectroscopic investigation, using the attenuated total reflection (ATR) technique, combined with UV-vis spectroscopy of the solution phase indicated that a monolayer of iron xanthate formed prior to dixanthogen deposition.

Yoon and co-workers studied the interaction of ethyl isopropyl thionocarbamate[71] and isobutyl ethoxycarbonyl thionocarbamate[71–73] with copper and chalcocite and the interaction of dicresyl monothiophosphate[71,74] and diisobutyl monothiophosphinate[71,75] with silver, gold, and silver–gold alloys. Adsorption in these systems could not be determined effectively by voltammetry because no currents could be unequivocally identified with reactions of these complex molecules. The authors overcame this problem by recording FTIR spectra at electrodes of the selected substrates, and this allowed the establishment of the extent of deposition as a function of potential. With the thionocarbamates, the band due to N-H vibration disappeared in the spectra of the adsorbed species, indicating that the adsorption process involves breakage of the N-H bond, that is,

$$Cu + ROCSNHR' \rightarrow Cu(ROCSNR') + H^+ + e^- \qquad (11)$$

The FTIR spectra were found to exhibit shifts in the characteristic absorption bands, indicating that both sulfur and oxygen are involved in the coordination with the copper on the surface. This was interpreted[71–73] in terms of the formation on the copper surface of a six-membered chelate as shown in Fig. 20.

Yoon and co-workers[71,73–75] concluded that thionocarbamate, thiophosphate, and thiophosphinate compounds adsorbed by an EC mechanism. The *electrochemical* (E) step was considered to be the formation of metal ions, and this was followed by the *chemical* (C) step, involving reaction with collector ions to precipitate the metal collector compound.

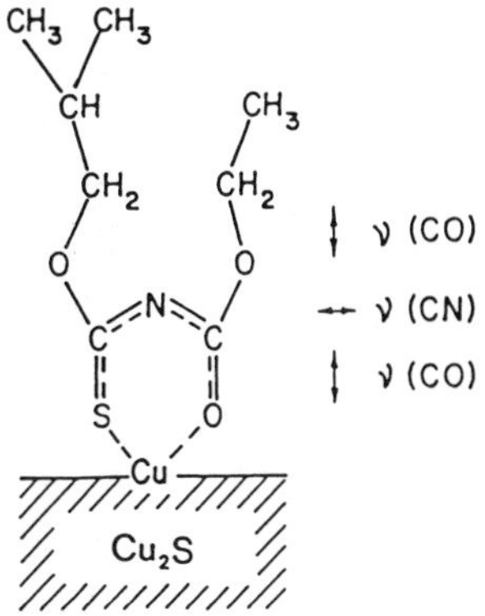

Figure 20. Molecular orientation of thionocarbamate on chalcocite proposed by Mielczarski and Yoon.[72]

It was suggested[71] that an EC mechanism could also be operative in the formation of metal xanthates. Consideration of the characteristics of the metal dissolution reaction indicates that an EC mechanism is not appropriate. For example, the equilibrium Cu^+ concentration at the potential at which the thionocarbamates were found to interact with copper (viz., -0.6 V) is only 10^{-19} mol dm^{-3}, and the rate of formation of Cu^+ must be at most $\sim 10^{-4} \mu A\, cm^{-2}$, taking the exchange current density, i_0 for 1 mol dm^{-3} to be 4 mA cm^{-2}.[76] These values are far too low for any meaningful interaction between thiol and metal to occur by an EC mechanism in a reasonable time frame. Similarly, silver ethyl xanthate forms on silver at -0.04 V,[39] and at this potential the equilibrium Ag^+ concentration is 10^{-14} mol dm^{-3} and its rate of formation is $\sim 10^{-3} \mu A\, cm^{-2}$ (taking i_0 for 1 mol dm^{-3} to be 20 mA cm^{-2}, as given in Ref. 76).

3. Electron Spectroscopy

Electron spectroscopy in its several modes has proved to be particularly powerful for determining the nature of species present on solid surfaces. These methods have been applied extensively to the characterization of SAMs, and they are finding an increasing application to mineral processing systems. Of the various available techniques, X-ray photoelectron spectroscopy (XPS) is particularly appropriate for the study of mineral surfaces because a knowledge of the chemical environment of atoms, in addition to elemental composition, is usually required. Such information is important in identifying surface thiol species on sulfide minerals because the sulfur in the thiol needs to be distinguished from that in the metal sulfide.

Clifford et al.[77] initiated the application of XPS to the characterization of sulfide mineral surfaces in flotation systems. These authors concluded that the adsorption of amyl (pentyl) xanthate and diethyl dithiophosphate on a range of binary and ternary metal sulfides "could be observed with difficulty by following the simultaneous increase in the C, S, and P molar ratios."

Mielczarski, Suoninen, and their co-workers utilized XPS to provide information on the structure of adsorbed xanthates[78–85] and dithiophosphate[86] on metals and metal sulfides. They interpreted their results in terms of the initial distribution of the adsorbed thiol without any special orientation, followed by island formation as the coverage increased and, eventually, a well-ordered monolayer. In subsequent multilayer formation by the metal thiol compound, the orientation of the thiol was considered to become more random. Interestingly, it was noted that other ions preadsorbed on the surface, such as hydroxyl and carbonate, were gradually removed as the xanthate monolayer developed. Only one sulfur environment was observed for surface xanthate species, supporting the conclusion that xanthate is bonded to metal atoms in the surface layer through both sulfur atoms (see Section VII.2). No differences were reported between the binding energies of the thiol in the initial monolayer and those of the bulk thiol compound.

In later studies, Mielczarski and co-workers[54,55] investigated surfaces after treatment with xanthate under potential control. This approach allows better definition of monolayer and multilayer conditions. The spectra observed for copper,[54] chalcocite,[54] and silver[55] electrodes in ethyl xanthate solutions after the potential was held in the prewave region indicated the presence of an additional species containing sulfur, carbon, and oxygen, but the authors considered this surface species not to be xanthate itself but an adsorbed impurity derived from the thiol collector. As pointed out in Sections IV, VII.1, and VII.2, this assignment is not consistent with the results of other studies.

Buckley, Woods, and co-workers have applied XPS to the study of the adsorption of thiols,[35,49,87,88] building on previous investigations of the surface oxidation of sulfide minerals using this technique. In a recent study[35] of the chalcocite/diethyl dithiophosphate system, coverages derived from XPS were compared with those derived by cathodic stripping voltammetry following immersion of chalcocite electrodes in dithiophosphate solutions in equilibrium with air under similar conditions. A good correlation was observed up to monolayer coverage. With multilayers, the XPS results indicated that the coverage was significantly underestimated by voltammetry; this was presumably due to incomplete stripping of the metal thiol compound before reduction of the chalcocite substrate occurred. Both XPS and the electrochemical technique indicated that a monolayer of dithiophosphate was adsorbed following immersion of chalcocite electrodes in 10^{-5} mol dm^{-3} solution for 10 min, whereas a multilayer developed after the same period at higher concentrations. Photoelectron peaks occurred at the same binding energy for the chemisorbed layer and multilayer copper diethyl dithiophosphate and were similar to those reported previously.[86] However, the X-ray-excited copper Auger electron spectrum of the monolayer was found to be similar to that for chalcocite itself and different from that for multilayer copper dithiophosphate (Fig. 21). This behavior indicates that the copper atom environment in the chemisorbed layer is similar to that in chalcocite and different from that in copper dithiophosphate. It was noted[35] that previous reports of Cu(LMM) Auger spectra for xanthate on copper[78] and chalcocite[85] also showed no change in kinetic energy when adsorption occurred, but the authors of those publications had not commented on this property. Also, the Ag(MNN) spectrum in the initial stages of the interaction of xanthate with silver was reported[55] to be the same as that of the substrate, but this was attributed to the involvement of only a small area of the surface. Such an explanation could not account for the results under all the conditions investigated. It must be concluded that the absence of a shift in the Auger kinetic energy is an intrinsic characteristic of these systems.

A similar correspondence of the electronic environment of the metal atom bonded to a chemisorbed thiol with that in the substrate, and different from that in the bulk compound, was observed for xanthate adsorption on galena.[49,87,88] In these investigations, galena crystals were cleaved under xanthate solution to ensure that no oxidation occurred prior to interaction between the mineral surface and the collector. For ethyl,[49,87] butyl,[88] and amyl xanthates,[87] the Pb($4f$) spectrum was indistinguishable from that of galena itself, whereas that for the bulk lead xanthate was shifted by 1 eV to higher binding energies.

The conclusion reached in Refs. 49 and 87 regarding the Pb($4f$) spectrum was disputed by Laajalehto et al.[84] These authors reported that a small additional Pb($4f$) component was observed when xanthate was adsorbed at the submonolayer level and considered that the adsorbed

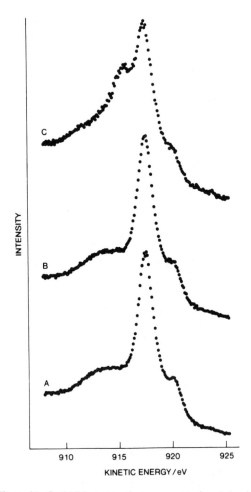

Figure 21. Cu(*LMM*) Auger electron spectra for chalcocite
abraded under nitrogen (A) and under 0.05 mol dm^{-3} sodium
tetraborate containing 10^{-5} mol dm^{-3} (B) or 5×10^{-5} mol dm^{-3}
(C) diethyl dithiophosphate and left immersed for 10 min. (From
Buckley and Woods.[35])

species was molecular lead xanthate rather than xanthate attached to lead atoms retained in the galena lattice. They considered that adsorption was facilitated by the conformation of the lead xanthate, with donor bonding between xanthate sulfur atoms and lead atoms and between xanthate lead atoms and sulfur atoms in the galena surface. Additional arguments in

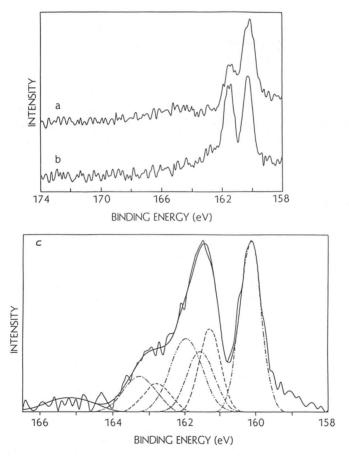

Figure 22. S(2p) photoelectron spectra from abraded galena surfaces maintained for 5 min in 0.05 mol dm^{-3} sodium tetraborate (pH 9.2) containing 10^{-3} mol dm^{-3} n-butyl xanthate at –0.010 V (a) and 0.2 V (b). The fitting of spectrum (b) is shown in (c). (From Shchukarev et al.[88])

favor of this model were presented by Nowak[47] and were discussed in Section V.

Such a model is not consistent with the observations[88] that xanthate forms a monolayer at underpotentials on lead as well as galena[24] and that the difference in potential between monolayer and multilayer formation, and hence the adsorption energy, is similar for both surfaces. Furthermore, subsequent XPS studies[88] on galena using an electron takeoff angle of 15° confirmed that no shifted $Pb(4f)$ component was discernible when the xanthate coverage was a monolayer or less.

The XPS study of the galena/butyl xanthate system[88] included investigations of surfaces treated under potential control. This approach more precisely defines the chemisorbed and metal xanthate states. At underpotentials to lead xanthate formation, the $S(2p)$ spectrum displayed an additional doublet shifted by ~1.5 eV from that of the substrate (Fig. 22) and no shifted $Pb(4f)$ component. After the potential was held above the reversible value of the $PbS/Pb(C_4H_9OCS_2)_2$ couple, two additional doublets were evident in the $S(2p)$ spectrum, shifted by 1.35 and 1.8 eV (Fig. 22), as well as a $Pb(4f)$ component shifted by ~1 eV. The two additional $S(2p)$ doublets were assigned to chemisorbed xanthate and lead xanthate, respectively.

VIII. CORRELATION OF ADSORPTION AND SURFACE HYDROPHOBICITY

The adsorption of thiols can be characterized by measurements of surface hydrophobicity. Wark and co-workers[5,8,10] refined methods for the determination of the three-phase contact angle at the gas–solid–solution interface and the application of contact angle measurements to the prediction of the flotation response of minerals when treated with collector solutions that simulate flotation conditions. The maximum contact angle observed for a particular collector was shown to be related to the length of the alkyl chain and to the presence or absence of chain branching, rather than to the identity of the head group responsible for attachment to the surface. Contact angle measurements have subsequently become established as an effective approach for the characterization of SAMs and to identify full monolayer coverage.[51,52]

The use of contact angle measurements to complement electrochemical techniques for the study of adsorption of thiols at metals and metal sulfides was first applied to the interaction of methyl, ethyl, and butyl xanthates with platinum,[89] gold,[89] pyrite,[23] and galena.[23] Dixanthogen had been identified electrochemically as the adsorbate on the first three surfaces, and the maximum contact angles observed were consistent with this assignment. The chemisorbed xanthate layer on galena was found to give rise to finite contact angles, with the value for a monolayer approaching that characteristic of the particular alkyl group. At the xanthate concentration employed, the angle observed when a multilayer was formed was indicative of the presence of dixanthogen in the surface layer.

Contact angle measurements were also correlated with voltammetry by Chander and Fuerstenau in their study of the interaction of diethyl dithiophosphate with copper and chalcocite surfaces.[33] Finite angles were observed when the potential was in the region of stability of copper dithiophosphate, but detailed information on the wettability of chemisorbed dithiophosphate was not reported. Recent studies[90] of this system established that the contact angle becomes finite when the chemisorbed species is present; an angle of ~60° was observed for a monolayer, and this is the characteristic value for the ethyl group.

Yoon and co-workers[38–40,54,71,73–75,90] have determined the contact angle as a function of potential for a range of systems. Figure 23 shows the values recorded for silver, gold, and silver–gold alloys.[40] Comparison with Fig. 11 shows that the development of hydrophobicity coincides with the chemisorption of xanthate onto silver sites in the pure metal and alloy surfaces. The potential at which the gold surface displays a finite contact angle is close to that at which dixanthogen is formed.

Contact angle measurements are particularly useful in situations in which it is not possible to unequivocally define adsorption using electrochemical techniques. For example, it is difficult to identify currents due to the interaction of dicresyl monothiophosphate (DCMTP) with silver on voltammograms, although there were indications of an anodic current rise above 100 mV versus SHE.[74] The contact angle measurements presented in Fig. 24 show that adsorption of DCMTP commenced just below 0 V.[74] It can also be seen from Fig. 24 that there is a good correlation between contact angle and FTIR relative signal intensities.

It has been demonstrated[90] that the potential at which the contact angle measurements indicate that abraded mineral surfaces become hydrophobic in the presence of a thiol can be somewhat higher than the value at which chemisorption commences. It was considered that this reflects the presence of a significant induction time in establishing a captive bubble on such a surface. Energy barriers in particle-to-bubble attachment

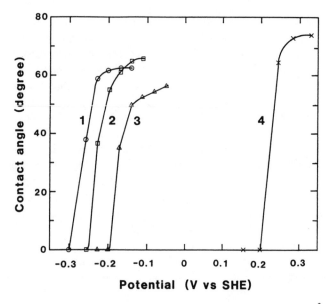

Figure 23. Dependence of contact angle on potential in 0.05 mol dm^{-3} sodium tetraborate (pH 9.2) containing 10^{-3} mol dm^{-3} ethyl xanthate for silver (1), 50:50 wt. % Ag–Au alloy (2), 20:80 wt. % Ag–Au alloy (3), and gold (4). (From Woods et al.[40])

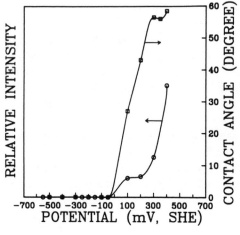

Figure 24. Potential dependence of the contact angle and FTIR relative signal intensity for a silver surface in a solution containing 10^{-4} mol dm^{-3} dicresyl monothiophosphate, 0.5 mol dm^{-3} acetic acid, and 0.5 mol dm^{-3} sodium acetate (pH 4.6). (From Basilio et al.[74])

are known to be important in flotation[91,92] and could inhibit interaction between a relatively large bubble and a flat mineral surface that is only partially covered with collector, since such a surface will contain both hydrophilic and hydrophobic sites. The situation is exacerbated by roughness of the surface. It was suggested that flotation recovery is a more meaningful measure of surface hydrophobicity than contact angle.

Flotation recovery is clearly the most appropriate measure of the efficacy of a flotation collector, since the purpose of the flotation process is the efficient recovery of minerals. The concept of combining floatability with electrochemical measurements was established with a particulate bed electrode composed of gold spheres.[89,93] These studies showed that, as the hydrocarbon chain length of the alkyl xanthate was lengthened, the coverage of xanthate required to float the gold decreased; it changed from about a monolayer for methyl xanthate to less than a fractional coverage of 0.1 for the amyl homolog. Analogous studies on galena particulate bed electrodes[23,93] indicated that less than a monolayer of chemisorbed ethyl or butyl xanthate was sufficient to induce effective flotation.

A recent FTIR study[66] has shown that ethyl xanthate is chemisorbed on gold at potentials slightly below that at which dixanthogen begins to deposit. Flotation of gold spheres with ethyl xanthate was found[89,93] to commence at about 0.02 V below the reversible potential of the xanthate/dixanthogen couple. This would indicate that the initiation of flotation results from the chemisorption process. For more rapid flotation, requiring a higher surface coverage, dixanthogen plays a supporting role.

The determination of flotation recovery using particulate sulfide mineral bed electrodes under potential control was refined by Richardson and co-workers and applied to the study of the interaction of ethyl xanthate with a range of sulfide minerals.[29,42,43,53] The approach introduced by these authors was to employ relatively large particles (590–840 μm) and to keep the bed under positive pressure during potential conditioning. With this technique, they were able to combine flotation recovery determinations with voltammetry and with UV-vis spectroscopy of the solution phase.

A different method for the determination of flotation recovery as a function of potential was conceived by Trahar and co-workers.[94–97] These authors applied a modified laboratory flotation cell in which flotation was carried out with nitrogen, and the potential, monitored with a platinum electrode inserted in the pulp, was adjusted by the controlled addition of oxidizing or reducing agents. This experimental technique had the disadvantage, compared with that of Richardson and co-workers, that complementary electrochemical investigations on the mineral bed were not available. On the other hand, it had the advantage that the particle size was the same as in conventional flotation recovery determinations, and hence the recovery data are more readily correlated with other flotation information and with plant practice. Yoon and co-workers[27,31,90] applied a variation on the Trahar concept; these authors employed a microflotation cell, with the potential controlled with redox reagents.

The flotation of chalcocite with ethyl xanthate has been determined by each of the above groups,[31,53,94] and there is good agreement among the reported potential dependences of flotation. It was noted[29,31] that the onset of flotation was in the potential region of xanthate chemisorption. In later studies[27] of the chalcocite/ethyl xanthate system, the flotation recovery was compared with the chemisorption isotherm (Fig. 25). It can be seen that significant flotation occurs at potentials at which the coverage is low, that 50% recovery corresponds to a fractional coverage of <0.2,

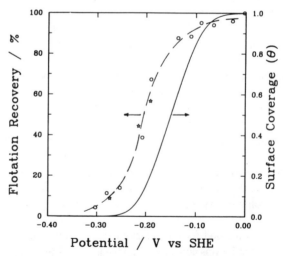

Figure 25. Comparison of flotation recovery after 1 min of chalcocite in 0.05 mol dm^{-3} sodium tetraborate (pH 9.2) containing 10^{-5} mol dm^{-3} ethyl xanthate with the adsorption isotherm for chemisorbed xanthate at this concentration. (From Woods *et al.*[27])

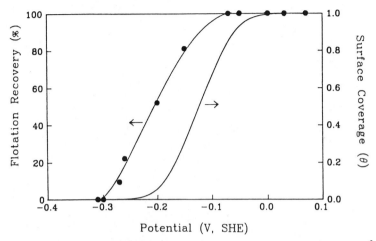

Figure 26. Comparison of flotation recovery after 1 min of chalcocite in 0.05 mol dm^{-3} sodium tetraborate (pH 9.2) containing 10^{-5} mol dm^{-3} diethyl dithiophosphate with the adsorption isotherm for chemisorbed xanthate at this concentration. (From Woods *et al.*[90])

and that 90% recovery corresponds to about half-coverage. Analogous investigations were carried out on the chalcocite/diethyl dithiophosphate system.[90] Figure 26 shows a comparison of flotation recovery with the adsorption isotherm. It can be seen that, as with ethyl xanthate, flotation commences at potentials corresponding to very low surface coverage, that 50% flotation corresponds to a fractional surface coverage of <0.1, and that ~90% recovery is observed at a fractional coverage of about one-half.

Figure 27 presents the flotation recovery curves of Richardson and Walker[43] for chalcocite, bornite, chalcopyrite, and pyrite. The results for chalcocite are similar to those shown in Fig. 25. The onset of flotation of bornite and chalcopyrite was found[43] to coincide with the potentials at which UV-vis spectroscopy showed xanthate to begin to be abstracted from solution. This indicated that attachment of xanthate to the surface was responsible for inducing flotation for both minerals. As pointed out in Section VII, these potentials are below the values at which copper xanthate or dixanthogen are formed but correspond to values at which chemisorption is expected.

A different conclusion regarding the flotation of chalcopyrite was reached by Leppinen *et al.*[63] based on *in situ* FTIR spectroscopic studies.

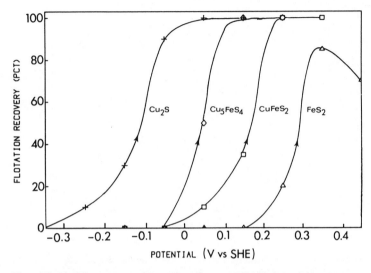

Figure 27. Flotation recovery after 2 min as a function of conditioning potential at which the bed was held for 10 min; 0.05 mol dm^{-3} sodium tetraborate (pH 9.2) containing 1.44 $\times 10^{-5}$ mol dm^{-3} ethyl xanthate for chalcocite and $2(\pm 0.1) \times 10^{-5}$ mol dm^{-3} ethyl xanthate for bornite, chalcopyrite, and pyrite. (From Richardson and Walker.[43])

These authors did not discern any spectra identifiable with xanthate at potentials below the reversible value of the xanthate/dixanthogen couple (see Fig. 19) whereas Fig. 27 indicates that flotation commences at ~0.2 V below this value. Leppinen *et al.* suggested that the flotation observed by Richardson and Walker[43] was induced by oxidation of the mineral surface itself rather than attachment of collector. Such *self-induced flotation* due to surface oxidation had been established for chalcopyrite[98-100] and found to commence at ~0.1 V. However, Richardson and Walker did not observe any flotation in the absence of collector. This can be explained by the relatively large particle size that they employed; Senior and Trahar[101] found self-induced flotation to diminish rapidly at pH 9 as the particle size exceeded 100 μm. In order for the mechanism put forward by Leppinen *et al.*[63] to be correct, the lack of self-induced flotation when large particles are employed would need to arise from inhibition by a surface species and the action of xanthate would have to be one of removing this species. Trahar and co-workers[101,102] have established that the presence of metal hydroxides inhibits self-induced flotation and suggested that this is the reason that chalcopyrite and some other sulfides do not float from complex ores in the absence of a collector under conditions in which the isolated minerals would respond efficiently. The addition of ethyl xanthate to such systems induced flotation, and it was concluded[101,102] that xanthate can have the dual role of increasing the hydrophobicity of the sulfide surface and of counteracting the hydrophilic effect of hydroxides. The latter was considered to occur through precipitation of a metal xanthate. If this is the case, then FTIR investigations of stationary chalcopyrite beds should still detect xanthate on the mineral surface at the potential of the flotation edge. An alternative explanation of the dual role is that the development of hydrophobicity through xanthate chemisorption results in the rejection of hydrophilic hydroxy species from the surface, but, again, xanthate should be present on the chalcopyrite surface at the flotation potential. Thus, it is suggested that the flotation of chalcopyrite with ethyl xanthate is induced by the chemisorption of xanthate and that the coverage required is below that detectable by FTIR spectroscopy at the sensitivity level of the technique employed by Leppinen *et al.*[63]

Guy and Trahar[96,97] found the potential dependence of the flotation of galena to vary with the mode of pretreatment of the mineral. When galena was ground under reducing conditions, the mineral was found to begin to float at potentials at which chemisorption commences. On the other hand, galena floated effectively at potentials as low as -0.5 V after grinding under conditions in which oxidation of the mineral could occur. It was concluded that sulfur had been lost from the galena surface during grinding under these conditions and that the excess lead had reacted with xanthate to form a lead xanthate surface. The flotation behavior was found to correlate with previously published voltammetry[19] for a preoxidized galena surface. The potential dependence of flotation of sulfidic lead ores in a pilot plant operation has been found[103,104] to correspond to that observed by Guy and Trahar for grinding under reducing environments. Thus, it would appear that this is the relationship more relevant to flotation practice.

The flotation recovery data of Guy and Trahar[96,97] have been correlated[50] with xanthate coverages determined for the same solution composition. It can be seen from Fig. 28 that significant flotation recoveries are realized at potentials at which the coverage of collector is very small and that maximum recovery occurs with little more than half-coverage.

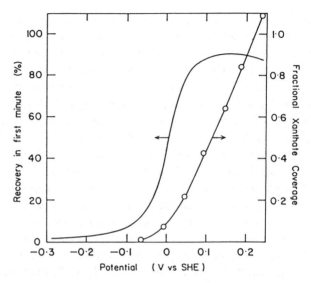

Figure 28. Comparison of the potential dependence of flotation recovery (data of Guy and Trahar[96]) and of xanthate coverage for galena in 2.3×10^{-5} mol dm^{-3} ethyl xanthate solution of pH 8. (From Buckley and Woods.[49])

IX. Eh–pH DIAGRAMS INCLUDING CHEMISORBED STATES

Electrochemical phase (Eh–pH) diagrams have proved to be a most convenient and valuable method for the presentation of free energy data in relation to mineral processing. This method of expressing thermodynamic information was conceived by Pourbaix[105] and first applied to mineral systems by Garrels and Christ.[106] The latter authors directed their attention to natural systems of geological interest and geochemical concern. The typical approach taken in constructing Eh–pH diagrams is appropriate to such applications but has a number of limitations when mineral processing is considered. First, the total system is usually taken to be essentially infinite whereas, in mineral processing, the system is closed, with mass balance being maintained during the various reactions that take place. Yoon and co-workers[27,31,60,107–110] have presented Eh–pH diagrams for systems relevant to flotation that take mass balance considerations into account. Second, for closed systems with multielement components, equilibria depend on elemental ratios, and then more than one diagram is required. It was shown[60] that separate diagrams were required to depict the possible reactions of chalcocite, djurleite, anilite, and covellite in the copper/sulfur/water system. Wadsley[111] calculated equilibria for the iron/sulfur/water system using the CSIRO-Monash Thermochemistry System and demonstrated that such systems may be described by multiple diagrams defined by stoichiometric constraints.

A third limitation of conventional diagrams is that they portray only stable phases whereas, in the relatively short time scales operative in mineral processing, kinetic limitations may allow the development of metastable phases. This situation is particularly relevant to the reactions of sulfide minerals owing to the irreversibility of sulfate formation. Peters[112,113] devised Eh–pH diagrams for sulfide mineral systems that took account of the irreversibility of sulfate formation by "destabilizing" sulfate by 300 kJ mol^{-1}. This shift in free energy for metastable sulfate was introduced to account for hydrometallurgical observations. Other workers[27,60,110] have adopted the approach of excluding oxygen–sulfur anions and hence allowing metastable sulfur species to occupy the sulfate stability region.

A fourth limitation of the conventional approach to constructing Eh–pH diagrams that is relevant to the flotation situation is that the formation of chemisorbed species is ignored. Clearly, since chemisorbed species form at underpotentials to bulk species and induce flotation, then Eh–pH diagrams that include only bulk species cannot predict the potential at which flotation commences. Fortunately, a stability region for the chemisorbed species can be included on the Eh–pH diagram; this is possible since the chemisorption isotherm is a thermodynamic expression. Eh–pH diagrams that include chemisorption coverages derived from the isotherm have been constructed for the copper/water/ethyl xanthate,[27] chalcocite/water/ethyl xanthate,[27] silver/water/ethyl xanthate,[38,39] and silver–gold alloy/water/ethyl xanthate[38] systems.

Figure 29 presents Eh–pH diagrams for the chalcocite/water/ethyl xanthate system.[27] In constructing these diagrams, a free energy for chemisorbed xanthate was used so that it corresponded to the onset of adsorption, which was defined as the potential at which the fractional coverage given by the appropriate isotherm was 0.01. This allowed the stability domains of the various other species to be determined. In defining the stability zone of chemisorbed xanthate, it was assumed that this species would only exist in regions in which copper or a copper sulfide is stable. Dashed lines have been included in Fig. 29 that correspond to fractional

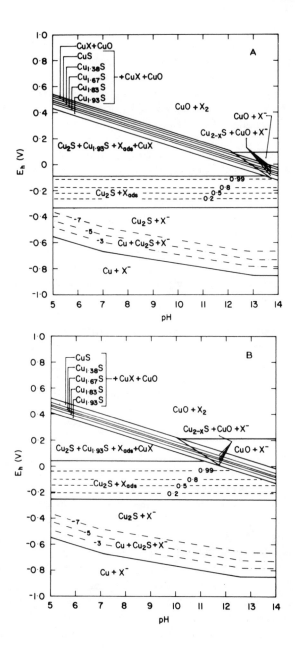

Figure 29. Eh–pH diagrams for the chalcocite/water/ethyl xanthate system for an initial xanthate concentration of 10^{-3} mol dm^{-3} (A) and 10^{-5} mol dm^{-3} (B). Sulfur–oxygen anions are not considered. Species designated $Cu_{2-x}S$ cover the same range of stoichiometries identified in the acid region. $---$, Fractional coverages of chemisorbed xanthate as indicated; $-\,-\,-$, copper coverages marked in log mol. (From Woods et al.[27])

coverages of chemisorbed xanthate of 0.2, 0.5, and 0.99. These coverages were calculated from the isotherm, with mass balance considerations taken into account.

The Eh–pH diagrams shown in Fig. 29 portray the region of chemisorption as extending nearly 0.3 V below the stability zone of copper(I) xanthate. At high Eh values, chalcocite is oxidized to copper sulfides of progressively lower copper content.[60] Chemisorbed xanthate will not coexist with these sulfides when CuO is formed since the oxide is expected to overlay the surface of the mineral. This implies that there is an upper potential limit to flotation of chalcocite with ethyl xanthate, and such behavior has been established for this system.[31,94]

X. CONCLUDING REMARKS

A significant degree of understanding has now been established on the nature and properties of chemisorbed layers formed by the interaction of the thiols used as flotation collectors with metal and metal sulfide surfaces. In the systems in which chemisorption has been shown to occur, the chemisorbed layer is the initial product, and its presence induces flotation. Hence, the chemisorbed species is the most important chemical entity in the interaction of flotation collectors with minerals in these systems. It must also be pointed out that chemisorption provides the most efficient distribution of collector on a mineral surface because it extends in a single molecular layer. On the other hand, metal thiol compounds and disulfides are expected to be formed by a nucleation and growth mechanism and could develop through multilayer islands. Thus, it would appear that the search for new collectors should seek to identify species that chemisorb at significant underpotentials to metal compound and dimer formation.

The work reviewed here also establishes that the interaction of collectors with minerals follows an electrochemical mechanism. One objective of research and development in flotation is to apply electrochemical measurements to the monitoring and control of flotation in practice. The advancement of understanding of the adsorption of thiol collectors and of how adsorption and the resultant wettability of the mineral surface are affected by a range of operational parameters is essential for success in this area.

Knowledge of the nature of adsorbed thiol species should become enhanced by the application of additional techniques to those discussed in this chapter. For example, static secondary-ion mass spectrometric (SIMS) imaging has begun to provide information[114] on the distribution of collector species adsorbed on different faces of mineral particles. Scanning tunneling microscopy and related techniques are now providing the opportunity of viewing atom positions in surface layers. These techniques are being applied to understanding mineral surfaces and have the capacity to provide information that elucidates the structure of chemisorbed thiol layers.

REFERENCES

[1] G. Blainey, in *Minerals of Broken Hill*, Ed. by H. K. Worner and R. W. Mitchell, AM&S Ltd., Melbourne, Australia, 1982, p. 12.
[2] C. H. Keller and C. P. Lewis, U.S. Patent 1,554,216; 1,554,220 (1925).
[3] M. G. Fleming and J. A. Kitchener, *Endeavour* **24** (1965) 101.
[4] A. M. Gaudin, AIME Technical Publication No. 4, 1927.
[5] I. W. Wark and A. B. Cox, *Trans. AIME* **112** (1934) 189.
[6] A. F. Taggart, T. C. Taylor, and A. F. Knoll, AIME Technical Publication No. 312, 1930.
[7] A. F. Taggart, G. R. M. del Giudice, and O. A. Ziehl, *Trans. AIME* **112** (1934) 348.
[8] I. W. Wark, Report of the Australia and New Zealand Association for the Advancement of Science, 25th Meeting, Adelaide, 1946, pp. 23–51.
[9] A. F. Taggart and M. D. Hassialis, *AIME, Min. Technol.* **10**(5), Technical Publication No. 2078 (1946).
[10] K. L. Sutherland and I. W. Wark, *Principles of Flotation*, Aus IMM, Melbourne, Australia, 1955.
[11] M. A. Cook and J. C. Nixon, *J. Phys. Colloid Chem.* **54** (1950) 445.
[12] J. C. Nixon, in *Proceedings of the 2nd International Congress on Surface Activity*, Vol. 3, Butterworths, London, 1957, p. 369.
[13] I. N. Plaksin and S. V. Bessonov, in *Proceedings of the 2nd International Congress on Surface Activity*, Vol. 3, Butterworths, London, 1957, p. 361.
[14] R. Woods, in *Principles of Mineral Flotation, The Wark Symposium*, Ed. by M. H. Jones and J. T. Woodcock, Aus IMM, Melbourne, Australia, 1984, pp. 91–115.
[15] R. Woods, *J. Phys. Chem.* **75** (1971) 354.
[16] P. E. Richardson and E. Maust, Jr., in *Flotation, A. M. Gaudin Memorial Volume*, Vol. 1, Ed. by M. C. Fuerstenau, AIME, New York, 1963, pp. 364–392.
[17] R. L. Paul, M. J. Nicol, J. W. Diggle, and A. P. Saunders, *Electrochim. Acta* **23** (1978) 625.
[18] J. R. Gardner and R. Woods, *J. Electroanal. Chem.* **100** (1979) 447.
[19] R. Woods, *Aust. J. Chem.* **25** (1972) 2329.
[20] A. N. Buckley and R. Woods, *Appl. Surf. Sci.* **17** (1984) 401.

[21] A. N. Buckley, I. M. Kravets, A. V. Shchukarev, and R. Woods, *J. Appl. Electrochem.* **24** (1994) 513.

[22] P. E. Richardson and C. S. O'Dell, *J. Electrochem. Soc.* **132** (1985) 1350.

[23] J. R. Gardner and R. Woods, *Aust. J. Chem.* **30** (1977) 981.

[24] M. D. Horne, M.Sc. Thesis, University of Melbourne, 1989.

[25] R. Woods, in *Copper 87*, Ed. by E. A. Mullar, G. Gonzalez, and C. Barama, Vol. 2, Mineral Processing and Control, University of Chile, Santiago, 1988, pp. 121–135.

[26] Z. Szeglowski, J. Czarnecki, A. Kowal, and A. Pomianowski, *Trans IMM* **86** (1977) C115.

[27] R. Woods, C. A. Young, and R.-H. Yoon, *Int. J. Miner. Process.* **30** (1990) 17.

[28] A. Kowal and A. Pomianowski, *J. Electroanal. Chem.* **46** (1973) 411.

[29] C. S. O'Dell, R. K. Dooley, G. W. Walker, and P. E. Richardson, in *Proceedings of the International Symposium on Electrochemistry in Mineral and Metal Processing*, Ed. by P. E. Richardson, S. Srinivasan, and R. Woods, Proceedings Vol. 84-10, The Electrochemical Society, Pennington, New Jersey, 1984, pp. 81–95.

[30] C. S. O'Dell, G. W. Walker, and P. E. Richardson, *J. Appl. Electrochem.* **16** (1986) 544.

[31] C. I. Basilio, M. D. Pritzker, and R.-H. Yoon, SME–AIME Annual Meeting, New York, 1985, Preprint No. 85-86, 10 pp.

[32] D. F. A. Koch and R. McIntyre, *J. Electroanal. Chem.* **71** (1976) 285.

[33] S. Chander and D. W. Fuerstenau, *J. Electroanal. Chem.* **56** (1974) 217.

[34] A. N. Buckley and R. Woods, in *Proceedings of the International Symposium on Electrochemistry in Mineral and Metal Processing III*, Ed. by R. Woods and P. E. Richardson, Proceedings Vol. 92-17, The Electrochemical Society, Pennington, New Jersey, 1992, pp. 269–285.

[35] A. N. Buckley and R. Woods, *J. Electroanal. Chem.* **357** (1993) 387.

[36] A. N. Buckley and R. Woods, *Colloids Surf.* **59** (1991) 307.

[37] P. Talonen, J. Rastas, and J. Leppinen, *Surf. Interface Anal.* **17** (1991) 669.

[38] R. Woods, C. I. Basilio, D. S. Kim, and R.-H. Yoon, in *Proceedings of the International Symposium on Electrochemistry in Mineral and Metal Processing III*, Ed. by R. Woods and P. E. Richardson, Proceedings Vol. 92-17, The Electrochemical Society, Pennington, New Jersey, 1992, pp. 129–145.

[39] R. Woods, C. I. Basilio, D. S. Kim, and R.-H. Yoon, *J. Electroanal. Chem.* **328** (1992) 179.

[40] R. Woods, C. I. Basilio, D. S. Kim, and R.-H. Yoon, *Colloids Surf.* **83** (1994) 1.

[41] R. Woods, in *Flotation, A. M. Gaudin Memorial Volume*, Vol. 1, Ed. by M. C. Fuerstenau, AIME, New York, 1963, pp. 298–333.

[42] J. B. Zachwieja, G. W. Walker, and P. E. Richardson, *Min. Metall. Process.* **4** (1987) 146.

[43] P. E. Richardson and G. W. Walker, in *XVth International Mineral Processing Congress*, Cannes, France, Vol. II, GEDIM, St. Etienne, France, pp. 198–210.

[44] A. N. Buckley, I. C. Hamilton, and R. Woods, *J. Appl. Electrochem.* **14** (1984) 63.

[45] J. W. Schultze, in *Proceedings of the Third Symposium on Electrode Processes*, 1979, Ed. by S. Bruckenstein, J. D. E. McIntyre, B. Miller, and E. Yeager, Proceedings Vol. 80-3, The Electrochemical Society, Princeton, New Jersey, 1980, pp. 167–189.

[46] J. L. White, R. L. Orr, and R. Hultgren, *Acta Metall.* **5** (1957) 747.

[47] P. Nowak, *Colloids Surf.* **76** (1993) 65.

[48] J. O. Leppinen, Ph.D. Thesis, University of Turku, Finland, 1986.

[49] A. N. Buckley and R. Woods, *Colloids Surf.* **53** (1991) 33.

[50] A. N. Buckley and R. Woods, *Colloids Surf.* **89** (1994) 71.

[51] A. Ulman, *Ultrathin Organic Films, from Langmuir Blodgett to Self-Assembly*, Academic Press, San Diego, 1991.

[52] L. H. Dubois and R. G. Nuzzo, *Annu. Rev. Phys. Chem.* **43** (1992) 437.

[53] P. E. Richardson, J. V. Stout III, C. L. Proctor, and G. W. Walker, *Int. J. Miner. Process.* **12** (1984) 73.

[54] J. A. Mielczarski, J. B. Zachwieja, and R.-H. Yoon, Society of Mining Engineers Annual Meeting, Salt Lake City, Utah, 1990, Preprint no. 90-174.

[55] J. O. Leppinen, R.-H. Yoon, and J. A. Mielczarski, *Colloids Surf.* **61** (1991) 189.

[56] R. Woods, C. I. Basilio, D. S. Kim, and R.-H. Yoon, *Int. J. Miner. Process.* **42** (1994) 215.

[57] M. H. Jones and J. T. Woodcock, *Int. J. Miner. Process.* **5** (1978) 285.

[58] J. R. Gardner and R. Woods, *Int. J. Miner. Process.* **6** (1979) 1.

[59] A. N. Buckley and R. Woods, *Aust. J. Chem.* **37** (1934) 2403.

[60] R. Woods, R.-H. Yoon, and C. A. Young, *Int. J. Miner. Process.* **20** (1987) 109.

[61] R. G. Greenler, *J. Phys. Chem.* **66** (1962) 879.

[62] J. Leja, L. H. Little, and G. W. Poling, *Trans. IMM* **72** (1963) 414.

[63] J. O. Leppinen, C. I. Basilio, and R.-H. Yoon, *Int. J. Miner. Process.* **26** (1989) 259.

[64] A. Ihs, K. Uvdal, and B. Liedberg, *Langmuir* **9** (1993) 733.

[65] N.-O. Persson, K. Udval, B. Liedberg, and M. Hellsten, *Prog. Colloid Polym. Sci.* **88** (1992) 100.

[66] R. Woods, D. S. Kim, C. I. Basilio, and R.-H. Yoon, *Colloids Surf.* **92** (1995) 67.

[67] J. O. Leppinen and J. K. Rastas, *Colloids Surf.* **20** (1986) 259.

[68] J. Leppinen and J. Mielczarski, *Int. J. Miner. Process.* **18** (1986) 3.

[69] J. O. Leppinen, Society of Mining Engineers Annual Meeting, Denver, Colorado, 1987, Preprint no. 87-73.

[70] J. Mielczarski, *Colloids Surf.* **17** (1986) 251.

[71] R.-H. Yoon and C. I. Basilio, in *XVIII International Mineral Processing Congress*, Sydney, Australia, 23–28 May 1993, Vol. 3, The Australian Institute of Mining and Metallurgy, Parkville, Australia (1993), pp. 611–617.

[72] J. A. Mielczarski and R.-H. Yoon, *Langmuir* **7** (1991) 101.

[73] C. I. Basilio and R.-H. Yoon, in *Proceedings of the International Symposium on Electrochemistry in Mineral and Metal Processing III*, Ed. by R. Woods and P. E. Richardson, Proceedings Vol. 92-17, The Electrochemical Society, Pennington, New Jersey, 1992, pp. 79–94.

[74] C. I. Basilio, D. S. Kim, R.-H. Yoon, and D. R. Nagaraj, *Miner. Eng.* **5** (1992) 397.

[75] C. I. Basilio, D. S. Kim, R.-H. Yoon, J. O. Leppinen, and D. R. Nagaraj, SME–AIME Annual Meeting, Phoenix, Arizona, Preprint 92-174.

[76] P. Cavallotti, D. Colombo, U. Ducati, and A. Piotti, in *Proceedings of the Symposium on Electrodeposition Technology, Theory and Practice*, Ed. by L. T. Romankiw and D. R. Turner, Proceedings Vol. 87-17, The Electrochemical Society, Pennington, New Jersey, 1987, pp. 429–448.

[77] R. K. Clifford, K. L. Purdy, and J. D. Miller, *AIChE Symp. Ser.* **71** (1975) 138.

[78] J. Mielczarski, F. Werfel, and E. Suoninen, *Appl. Surf. Sci.* **17** (1983) 160.

[79] J. Mielczarski and E. Suoninen, *Surf. Interface Anal.* **6** (1984) 34.

[80] L.-S. Johansson, J. Juhanoja, K. Laajalehto, E. Suoninen, and J. Mielczarski, *Surf. Interface Anal.* **9** (1986) 501.

[81] J. Mielczarski, *J. Colloid Interface Sci.* **120** (1987) 201.

[82] J. Mielczarski, E. Suoninen, L.-S. Johansson, and K. Laajalehto, *Int. J. Miner. Process.* **26** (1989) 181.

[83] K. Laajalehto, P. Nowak, A. Pomianowski, and E. Suoninen, *Colloids Surf.* **57** (1991) 319.

[84] K. Laajalehto, P. Nowak, and E. Suoninen, *Int. J. Miner. Process.* **37** (1993) 123.

[85] J. Mielczarski and R.-H. Yoon, in *Proceedings of the Engineering Foundation Conference on Flocculation and Dewatering*, Palm Coast, Florida, January 10–15, 1988.

[86] J. Mielczarski and E. Minni, *Surf. Interface Anal.* **6** (1984) 221.

[87] A. N. Buckley and R. Woods, *Int. J. Miner. Process.* **28** (1990) 301.

[88] A. V. Shchukarev, I. M. Kravets, A. N. Buckley, and R. Woods, *Int. J. Miner. Process.* **44** (1994) 99.

[89] J. R. Gardner and R. Woods, *Aust. J. Chem.* **27** (1974) 2139.

[90]R. Woods, C. I. Basilio, D. S. Kim, and R.-H. Yoon, *Int. J. Miner. Process.* **39** (1993) 101.

[91]R.-H. Yoon, in *XVII International Mineral Processing Congress*, Dresden, 1991, Vol II, pp. 17–31.

[92]J. S. Laskowski, Z. Xu, and R.-H. Yoon, in *XVII International Mineral Processing Congress*, Dresden, 1991, Vol II, pp. 237–249.

[93]J. R. Gardner and R. Woods, *Aust. J. Chem.* **26** (1973) 1635.

[94]G. W. Heyes and W. J. Trahar, *Int. J. Miner. Process.* **6** (1979) 229.

[95]W. J. Trahar, in *Principles of Mineral Flotation, The Wark Symposium*, Ed. by M. H. Jones and J. T. Woodcock, Aus IMM, Melbourne, Australia, 1984, pp. 117–136.

[96]P. J. Guy and W. J. Trahar, *Int. J. Miner. Process.* **12** (1984) 15.

[97]P. J. Guy and W. J. Trahar, in *Flotation of Sulphide Minerals, Developments in Mineral Processing*, No. 6, Ed. by K. S. E. Forssberg, Elsevier, Amsterdam, 1985, pp. 91–110.

[98]G. W. Heyes and W. J. Trahar, *Int. J. Miner. Process.* **4** (1977) 317.

[99]J. R. Gardner and R. Woods, *Int. J. Miner. Process.* **6** (1979) 1.

[100]G. H. Luttrell and R.-H. Yoon, *Colloids Surf.* **12** (1984) 239.

[101]G. D. Senior and W. J. Trahar, *Int. J. Miner. Process.* **33** (1991) 321.

[102]L. K. Shannon and W. J. Trahar, in *Advances in Mineral Processing, 2* Ed. by P. Somasundaran, SME/AIME, New York, 1986, pp. 408–425.

[103]N. W. Johnson, A. Jowett, and G. W. Heyes, *Trans. IMM* **91** (1982) C32.

[104]S. R. Grano, J. Ralston, and R. Smart, *Int. J. Miner. Process.* **30** (1990) 69.

[105]M. Pourbaix, *Atlas d'Equilibres Electrochimiques*, Gauthier-Villars, Paris, 1963.

[106]R. M. Garrels and C. L. Christ, *Solutions, Minerals and Equilibria*, Harper and Row, New York, 1965.

[107]M. D. Pritzker and R.-H. Yoon, in *Proceedings of the International Symposium on Electrochemistry in Mineral and Metal Processing*, Ed. by P. E. Richardson, S. Srinivasan, and R. Woods, Proceedings Vol. 84-10, The Electrochemical Society, Pennington, New Jersey, 1984, pp. 26–53.

[108]M. D. Pritzker and R.-H. Yoon, *Int. J. Miner. Process.* **12** (1984) 95.

[109]M. D. Pritzker, R.-H. Yoon, C. I. Basilio, and W. Z. Choi, *Can. Metall. Q.* **24** (1985) 27.

[110]C. A. Young, R. Woods, and R.-H. Yoon, in *Proceedings of the International Symposium on Electrochemistry in Mineral and Metal Processing II*, Ed. by P. E. Richardson and R. Woods, Proceedings Vol. 88-21, The Electrochemical Society, Pennington, New Jersey, 1988, pp. 3–17.

[111]M. Wadsley, *Hydrometallurgy* **29** (1992) 91.

[112]E. Peters, in *Proceedings of the International Symposium on Electrochemistry in Mineral and Metal Processing*, Ed. by P. E. Richardson, S. Srinivasan, and R. Woods, Proceedings Vol. 84-10, The Electrochemical Society, Pennington, New Jersey, 1984, pp. 343–361.

[113]E. Peters, in *Advances in Mineral Processing*, Ed. by P. Somasundaran, SME/AIME, New York, 1986, pp. 445–462.

[114]J. S. Brinen and F. Reich, *Surf. Interface Anal.* **18** (1992) 448.

Cumulative Author Index for Numbers 1–29

Author	Title	Number
Bauer, H. H.	Critical Observations on the Measurement of Adsorption at Electrodes	7
Bebelis, S. I.	The Electrochemical Activation of Catalytic Reactions	29
Becker, R. O.	Electrochemical Mechanisms and the Control of Biological Growth Processes	10
Beden, B.	Electrocatalytic Oxidation of Oxygenated Aliphatic Organic Compounds at Noble Metal Electrodes	22
Benderskii, V. A.	Phase Transitions in the Double Layer at Electrodes	26
Berg, H.	Bioelectrochemical Field Effects: Electrostimulation of Biological Cells by Low Frequencies	24
Berwick, A.	The Study of Simple Consecutive Processes in Electrochemical Reactions	5
Blank, M.	Electrochemistry in Nerve Excitation	24
Bloom, H.	Models for Molten Salts	9
Bloom, H.	Molten Electrolytes	2
Blyholder, G.	Quantum Chemical Treatment of Adsorbed Species	8
Bockris, J. O'M.	Electrode Kinetics	1
Bockris, J. O'M.	Ionic Solvation	1
Bockris, J. O'M.	The Mechanism of Charge Transfer from Metal Electrodes to Ions in Solution	6
Bockris, J. O'M.	The Mechanism of the Electrode Position of Metals	3
Bockris, J. O'M.	Molten Electrolytes	2
Bockris, J. O'M.	Photoelectrochemical Kinetics and Related Devices	14
Boguslavsky, L. I.	Electron Transfer Effects and the Mechanism of the Membrane Potential	18
Breiter, M. W.	Adsorption of Organic Species on Platinum Metal Electrodes	10
Breiter, M. W.	Low-Temperature Electrochemistry at High-T_2 Superconductor/Ionic Conductor Interfaces	28
Brodskii, A. N.	Phase Transitions in the Double Layer at Electrodes	26
Burke, L. D.	Electrochemistry of Hydrous Oxide Films	18
Burney, H. S.	Membrane Chlor-Alkali Process	24

Author	Title	Number
MacDonald, D. D.	Impedance Measurements in Electrochemical Systems	14
Maksimović, M. D.	Theory of the Effect of Electrodeposition at a Periodically Changing Rate on the Morphology of Metal Deposits	19
Mandel, L. J.	Electrochemical Processes at Biological Interfaces	8
Marchiano, S. L.	Transport Phenomena in Electrochemical Kinetics	6
Marincic, N.	Lithium Batteries with Liquid Depolarizers	15
Markin, V. S.	Thermodynamics of Membrane Energy Transduction in an Oscillating Field	24
Martinez-Duart, J. M.	Electric Breakdown in Anodic Oxide Films	23
Matthews, D. B.	The Mechanism of Charge Transfer from Metal Electrodes to Ions in Solution	6
Mauritz, K. A.	Structural Properties of Membrane Ionomers	14
McBreen, J.	The Nickel Oxide Electrode	21
McKinnon, W. R.	Physical Mechanisms of Intercalation	15
McKubre, M. C. H.	Impedance Measurements in Electrochemical Systems	14
Murphy, O. J.	The Electrochemical Splitting of Water	15
Nagarkan, P. V.	Electrochemistry of Metallic Glasses	21
Nágy, Z.	DC Electrochemical Techniques for the Measurement of Corrosion Rates	25
Nágy, Z.	DC Relaxation Techniques for the Investigation of Fast Electrode Reactions	21
Neophytides, S. G.	The Electrochemical Activation of Catalytic Reactions	29
Newman, J.	Photoelectrochemical Devices for Solar Energy Conversion	18
Newman, J.	Determination of Current Distributions Governed by Laplace's Equation	23
Newman, J.	Metal Hydride Electrodes	27
Newman, K. E.	NMR Studies of the Structure of Electrolyte Solutions	12
Nişanciağlu, K.	Design Techniques in Cathodic Protection Engineering	23
Novak, D. M.	Fundamental and Applied Aspects of Anodic Chlorine Production	14

Author	Title	Number
Pound, B. G.	Electrochemical Techniques to Study Hydrogen Ingress in Metals	25
Power, G. P.	Metal Displacement Reactions	11
Reeves, R. M.	The Electrical Double Layer: The Current States of Data and Models, with Particular Emphasis on the Solvent	9
Revie, R. W.	Environmental Cracking of Metals: Electrochemical Aspects	26
Ritchie, I. M.	Metal Displacement Reactions	11
Rohland, B.	Advanced Electrochemical Hydrogen Technologies: Water Electrolyzers and Fuel Cells	26
Roscoe, S. G.	Electrochemical Investigations of the Interfacial Behavior of Proteins	29
Rusling, J. F.	Electrochemistry and Electrochemical Catalysis in Microemulsions	26
Russell, J.	Interfacial Infrared Vibrational Spectroscopy	17
Rysselberghe, P. Van	Some Aspects of the Thermodynamic Structure of Electrochemistry	4
Sacher, E.	Theories of Elementary Homogeneous Electron-Transfer Reactions	3
Saemann-Ischenko, G.	Low-Temperature Electrochemistry at High-T_2 Superconductor/Ionic Conductor Interfaces	28
Salvarezza, R. C.	A Modern Approach to Surface Roughness Applied to Electrochemical Systems	28
Sandstede, G. S.	Water Electrolysis and Solar Hydrogen Demonstration Projects	27
Savenko, V. I.	Electric Surface Effects in Solid Plasticity and Strength	24
Scharifker, B. R.	Microelectrode Techniques in Electrochemistry	22
Schmickler, W.	Electron Transfer Reactions on Oxide-Covered Metal Electrodes	17
Schneir, J.	Scanning Tunneling Microscopy: A Natural for Electrochemistry	21
Schultze, J. W.	Electron Transfer Reactions on Oxide-Covered Metal Electrodes	17
Scott, K.	Reaction Engineering and Digital Simulation in Electrochemical Processes	27

Cumulative Title Index for Numbers 1–29

INDEX